Sustainability and the Rights of Nature

An Introduction

Social-Environmental Sustainability Series

Series Editor

Chris Maser

Published Titles

Sustainability and the Rights of Nature: An Introduction
Cameron La Follette, Chris Maser

Interactions of Land, Ocean and Humans: A Global Perspective
Chris Maser

Land-Use Planning for Sustainable Development, Second Edition
Jane Silberstein and Chris Maser

Insects and Sustainability of Ecosystem Services
Timothy D. Schowalter

Biosequestration and Ecological Diversity: Mitigating and Adapting to Climate Change and Environmental Degradation
Wayne A. White

Decision-Making for a Sustainable Environment: A Systemic Approach
Chris Maser

Economics and Ecology: United for a Sustainable World
Russ Beaton and Chris Maser

Resolving Environmental Conflicts, Second Edition
Chris Maser and Carol A. Pollio

Fundamentals of Practical Environmentalism
Mark B. Weldon

Sustainable Development: Principles, Frameworks, and Case Studies
Okechukwu Ukaga, Chris Maser, and Michael Reichenbach

Social-Environmental Planning: The Design Interface Between Everyforest and Everycity
Chris Maser

Sustainability and the Rights of Nature

An Introduction

Cameron La Follette
Chris Maser

CRC Press
Taylor & Francis Group
Boca Raton London New York

CRC Press is an imprint of the
Taylor & Francis Group, an **informa** business

Cover photo of Monterey, VA © Tim Palmer.

CRC Press
Taylor & Francis Group
6000 Broken Sound Parkway NW, Suite 300
Boca Raton, FL 33487-2742

First issued in paperback 2019

ISBN-13: 978-1-4987-8844-1 (hbk)
ISBN-13: 978-0-367-88435-2 (pbk)

Library of Congress Cataloging-in-Publication Data

Names: La Follette, Cameron, author. | Maser, Chris, author.
Title: Sustainability and the rights of nature : an introduction / Cameron La Follette and Chris Maser.
Description: Boca Raton, FL : Taylor & Francis, 2017. | Includes bibliographical references.
Identifiers: LCCN 2016045225| ISBN 9781498788441 (hardback : alk. paper) | ISBN 9781498788465 (ebook)
Subjects: LCSH: Environmental law, International. | Sustainable development--Law and legislation. | Nature conservation--Law and legislation. | Environmental protection--International cooperation. | Environmentalism--International cooperation. | Sustainability--International cooperation. | Nature--Effect of human beings on.
Classification: LCC K3585 .L33 2017 | DDC 344.04/6--dc23
LC record available at https://lccn.loc.gov/2016045225

Visit the Taylor & Francis Web site at
http://www.taylorandfrancis.com

and the CRC Press Web site at
http://www.crcpress.com

We dedicate this book to Ecuador, Bolivia, and their indigenous peoples. Building on their indigenous philosophies, these countries, with foresight and courage, have led the world by first placing the Rights of Nature in their legal systems. They are the pioneers in creating environmental sustainability for all generations.

Contents

Section II Building Blocks for a Rights of Nature System

Section III Stumbling Blocks to a Rights of Nature System

Section IV Rights of Nature for Land, Water, and Air

Foreword

The crises of climate change and ecosystem disruption are true planetary emergencies. Given the magnitude of these problems (and many others), old thinking will not be enough. Only bold new approaches will suffice.

This wonderful book proposes a new framing that provides some questions that we need to ask—and some answers that may be the ones we need. It is not surprising that such new ideas would come from Cameron La Follette and Chris Maser.

Nearly 40 years ago, Chris Maser (a Bureau of Land Management wildlife biologist at that time) helped open my eyes at a weeklong "short course" in the ecology of old-growth (ancient) forests. The course was held at the H.J. Andrews Experimental Forest in the summer of 1978. He and other scientists associated with Oregon State University's Department of Forestry had started to untangle the puzzles of the ancient forests, which the timber industry viewed as simply a source to be exploited and which the industry called a "biological desert." Maser and his colleagues showed this was far from the truth.

A few months later, the Lane County Audubon Society organized a series of evening lectures on the ecology of old growth forests. One of those lecturing was a college student named Cameron La Follette, wise beyond her years, who was in the process of publishing a small book titled *Saving All the Pieces: Old Growth Forest in Oregon*. She pointed out that the "first rule of intelligent tinkering is saving all the pieces," a paraphrase of the advice of conservationist Aldo Leopold. Cameron later went to law school, but decided to devote her life to other ways of saving all the pieces.

Cameron and Chris have long challenged us to think in new ways. This book is their latest effort, and it is one of stunning scope. Both of them have always exhibited passionate creativity. Their thinking was indispensable to the reframing of lawsuits and politics that led to new visions of the disappearing ancient forests of the Pacific Northwest of the United States. They are now embarking on a new journey with this book—one that asks us to redesign legal approaches and reimagine the tools that we need to craft solutions. The heart of their vision involves ecosystem rights—rights for Nature itself.

In recent years, the linkages between human rights and environmental protection have been increasingly recognized around the world. More than 100 countries have embedded a right to a safe and healthy environment in their national constitutions—as have five states of the United States in their state constitutions. Books, articles, and journals are being published on "human rights and the environment." The organization that I co-founded, the Environmental Law Alliance Worldwide, has lawyers in countries all around the world who litigate and advocate on behalf of human environmental rights. But Chris and Cameron ask us to look beyond "human" rights. They ask us to look even more deeply at Nature, ecology, and law.

Laws in numerous countries and international agreements have evolved to give citizens the right to file lawsuits against government bodies or polluters for causing environmental harm. United States laws since 1970 have provided for "citizen suits" in the fields of air pollution, water pollution, endangered species, and many other areas. In Europe, the Aarhus Public Participation Convention, now nearly 20 years old, states explicitly that a right to a healthy environment exists and requires that the 45 countries that ratified or acceded to it must open their courts to citizen groups that are defending the environment, without having to prove an economic or personal interest will be affected before suing.

Various experts have talked about "legal standing" in court for natural places, about the glacial pace of environmental regulation, about the difficulty of getting political bodies to react with urgency to the loss of biodiversity and the looming tragedy of global warming. What La Follette and Maser do in this exciting book is bring these problems together in a new synthesis and describe a radically new solution that is already forming in a few scattered countries and cities—recognizing that Nature itself could have legal rights. They also demand a radical (that is, fundamental) understanding of true "sustainability"—not as a political or economic magic phrase, but as a truly ecological concept. They take their vision and work through its implications in a multitude of contexts. One need not subscribe to every argument they offer in this book. But one cannot read this book without opening one's eyes to new insights and new possibilities.

When the Universal Declaration of Human Rights of 1948 was drafted, there was little thought that it would apply to issues of clean water, clean air, and other environmental matters. Yet decisions of the European Court of Human Rights, Inter-American Court of Human Rights, and other bodies have interpreted older human rights language to extend to environmental rights for human beings. Who is to say that this process is complete—or that it must stop with "human" rights? Reading this book will raise new questions. Will Nature's own rights—to be free from excessive and debilitating human contamination and pollution, for example—be next in the evolution of ideas? Is it possible to create a legal and sustainable order in which Nature can continue its own natural processes without its protectors being restricted to narrow, 17th or even 20th-century conceptions?

If Nature's rights are to be recognized, who would have the status to defend Nature's rights in courts or other bodies? Will it be government entities, conservation groups, or individuals? Will development in law and consciousness be limited to litigation in the courts set in place, or will there arise something new, like Nature Tribunals? And who will be the lawyers, scientists, politicians, and citizens who drive us toward a new generation of rights?

Read, admire, argue, challenge, and enjoy this statement of a new vision for Earth. And consider ways to make this kind of vision real. The life of Planet Earth depends on new thinking such as this.

John E. Bonine
Founder, Environmental Law Alliance Worldwide (ELAW)
B.B. Kliks Professor of Law, University of Oregon

Series Preface

In reading this book, keep in mind that our earthscape is composed of three interactive spheres: the atmosphere (air), the litho-hydrosphere (the rock that constitutes the restless continents and the water that surrounds them), and the biosphere (the life-forms that exist within and between the other two spheres). We humans have, however, arbitrarily delineated our seamless world into discrete ecosystems since the advent of agriculture, as we try to control—"manage"—the fluid interactions among the nonliving and living components of planet Earth for our material benefits. If you picture the interconnectivity of the three spheres as being analogous to the motion of a filled waterbed, you will see how patently impossible such divisions are because you cannot touch any part of a filled waterbed without affecting the whole of it.

Together, these three spheres form myriad interactive, self-reinforcing feedback loops that affect all life on Earth. It is the degree to which we humans either honor or defy the reciprocity of these feedback loops that form the legacy we leave—one that either liberates or progressively binds all generations. The choice of how we, today's adults, behave is ours—either with psychological maturity and our respectful treatment of Nature through sacred humility *or* continued self-indulgence through materialism, profit seeking, and its ensuing environmental violence.

Where in the United States today are there unequivocal voices that speak for protecting the sustainability and productive capacity of Nature, as our bequest to all children—present and future? Without such voices of consciousness, courage, and unconditional commitment to the Rights of Nature through the present into the future in *all* ranks of leadership, we, the adults, are increasingly condemning our children, grandchildren—and every generation yet unborn—to pay a progressively awful price rather than accept the sometimes-difficult choices of our adult responsibilities as trustees of our home planet's social-environmental integrity.

The psychologically mature choice to become a truly peaceful society for the rest of the world to emulate requires that we, in the United States, transcend the environmental violence of our often-declared "war" on Nature. To rise above this violence, we must focus first and foremost on the "Rights of Nature" by accepting and honoring Nature's Laws of Reciprocity with total concentration, dedication, and persistence, in the legal system and in our social systems.

Peace will reign only when there is no longer any thought of abusing Nature for personal gain. But to eliminate this linear, economically oriented mentality, its environmental violence, and its waste of dwindling resources, we must shift the legal system from its current focus on flagrant overexploitation of Nature to a legal system that fully protects and prioritizes Nature's existence and its rights to flourish as an entity.

Finally, this CRC series of books on the various facets of social-environmental sustainability is a forum wherein those who dare to seek harmony and wholeness can struggle to integrate disciplines and balance the material world with the spiritual, the scientific with the social, and in so doing expose their vulnerabilities, human frailties, and hope, as well as their visions for a viable future.

Chris Maser
Series Editor

Author's Preface

Chris Maser approached me with the idea of collaborating on a book focusing on change needed in the American legal systems to meet the challenge of environmental sustainability. As we began the initial research, it quickly became evident that an entirely new paradigm had already emerged, thanks to Ecuador, Bolivia, and a dedicated group of what is now called "earth jurisprudence" activists: the Rights of Nature. I had heard of the courageous and monumental change initiated by Ecuador and Bolivia when they placed Rights of Nature in their Constitution and laws but knew little about it.

As I researched it further, learning of New Zealand's ground-breaking Rights of Nature activity in addition, it became increasingly clear that the deep paradigm shift to a Rights of Nature system is the missing puzzle piece that, if implemented with foresight and courage, would provide the desperately needed framework to create lasting environmental sustainability. Only Rights of Nature expands relationship and the *responsibility for relationship* to all ecosystems and their components, which collectively sustain all life. Only this system would break the cycle of profit-seeking and plundering the world.

I was reluctant to commit to such a project, mainly because Rights of Nature is a marginalized, derided concept in the public discussion of natural resources issues. I work as an environmental activist for a nonprofit organization that focuses on protecting the natural resources and livable communities of the Oregon coast. The current legal framework for land and water protection in Oregon—and elsewhere in the United States, with a few local exceptions—offers but narrow opportunities to create true sustainability. The legal system forces the conversation to take place only in terms of what would benefit human needs. No one mentions what Nature itself needs.

Ultimately, I agreed to tackle this project because the need is so great. My greatest hope is that this book will catalyze Americans' courage to mend our broken relationship with Nature. The need is desperate: unending news of near-extinction of species, uncontrollable global warming, increased fires and floods, and many other disasters pours in daily. Implementing a Rights of Nature system is the most positive, life-affirming step every person, community, and state can take in the face of all the grim news.

I extend my deepest admiration and thanks to all those in the environmental and conservation movements worldwide, staff and volunteers, who work against incredible odds to stop destructive mining, restore watersheds, enlarge recycling opportunities, oppose damaging resort and subdivision projects, protect forests, change transportation patterns, create livable communities, renew organic agriculture, and assume a thousand other tasks. They are the vanguard of a renewed world. May the Rights of Nature, providing the legal groundwork for people to expand our relationships with Nature, bring to fruition all that they strive for.

I researched and wrote my share of *Sustainability and the Rights of Nature: An Introduction* in my personal capacity. The opinions expressed in this work are my own and do not reflect the view of any past, present, or future employers or organizations with which I am, have been or may in future be affiliated.

Cameron La Follette
Co-author

Authors

Cameron La Follette has a law degree from Columbia University School of Law, a master's degree in psychology from New York University, and a bachelor's degree in journalism from the University of Oregon. Her initial environmental activism (1978–1982) was with Oregon nonprofit organizations that focused on preserving ancient forests on federal public lands managed by the U.S. Forest Service and U.S. Bureau of Land Management to protect salmon habitat, clean drinking water, and forest ecosystems. She served on the Salem, Oregon, Planning Commission for three years (2002–2005), applying the City of Salem's land use and zoning ordinances to many situations ranging from residential housing to industrial and commercial properties. As director of a nonprofit project (2004–2006), she focused on bringing people together to collaborate on coastal environmental problems. Since 2010, she has been executive director of an environmental and land use activist nonprofit that focuses on protecting the natural resources of the Oregon coast, working with residents to oppose ill-advised land use projects, and helping maintain livable coastal communities.

Chris Maser spent over 25 years as a research scientist in natural history and ecology in forest, shrub steppe, subarctic, desert, coastal, and agricultural settings. Trained primarily as a vertebrate zoologist, he was a research mammalogist in Nubia, Egypt (1963–1964) with the Yale University Peabody Museum Prehistoric Expedition and a research mammalogist in Nepal (1966–1967), where he participated in a study of tick-borne diseases for the U.S. Naval Medical Research Unit #3 based in Cairo, Egypt. He conducted a three-year (1970–1973) ecological survey of the Oregon Coast for the University of Puget Sound, Tacoma, Washington. He was a research ecologist with the U.S. Bureau of Land Management for 13 years, with the first seven years (1974–1981) spent studying the biophysical relationships in rangelands in southeastern Oregon and the last six years (1982–1987) spent studying old-growth forests in western Oregon. He also spent a year as a landscape ecologist with the U.S. Environmental Protection Agency (1990–1991).

Today, he is an independent author as well as an international lecturer, facilitator in resolving environmental conflicts, vision statements, and sustainable community development. He is also an international consultant in forest ecology and sustainable forestry practices. He has written or edited over 290 publications, including 40 books.

Introduction

What would it really take for us, in the United States, to create and maintain a sustainable lifestyle? How would we even know where to begin, since "sustainability" is impossible to define—albeit the word is increasingly used? One sees it everywhere. But few of the many definitions have anything in common, because sustainability is not a fixed endpoint. Rather, it is a lifelong journey toward humility and reciprocity of caring in our relationship with Planet Earth.[1]

Beyond that, sustainability must mean creating and living in a society in which the integrity, resilience, and productivity of Nature's Laws of Reciprocity come first. Sustainability is so frequently mentioned that we forget there are no "levels of sustainability." It is not systemically sustainable, for example, to build a gigantic "sun farm" out in the desert, thereby degrading thousands of acres to power cities and industrial systems with "clean" solar energy. We might call such a thing "shallow sustainability," but nothing more. Only by setting human use of resources within the Rights of Nature paradigm, in which Earth's integrity comes first, can human societies flourish, living in harmony with Nature's Laws of Reciprocity. That is a critical premise of this book.

Human beings depend on Nature's systems for life. If the global ecosystem becomes less able to support a good quality of life, humans will suffer accordingly. Thus, the central focus of this book is creating a blueprint for changing our relationship with the Earth by placing a Rights of Nature framework at the center of the American legal system, both at the federal and state levels. Furthermore, the Rights of Nature must be "first among rights" in our rights-based system, so Nature's resilience is the basis from which all human society springs. We then explore ways in which placing the Earth's ability to flourish first and foremost will change our relationship with the environment and our current land-management strategies. Finally, we offer suggestions and guidelines on how to bring those changes to fruition.

We take this approach because true sustainability can only be achieved by thinking systemically—not by merely tackling one symptom after another without addressing the system as a whole. The blueprint is in our hands, thanks to the courageous and farsighted decisions by the nations of Ecuador and Bolivia, which have already placed the Rights of Nature at the heart of their legal systems. Ecuador, in 2008, became the first nation in the world to have the Rights of Nature in their Constitution, which has been accepted by a nationwide vote. Bolivia followed suit in 2010 with an even more comprehensive Rights of Nature statute, which the country is in the process of fully implementing. These two nations are leading the world in reorienting the human relationship with the Earth to a Nature-based legal and economic system. We acknowledge their courage, strength, and visionary action to protect the Earth for the future of all living beings.

We begin by laying out the Nature's 15 Laws of Reciprocity. These laws constitute the biophysical foundation of all living systems—the often-invisible laws by which they operate. Human economic and social activity must fit within these laws if the Earth's biological resilience is to continue. Laying these laws open to scrutiny helps us better understand the Earth's systems in which we are embedded, and of which we are an inseparable part.

Following this, we discuss the emerging paradigm of Nature's Rights, led by Ecuador, Bolivia, and the Global Alliance for the Rights of Nature. This new paradigm was formed in the wake of those two countries' extraordinary, courageous leap in rewriting their laws to place Nature's Rights first. Although still entirely outside the framework of existing international law, the Global Alliance has already created an international voice for Nature's Rights. Innovative partnerships involving the cultural principles of indigenous peoples are beginning to appear in other places, such as New Zealand. In addition, new voices—such as the Catholic Papacy—are speaking out on the importance of protecting Nature's integrity.

Although no country, with the exception of Ecuador and Bolivia, has ensconced the Rights of Nature in their Constitutions or national laws, the concept is now being discussed, studied, and seriously considered, even in the United States. Although the heavily corporatized economic system in the United States might seem like the last bastion against Nature's Rights, a growing number of American towns have adopted Rights of Nature clauses. As environmental degradation expands, so too does the urgent discussion of the need for the Rights of Nature as the central operating principle within the legal system. It is the only way humans' relationship with Nature can be fully repaired.

Our first contribution to this discussion is an examination of a time in Western culture when a form of reciprocity and sustainability was actually practiced: the feudal system of the High Middle Ages. The feudal system created only a partial system of reciprocity that was highly formalized and only extended among humans—and then only among certain classes of people. Nevertheless, it is the one time in the Western tradition when European culture sought to model itself on a sense of obligation and reciprocity to others. Using these ideas as a seedbed, we can expand and shape a Rights of Nature system for the future. As part of this survey, we also explore the food and ritual systems of the indigenous peoples of the Northwestern United States, which centered on expanding the local habitats for culturally important species to strengthen both the human culture and the animal populations the culture depended on. This is a vital angle of vision missing from the feudal system, but which can be incorporated in a manner that dovetails with current society.

Next, we examine two of the largest stumbling blocks to a Rights of Nature-based sustainable environment—the misuse of technology, and the corporate legal structure. Western societies tend to rely on technology for quick fix, symptomatic solutions to looming problems, but often inappropriately apply the technology. A Rights of Nature–based legal system can correct for this by using a systemic approach to sustainability with the wise use of responsible technology, when and where appropriate. We might call such technologies "biomimicry," an approach to innovation where people consciously seek sustainable solutions to human challenges by emulating Nature's time-tested patterns and processes.[2] True biomimicry harmonizes with the integrity of Nature's Laws of Reciprocity, unlike profit-oriented, degrading technologies, such as genetically modified organisms (GMOs) and their dangerously unpredictable pesticide tolerance.

A potent instrument in the decimation of the global ecosystem is a relatively modern invention—the corporation. The United States Supreme Court's decision to grant corporations personhood has created a mindless, faceless entity shorn of all accountability that thrives on profit. Corporations, which pay attention only to their bottom line, cross borders seeking natural resources without regard for the social-environmental costs. This fiction of legal personhood must be dismantled, both in the United States and internationally, before we can place Rights of Nature at the center of the legal system and create a sustainable lifestyle.

The second half of the book explores the directions needed to create a Rights of Nature-based society. We begin with a discussion of existing legal principles that build the entire structure, especially the concept of the "Commons" in Western thought: how it has been used, abused, and ignored. Then we focus on the hard questions: How must society treat the land, water, and air to be sustainable? What choices could the legal system make that would allow humans to use natural resources, while at the same time allow ecosystems to retain their biophysical integrity? How would a sustainable energy system work? What must be done about the vast piles of recyclable materials rotting on land and polluting the seas?

Two main principles guide us in laying out the Rights of Nature legal framework: the first is the *precautionary principle*, which reorients our decision-making toward being cautious about the effects of a project—not just the principle of "look before you leap," but not leaping at all if there are known or suspected ill effects, or even the possibility of them, in a proposal. The second principle is the creation of a legal *biophysical living trust*. Our ways of living must be focused on the future generations who will be the beneficiaries of the decisions we make today. Acting in ways that further degrade the global ecosystem is directly contrary to the essence of a biophysical living trust, and would no longer be permitted.

The goal of both of these principles, which shape the Rights of Nature legal system we propose, is to allow the natural resources, on which we depend, to "speak for themselves." How can they do that, since they do not have language as we understand it? Nevertheless, they "speak" all the time. Through their biophysical integrity, resilience, and productivity—or the lack thereof—ecosystems tell us what effect our human actions are having, and offer indisputable course corrections. We have the means to understand Nature's speech through scientific data, direct observation, historical research, and similar tools. Nature's Laws of Reciprocity create the means by which humans can understand the speech of the world's ecosystems.

Although many of the current problems appear insoluble, we can mitigate or repair many of the problems humans have created. Pouring skill, ingenuity, funding, and priority into such restoration is the critical first step to shifting and beginning to live within a Rights of Nature system. We treat these necessary restoration activities not as "moving backward," but as moving forward into sustainability. First of all, human societies need to stop doing further harm. Although this ought to be the most easily taken step, it is often the most difficult because, since the Industrial Revolution, our society has depended on resource exploitation far more than can be sustained by the world's ecosystems with their burgeoning human populations.

The surprising and heartening news is that the pieces needed to fashion a true Rights of Nature society are already available in American law and policy as well as worldwide; they are just not yet connected into a whole and set into the ruling frameworks. Rights of Nature is neither radical nor impossible. Its components are already in use. In this book, we use examples from around the world, but our immediate focus is changes needed in American legal systems and lifeways.

Examining on-the-ground management problems in the second half of the book, we assume that the United States has already taken the critically important initial step of placing Rights of Nature language in the federal and state Constitutions. This has not yet happened, of course, and the struggles to realize this vision will be many and vigorous. But in order to create a template for future action, we assume this initial victory has already occurred, so that we can explore the options over the near horizon.

American laws, both federal and local, often allow some of the necessary activities to protect Nature or encourage its flourishing, but they are only regulatory and/or permissive. Most fundamentally, they always have human needs in mind first, not Nature's.

There are, however, several things this book is not. It is not a comprehensive legal handbook designed to provide templates on how to incorporate the Rights of Nature into the

American legal framework at every level. Basic language is already available from Ecuador, Bolivia, New Zealand, many American local ordinances, and the Global Alliance for the Rights of Nature. Detailed discussions of which nuance to choose, and how to reorganize the law to place Nature first, must occur at every level of American society. Each region of the country differs in ecology, economy, and history. It is thus fitting that the details of the change be grappled with locally.

This book is also not a workbook on how to apply the Rights of Nature to every ecosystem and human community. The United States, not to mention the rest of the world, contains many ecosystems, each of staggering complexity and local variation. Human cultures have adapted to the land and weather. Thus, the details of how best to fit Nature first in human activities must be worked out in the context of local, land-based knowledge by residents of that place, because the solution in one area may well differ significantly from that in another. The most we can do is survey three major ecosystems in the United States: forests, grasslands, and wetlands; discuss the problems; and show how the Rights of Nature create the new center from which human activities take place in ways that prioritize the integrity of Nature's right to flourish.

Many will dismiss the central thesis of this book out of hand as impossible, destructive, unrealistic, and thus unworthy of serious attention. Some will argue that a Rights of Nature is "anti-people," and cares northing for human lives and needs. Nothing could be farther from the truth. "There are no jobs on a dead planet," as West Virginia mountaintop-removal mining activist Judy Bonds often said. We argue that no economic system can flourish—and human life cannot flourish—if Nature does not flourish *first*.[3] This simple truth has been ignored for a long time, as the consequences of human overreach have been unclear. Now, however, they are becoming starkly obvious.

Placing the Rights of Nature first will lead to setting us on a course of true sustainability, although some would deride it as "backtracking." It is not backtracking, however; it is moving forward in a new direction under a Rights of Nature legal and economic system. Changing the laws gives us the framework in which to respect the natural environment, repair its damaged processes, revitalize its resilience, and perpetuate its productive capacity.

"Sustainability" in our vision simply means the ability of a natural system to maintain its critical functions. By analogy, if you commence cutting the strands of a net, it will retain the capacity to act as a net only so long as enough of the strands remain interconnected. But the net's overall effectiveness is reduced with the severing of each strand. When enough of the apparently useless strands have been slashed, the net will develop such large holes that its ability to function as it was designed to will end.

Linear-thinking, economically competitive, industrialized nations such as the United States often extract Nature's commodities to the "bare bones" of ecosystem function. It will require political determination, courage, humility, and cooperation to go down the road we describe in this book and begin living in a Rights of Nature legal system. Can it be done? Yes. We are today at a crossroads in terms of whether humanity's future is reasonably assured because ecosystem resilience remains strong, or dangerously unstable because our intervention has sufficiently degraded them.

The choice is ours. This choice does not mean we shall be thrown into a dark age, but it does mean that people must be more personally responsible, mindful, humble, and caring—both locally and globally. Put differently: Act locally and affect the whole world. We must repair our relationship with the Earth that sustains all, and live in a Rights of Nature culture. Diverse peoples throughout history have proven this is possible.

Endnotes

1. Chris Maser. *Decision Making for a Sustainable Environment: A Systemic Approach*. CRC Press, Boca Raton, FL. (2013) 304 pp.
2. Biomimicry Institute. http://biomimicry.org/what-is-biomimicry/ (accessed June 1, 2015).
3. Russ Beaton and Chris Maser. *Economics and Ecology: United for a Sustainable World*. CRC Press, Boca Raton, FL. (2012) 191 pp.

Section I

Nature's Laws of Reciprocity

A Beginning

No person, institution, or nation has the right to participate in activities that contribute to large-scale, irreversible changes of the Earth's biogeochemical cycles or undermine the integrity, stability, and beauty of the Earth's ecologies—the consequences of which would fall on succeeding generations as an irrevocable form of remote tyranny.

David Orr[1]

Endnote

1. David Orr. 2020: A proposal. *Conservation Biology*, 14 (2000):338–341.

1

Nature's Laws of Reciprocity

Every crisis in today's world—whether social or environmental—is a historical archive of human choices, decisions, legal opinions, subsequent actions, and compounding consequences, including those of yesterday. Despite all the evidence before us, we keep making the same types of unwise social-environmental decisions, staunchly defended by legal opinions that ignore the myriad volumes of historical evidence depicting their dire consequences. Each time we expect a new and different outcome. We humans face a growing cataclysm of suffering, not the least of which is caused by ideological strife, with its wanton destruction of irreplaceable resources, and the growing threat of global warming. Yet, those rare individuals who make decisions that would, in fact, further the social-environmental well-being of humanity and the Earth are too often thwarted through the political/legal system by the socially powerful minority who are afraid of losing their economic advantages.

Change often occurs on the brink of disaster between need and fear. On the one hand, we know we need to do things differently. On the other hand, we are terrified of facing the unknown, unfamiliar, and uncertain. Many people prefer to err again and again rather than let go of some cherished belief, assumption, or staunchly defended position. Others err because they are pessimistic in their outlook and are thus blind to viable options.[1]

Whoever makes a social decision is simultaneously making an environmental decision—and vice versa. This is an inviolable relationship because human society is an inseparable part of the environment, just as the environment is an indivisible part of human society. Thus, every leader, regardless of his or her hierarchical level in government—local, national, or international—is a social-environmental decision maker, despite whether the leader understands and accepts it or not.[2]

Social-environmental sustainability demands that decision-making within the legal system go beyond the immediate human valuation of a resource to examine—*and disclose*—how its use will affect the long-term, productive sustainability of the ecosystem of which it is a component. Decision makers must also recognize and disclose the long-term, social-environmental issues embedded in the resource extraction method. This is necessary because the overall integrity of an ecosystem will determine the array of options available to all generations—for which individuals in the legal profession are setting the long-term precedent.

To sustain the environment's ecological integrity for the common good of humanity, each person in the legal profession must integrate Nature's inviolable Laws of Reciprocity given in this chapter as a condition of wise decision-making. They are the foundation of Nature's Rights we discuss in the forthcoming chapters. These laws form the underpinnings through which Nature operates and the social limitations we must understand and accept if we are to participate with Nature in a sustainable fashion.

Is it really so imperative for individuals to change their behavior, if it infringes on their sense of personal *rights*? But in making a choice, each person bequeaths the consequences of that choice to *all* generations. Therefore, it would be well to consider the counsel of professor Johan Rockström and his interdisciplinary team of 29 scientists:

> Although Earth has undergone many periods of significant environmental change, the planet's environment has been unusually stable for the past 10,000 years. This period of stability—known to geologists as the Holocene—has seen human civilizations arise, develop and thrive. Such stability may now be under threat. Since the Industrial Revolution, a new era has arisen, the Anthropocene, in which human actions have become the main driver of global environmental change. This could see human activities push the Earth system outside the stable environmental state of the Holocene, with consequences that are detrimental or even catastrophic for large parts of the world.[3] [Holocene comes from the Greek *holos*, ("whole") and *cene* ("new"). Anthropocene[4] comes from the Greek *anthropo* ("human") and *cene* ("new").]

The dawning of the Anthropocene Epoch represents "a new phase in the history of both humankind and of the Earth, when natural forces and human forces became intertwined, so the fate of one determines the fate of the other. Geologically, this is a remarkable episode in the history of this planet."[5] Hiking in a wilderness area or wandering through a national park, no matter how far removed from the center of society, we are still breathing pollution. It is everywhere and will worsen, as long as legal decisions to placate big industry continually trump a global pursuit of dramatically cleaning the world's air. Here it must be stated in fairness that our material appetites feed the corporate drive for more, and the corporate drive for more stimulates our material appetite for more—always more—through advertising, in a self-reinforcing feedback loop.

Yet we, as a society, listen to the world's traditional economists, corporate lawyers, and the political elite and *assume* they are correct when they take such ecological variables as air, soil, water, sunlight, biodiversity, genetic diversity, and climate and convert them—in theory at least—into economic/political constants whose values are unchanging, or discount them altogether as "externalities." Ecological variables are therefore omitted from consideration in most economic and planning models and even from our thinking—to say nothing of the decisions rendered by today's legal professionals. Biodiversity and genetic diversity, on the other hand, are euphemistically discounted as "externalities" when their consideration interferes with monetary profits.[6]

Nevertheless, the relationships among things are in constant flux, as complex systems arise from subatomic and atomic particles in the giant process of evolution on Earth. Moreover, a system's functional dynamics are characterized by their diversity, as well as by the constraints of the overarching laws and subordinate laws that govern them.

These laws can be said to govern the world and our place in it because they form the behavioral constraints without which nothing could function in an orderly manner. In this sense, Nature's Laws of Reciprocity inform society of the latitude whereby it can cooperate with Nature and survive in a sustainable manner. Beyond that, the global ecosystem will always function in a manner sustainable unto itself based on Nature's inviolable Laws of Reciprocity, but not necessarily in a way that is favorable to human survival, let alone a human life of well-being and dignity—despite the assumed sanctity of the legal system. "Inviolable" means that we can manipulate the effects of a law through our decisions and subsequent actions, but we do not—and cannot—control the functional effect of the law itself.

Nature's Laws of Reciprocity

Although we have done our best to present Nature's Laws of Reciprocity in a seemingly logical order, it is impossible to be definitive because each law is an interactive strand in the multi-dimensional web of energy interchange that constitutes the universe, our world within it, and how we affect our world. Moreover, a different order can be found each time the laws are read, and each arrangement seems logical. Because each law affects all laws (like touching a filled waterbed), every arrangement is equally correct in its overall function.

Law 1—Everything Is a Relationship

The universe is a single, all-inclusive relationship constituted of an ever-expanding web of biophysical feedback loops, each of which is novel and self-reinforcing. Each feedback loop is a conduit whereby energy is moved from one place, one dimension, and one scale to another. And, all we humans do—ever—is practice relationships with the flow of energy within this web, because the existence of everything in the universe is an expression of its relationship to everything else within the web through the continual exchange of energy. Moreover, all relationships are forever dynamic. Herein lies one of the foremost paradoxes of life: The ongoing process of change is a universal constant over which, much to our dismay, we have *absolutely* no control.

Law 2—All Relationships Are All Inclusive and Productive of an Outcome

It is often said that a particular piece of land is "unproductive"—a strictly human concept based on commodity-oriented, monetary outcomes—and needs to be "brought under management," and management decisions are made accordingly. Here it must be rendered clear that every relationship is productive of a cause that has an effect, and the effect, which is the cause of another effect, *is* the product—but often contrary to human-desired, commodity-oriented, monetary outcomes. For example, the on and off storms in Oregon and Washington during the first two weeks of December 2015 "sent rivers bursting from their banks, spilled boulders and trees into [a] major highway and spawned a rare tornado that snapped power poles and battered homes [all negative outcomes by human valuation]. They've also had one positive effect—easing drought concerns after an unusually dry summer."[7]

Therefore, the notion of an unproductive parcel of ground or an unproductive political meeting is an illustration of the narrowness of human valuation because such judgment is viewed strictly within the *extrinsic* realm of personal values, usually economics—not the *intrinsic* realm of Nature's dynamics. This narrow valuation not only illuminates how little we understand Nature's inviolable dynamics but also defies the certainty within which we hallow the validity of our knowledge and thus the ability of our economic assessments—and the legal decisions based on them—to address these profound issues.[8]

We humans are not so powerful a natural force that we can "destroy" an ecosystem, because it still obeys Nature's Laws of Reciprocity that determine how it functions at a given point in time. Nevertheless, we can, we have, and we continue to so severely alter ecosystems to the point they are incapable of providing—for many decades and even for all time—those goods and services they once provided and we require to sustain human life and society.

Law 3—The Only True Investment in Our Global Ecosystem Is Energy from Sunlight

The *only true investment* in the global ecosystem is energy from solar radiation (materialized sunlight). Everything else is merely the recycling of already-existing energy. In a business sense, for example, one makes money ("economic capital") and then takes a percentage of those earnings and recycles them by putting them back into the infrastructure of the enterprise for maintenance of buildings and equipment to facilitate making a profit by protecting the integrity of the initial outlay of capital over time. In a business, one recycles economic capital after the profits have been earned.

"Biological capital," on the other hand, must be "recycled" *before* the profits are earned to stay even. This means forgoing some potential monetary gain by leaving enough of an ecosystem intact for it to function in a productively sustainable manner. In a forest, for instance, one leaves some proportion of the merchantable dead and live trees, the latter to gather solar radiation and produce living tissue that will subsequently die—and together with the already-dead trees—rot, recycle into the soil, and thereby replenish the fabric of the living system. In rangelands, one leaves the forage plants in a viable condition so they can capture solar radiation, grow, seed, die, decompose, and protect the soil from erosion as well as add organic material to the soil's long-term ecological integrity and productive capacity.

With respect to biological capital, it has long been understood that green plants use chlorophyll molecules to absorb sunlight and use its energy to synthesize carbohydrates from carbon dioxide and water. These carbohydrates, in turn, are partly stored energy from the sun—a new input of energy into the global ecosystem—and partly the storage of existing energy from the amalgam of carbon dioxide and water, a process sustained in part by chemical nutrients from the soil.[9]

This process is known as "photosynthesis," from the Greek *photo* ("light"), *syn* ("with, together"), and *thesis* ("putting, placing"). In other words, photosynthesis is the fusion of energy embodied in sunlight with the recycled energy stored in earthly materials. The combination is the basis for sustaining the life processes of virtually all plants, and, in turn, our human well-being. The energy is derived from the sun (an original input) and combined with carbon dioxide and water, as well as nutrients from the soil in which the plants grow (from existing recycled chemicals), to create a renewable source of usable energy that a decomposing plant adds to and recycles through the soil.

Law 4—All Systems Are Defined by Their Function

The behavior of every system—including the legal system—depends on how its individual parts interact as functional components of the whole, not on what an isolated part is doing. The whole, in turn, can only be understood through the relationships, the interaction of its parts. The only way anything can exist is via its interdependent relationship to everything else, which means an isolated fragment or an "independent variable" can exist only on paper as a figment of the human imagination—of which it is, nevertheless, an interactive part.

Put differently, it is a false assumption that an independent variable of one's choosing can exist in a system of one's choice, and that it will indeed act as an independent variable of one's choosing. In reality, all systems are interdependent and thus rely on their pieces to act in concert as a functioning whole. This being the case, no individual piece can stand on its own *and* simultaneously be part of an interactive system. Thus, there neither is nor can there be an *independent* variable in any system, be it biophysical, mechanical, or even intellectual, because every system is interactive by its very definition as a system.

What is more, every relationship is constantly adjusting itself to fit precisely into other relationships that, in turn, are consequently adjusting themselves to fit precisely into all relationships, a dynamic that precludes the existence of an independent variable. Therefore, *no given thing can be held as a constant value beyond the number one* (the universal common denominator) because to do so would necessitate the detachment of the thing in question from the system as an independent variable. This being the case, all relationships are constituted by additions of *one* in all its myriad forms, from quarks, atoms, molecules, and proteins, which comprise the building blocks of life, to the living organisms, which collectively form the species and communities.

So, to understand a system as a functional whole, we need to understand how it fits into the larger system of which it is a part and so gives us a view of systems (a municipal court) supporting systems (state supreme court) supporting systems (United States Supreme Court) supporting systems (United Nations), ad infinitum.

Law 5—All Relationships Result in a Transfer of Energy

Although a "conduit" is technically a hollow tube of some sort, the term is used here to connote any system employed specifically for the transfer of energy from one place to another. Every living thing, from a virus to a bacterium, fungus, plant, insect, fish, amphibian, reptile, bird, mammal, and every cell in our body, is a conduit for the collection, absorption, transformation, storage, transfer, and expulsion of energy. In fact, the function of the entire biophysical system is tied up in the collection, absorption, transformation, storage, transfer, and expulsion of energy—one gigantic, energy-balancing act, or perhaps more correctly, energy-juggling act.

Human activities may be constructive (caring for a wounded animal), benign (swimming in the ocean), or destructive (mountaintop-removal mining), all of which are subjective concepts based on human values, but change—the dance of energy—is continual, albeit at various rates and in various directions. All changes are, in addition, cumulative. Even mild, slow change can show dramatic effects over the long term. Although there was, of course, some localized European impact before 1750, thereafter (but especially after 1850) populations of European Americans expanded tremendously; they severely exploited the resources, greatly increasing the process of environmental modification—a dynamic that today not only is accelerating exponentially but also is promoted by the legal system. This is one reason why changing the legal system to the Rights of Nature is essential to enable humans to live within the limitations of Nature's Laws of Reciprocity rather than seeking to overrule them.

Law 6—All Relationships Are Self-Reinforcing Feedback Loops

Everything in the universe is connected to everything else in a cosmic web of interactive feedback loops, all entrained in self-reinforcing relationships that continually create novel, never-ending stories of cause and effect, stories that began with the original story, the original cause. Everything, from a microbe to a galaxy, is defined by its ever-shifting relationship to every other component of the cosmos. Thus, "freedom" (perceived as the lack of constraints) is merely a continuum of fluid relativity. In contraposition, every relationship is the embodiment of interactive constraints to the flow of energy—the very dynamic that perpetuates the relativity of freedom and thus all relationships, which, by definition, *precludes absolute freedom*, such as a "free market," despite political/legal proclamations.

Hence, every change (no matter how minute or how vast) constitutes a systemic modification that produces novel outcomes. A feedback loop, in this sense, is a reciprocal relationship among countless bursts of energy moving through specific strands in the cosmic web that causes forever-new, compounding changes in Nature at either end of the strand, as well as every connecting strand.[10] And here, we humans often face a dichotomy with respect to our interests.

On the one hand, while all feedback loops are self-reinforcing, their effects in Nature are neutral because Nature is impartial with respect to consequences. We, on the other hand, have definite desires as far as outcomes are concerned, and thus assign a preconceived value to what we think of as the outcome or product of our manipulation of Nature's biophysical feedback loops.

Law 7—All Relationships Have One or More Trade-Offs

All relationships have trade-offs that may or may not be readily apparent or immediately understood. In the end, however, each trade-off is couched in terms of a decision to change or not to change, based largely on a personal value that blends naturally into an emotional criterion of choice, and thus a decision to choose to act one way or another—or not to act, which is still by choice an action.

Other relationships have much more discernible trade-offs. To illustrate, the springtime ozone hole over Antarctica is finally shrinking after years of growing. As the hole grew in size due to human-induced, ozone-destroying chemicals in the stratosphere, the risk of skin cancer increased because more ultraviolet radiation reached Earth. Although today the "good news"—from a cancer patient's point of view—may be that the ozone hole is now shrinking and, through a complicated cascade of effects, could fully close within this century, what about tomorrow? (The *stratosphere* is the atmospheric layer immediately above the aforementioned troposphere and contains most of the Earth's ozone.)

Because the hole in the stratospheric ozone layer does not absorb much ultraviolet radiation, it keeps the temperature of Antarctica much cooler than normal, helping to minimize its influence on the rising sea level that is beginning to displace island-dwelling nations. As much as we might applaud the completely recovered ozone layer and its increased protection from skin cancer, on the one hand, the recovery could significantly boost atmospheric warming over the icy continent on the other hand, possibly augmenting its melting,[11] and thus a rise in sea level, to the detriment of island dwellers. In this case, what is good for humans in one sense may not be good for Antarctica or the island nations, and vice versa. So it is, by analogy, with all legal decisions between the humans of today and the natural environment that supports them, on the one hand, and the consequences of today's legal decisions passed forward, as an irreversible legacy, to all the generations of the future, on the other.

Law 8—Change Is a Process of Eternal Becoming

Change, as a universal constant, is a continual process of inexorable novelty. It is a condition along a continuum that may reach a momentary pinnacle of harmony. Then, the very process that created the harmony takes it away and replaces it with something else.

We all cause change of some kind every day. I (Chris) remember a rather dramatic one I made inadvertently along a small stream flowing across a beach on its way to the sea. The stream, having eroded its way into the sand, created a small undercut that could not be seen from the top. Something captured my attention in the middle of the stream, and I stepped on the overhang for a better look, causing the bank to cave in and me to get a really close-up view of the water. As a consequence of my misstep, I had both altered the configuration of the bank and caused innumerable grains of sand to be washed back into the sea from whence they had come several years earlier riding the crest of a storm wave.

So it is that increasing global warming and the potential for shortages of water is a portent of growing civil conflicts to come, as the exploding human population is increasingly confronted with and stressed by a progressively unpredictable environment.[12]

Law 9—All Relationships Are Irreversible

Because change is a constant process orchestrated along the interactive web of universal relationships, it produces infinite novelty that precludes anything in the cosmos from ever being reversible. Take mountaintop-removal mining. Can one of the destroyed mountaintops be restored to its original condition? No. No one can go back in time and make a different decision of whether or not to remove the mountaintop for whatever reason the idea was conceived in the first place. And, because we cannot go back in time, nothing can be restored to its former condition. All we can ever do is repair a process that is broken so it can continue to function, albeit differently than in its original form. (For a detailed discussion of relationship irreversibility, read *Earth in Our Care*.[13])

Law 10—All Systems Are Based on Composition, Structure, and Function

We perceive objects by means of their obvious structures or functions. Structure is the configuration of elements and parts (composition of constituents), be it simple or complex. The structure can be thought of as the organization, arrangement, or make-up of a thing. Function, on the other hand, is what a particular structure either can do or allows to be done to it, with it, or through it.

To maintain biophysical functions means that we humans must maintain the characteristics of an ecosystem in such a way that its processes are sustainable. The characteristics we must be concerned with are (1) composition, (2) structure, (3) function, and (4) Nature's disturbance regimes, which periodically alter an ecosystem's composition, structure, and function.

We can, for example, change the composition of an ecosystem, such as the kinds and arrangement of plants in a forest, grassland, or agricultural crop. This alteration means that composition is malleable to human desire and thus negotiable within the context of cause and effect. In this case, composition is the determiner of the structure and function in that composition is the cause, rather than the effect.

Composition determines the structure, and structure determines the function. Thus, by negotiating the composition, we simultaneously negotiate both the structure and the function. On the other hand, once the composition is in place, the structure and function are set—unless, of course, the composition is manipulated for a particular purpose, at which time both the structure and function are altered accordingly.

The kinds of plants and their age classes within a plant community create a structure that is characteristic of the community at any given age. The structure of the plant community creates and maintains certain functions. In turn, the composition, structure, and function of a plant community determine what kinds of animals can live there, how many, in what type of dynamic relationship, and for how long.

If, for example, an owner of forestland changes the composition of the plants (clear-cutting the mixed forest and planting a single-species, monocultural tree farm), he or she changes the structure, hence the function, and thus affects the animals. The animals living in the forest are not just a reflection of its composition at any given point in time but are ultimately constrained by it.

People and Nature are continually altering the structure and function of ecosystems by manipulating the composition of its plants, an act that subsequently changes the composition of the animals dependent on the resultant habitat—including humans worldwide. By altering the composition of plants within an ecosystem, people and Nature alter its structure and, in turn, affect how it functions, which in turn determines not only its potential ecosystem services but also what uses humans can derive from those services.

By analogy, a legal system is constantly being altered by the influence of whichever political party or ruler is presently in control of the government and, thus, the courts. The composition (kinds of laws and statutes), structure (how the laws and statutes are configured into a legal system), and function of the legal system (how its administration affects both present and future generations) are therefore continually swayed by people's struggle for political power. Consequently, the outcome of virtually every legal decision is a reflection of today's power struggle based on the composition, structure, and function of the current legal system.

Law 11—All Systems Have Cumulative Effects, Lag Periods, and Thresholds

Nature, as previously stated, has intrinsic value only and so allows each component of an ecosystem to develop its prescribed structure, carry out its biophysical function, and interact with other components through their evolved, interdependent processes and self-reinforcing feedback loops. No component is more or less important than another—except in human valuation based on personal desire for a particular outcome. Though components may differ from one another in form, all are complementary in function.

Our intellectual challenge in decision-making (legal and otherwise) is to recognize that no given factor can be singled out as the sole cause of anything. All things operate synergistically as cumulative effects that exhibit a lag period before manifesting a clearly noticeable outcome of some type. Cumulative effects, which encompass many little, inherent novelties, cannot be understood statistically because ecological relationships are far more complex and far less predictable than our statistical models lead us to believe—a circumstance Francis Bacon may have been alluding to when he said, "The subtlety of Nature is greater many times over than the subtlety of the senses and understanding."[14] In essence, Bacon's observation recognizes that we live in the "invisible present" and thus are seldom able to recognize cumulative effects.[15]

The invisible present is our inability to stand at a given point in time and see the small, seemingly innocuous effects of our actions as they accumulate over weeks, months, and years. Obviously, we can all sense change—day becoming night, night turning into day, a hot summer changing into a cold winter, and so on. There are, however, some people who live for a long time in one place who can see generalized, longer-term events and remember a winter of exceptionally deep snow or a summer of deadly heat.

Despite such a gift, it is a rare individual who can sense, with any degree of precision, the changes that occur over the decades of their lives. At this scale of time, we tend to think of the world as being in some sort of ongoing, relatively steady state, with the exception of technology, wars, or periodic natural disasters. Moreover, we typically underestimate the degree to which slow, seemingly innocuous change has occurred—such as global warming. We are unable to directly sense slow changes, and we are even more limited in our abilities to interpret the relationships of cause and effect in these changes.

Nevertheless, these subtle processes, acting quietly and unobtrusively over decades, reside cloaked in the invisible present, such as gradual declines or improvements in habitat quality. Thus, from a legal point of view, the invisible present can be fraught with the tyranny of many, little, seemingly innocuous and unrelated social-environmental decisions that invite disaster—especially when abetted by informed denial in the struggle for political power, despite the warnings of history and people with unimpaired foresight, such as Winston Churchill who, with clear foreboding, saw the onrushing threat of Nazi Germany. He said:

> When the situation was manageable it was neglected, and now that it is thoroughly out of hand we apply too late the remedies which then might have effected a cure. There is nothing new in the story. ... It falls into that long, dismal catalogue of the fruitlessness of experience and the confirmed unteachability of mankind.
>
> Want of foresight, unwillingness to act when action would be simple and effective, lack of clear thinking, confusion of counsel until the emergency comes, until self-preservation strikes its jarring gong—these are the features which constitute the endless repetition of history.[16]

And, today, for example, virtually all the world's roughly 3,000 professional climate scientists (and other professionals who work closely with them) foresee a more inhospitable world for the generations of young children already born and those unborn, including higher prices for food; growing numbers of plants and animals becoming extinct, along with their ecological functions and services; as well as other biophysical disruptions linked directly to the warming climate. Yet, despite all the evidence observed and irrefutably measured worldwide, such as melting glaciers and rising ocean levels, the fossil fuel industry (among others) and ideological groups are trying to delay regulation of greenhouse emissions by funding campaigns that create confusion and intimidation in an effort to discount the data and win agreement with their points of view through the legal system.[17]

At length, however, cumulative effects, gathering themselves below the level of our conscious awareness, seem to suddenly become visible. By then, it is too late to retract our decisions and actions, even if the outcome they cause is decidedly negative with respect to our intentions. So it is that cumulative effects from legal decisions favoring corporate exploitation of natural resources multiply unnoticed until something in the environment shifts dramatically enough—even for people without social-environmental foresight—to see the outcome through casual observation. A threshold of tolerance defines this shift in the system, beyond which the system, as we knew it, appears to suddenly, visibly, irreversibly become something else. Within our world, this same dynamic takes place in a vast array of scales in all natural *and* artificial systems, from the infinitesimal to the gigantic, from the personal, to the corporate, to the national, to the global—despite legal rulings to the contrary.

Law 12—All Systems Are Cyclical, but None Are Perfect Circles

While all processes in Nature are cyclical, no cycle is a perfect circle, despite such depictions in the scientific literature and textbooks. They are, instead, a coming together in time and space at a specific point, where one "end" of a cycle approximates—*but only approximates*—its "beginning" in a particular time and place. Between its beginning and its ending, a cycle can have any configuration of cosmic happenstance. Clearly, there will be change, but the degree and form of the change will be determined by the "original" conditions. Biophysical cycles can thus be likened to a coiled spring insofar as every coil approximates the curvature of its neighbor but always on a different spatial level (temporal level in Nature), thus never touching.

The size and relative flexibility of a metal spring determine how closely one coil approaches another—such as the small, flexible, loosely coiled spring in a ballpoint pen juxtaposed to the large, stiff, coiled spring on the front axle of an 18-wheel truck. In this sense, the cycles of annual plants in a backyard garden or a mountain meadow are relatively rapid and thus close together in time. Conversely, the millennial cycles of Great Basin bristlecone pines growing on rocky slopes in the mountains of Nevada, where they are largely protected from fire, or a Norway spruce growing on a rocky promontory in the Alps of Switzerland have cycles that extend over centuries to millennia.

Regardless of its size or flexibility, a spring's coils are forever reaching outward. In Nature's Laws of Reciprocity, they are cycles that are forever moving toward the next level of novelty in the creative process and so are perpetually embracing the uncertainty of future conditions—never to repeat the exact outcome of an event as it once happened. This phenomenon occurs even in times of relative climatic stability. Today's progressive global warming, however, will only intensify tomorrow's uncertainties—such as the increasingly quick loss of groundwater beneath the Central Valley of California, Argentine Patagonia, the Middle East and Russia, northeastern China, northern India, and the Canning Basin of Australia because it is being pumped out for agriculture faster than it can be replenished[18]; reduced production of global crops of grain[19]; rising sea levels[20]; and the collapsing coastlines in the Arctic.[21] Each of these outcomes of human behavior are in one way or another influenced by legal systems in the various countries worldwide, and each legal decision will affect all generations to come, because none of the consequences are reversible.

Law 13—Systemic Change Is Based on Self-Organized Criticality

When dealing with scale (a small, mountain lake as opposed to the drainage basin of a large river, such as the Mississippi in the United States or the Ganges in India), scientists have traditionally analyzed large, interactive systems in the same way that they have studied small, orderly systems, mainly because their methods of study had proven so successful to their limited sense of understanding. The prevailing wisdom is that the behavior of a large, complicated system, such as the effects of a country's legal system, can be predicted by studying its elements separately and by analyzing its micro-mechanisms individually. Such reductionist-mechanical thinking is predominant in Western society and tends to view the world and all it contains through a lens of intellectual isolation embedded in symptomatology. During the last few decades, however, it has become increasingly clear that many complicated systems, like forests, oceans, cities, and legal systems do not yield to such traditional analysis.

Instead, large, complicated, interactive systems evolve naturally to a critical state in which even a minor event starts a chain reaction that can affect any number of internal elements and can lead to a dramatic alteration in the system—such as listening to and accepting the plea of children to act in a way that helps curb global warming.[22] Although such systems produce more minor events than catastrophic ones, chain reactions of all sizes are an integral part of system dynamics. According to the theory called "self-organized criticality," the mechanism that leads to minor events (a pebble falling from a cliff) is the same mechanism that leads to major events (an earthquake).[23] With respect to earthquakes, for example, the weight of seasonal rains or snow on the soil and floodwaters, such as might collect in a lake in areas of geological faults, can be heavy enough to slightly depress a fault zone and thus cause initial movement along the fault.[24]

Not understanding this, analysts have typically and erroneously blamed some rare set of circumstances (some exception to the rule) or some powerful combination of mechanisms when catastrophe strikes—a common reason given for continually engaging the legal system in the interest of corporate, monetary gain.

Law 14—Dynamic Disequilibrium Rules All Systems

If change is a universal constant in which nothing is static, what, then, is a natural state? In answering this question, it becomes apparent that the "balance of Nature" in the classical sense (disturb Nature and Nature will return to its former state after the disturbance is removed) is an illusion. In fact, the so-called balance of Nature is a romanticized figment of the human imagination, a snapshot image of the world in which we live and hope to have control. In reality, Nature exists in a continual state of ever-shifting "dis-equilibrium."

Ecosystems are always in an irreversible process of change and novelty, thereby altering their composition, structure, function, and the resultant interactive feedback loops—irrespective of human desires and a false sense of control. Thus, despite how closely an ecosystem might approximate its former state following a disturbance, the existence of every ecosystem is a tenuous balancing act because every system is in a continual state of reorganization that occurs over various scales of time, from the cycle of an old forest to a geological phenomenon, such as Mauna Loa, the active volcanic mountain in Hawaii. Whereas people can manipulate a forest to some extent, Mauna Loa is in an eternal flux over which no human has any control.

Perhaps the most outstanding evidence that an ecosystem is subject to constant change and disruption, rather than remaining in a static balance, comes from studies of naturally occurring *external factors* that dislocate ecosystems, primarily human influence through such things as chemical pollution of the air, soil and water; overfishing the oceans; and mountaintop-removal mining—all of which help change the worldwide climate. In turn, each of these human influences is, individually and collectively, altering the quality of human life,[25] through their compounding, social-environmental degradation. Moreover, the long-term effects of each of these external factors have been exacerbated by court decisions that, in one way or another, have favored corporate exploitation worldwide.

Finally, every decision has its antithesis, as clearly stated in the Chinese proverb: To every man is given the key to the gates of heaven, and the same key opens the gates of hell.

Law 15—This Present Moment, the Here and Now, Is All We Ever Have

This eternal, present moment is all we ever have in which to act. The past is a memory, and the future never comes. Now is the eternal moment. These inviolable rules of Nature are enshrined in every moment of everyday life, whether we recognize them or not. As such, the person who honors them will move forward unafraid.

Perhaps the toughest decision a leader is confronted with is to bear, unflinchingly, all the abuses that a person or persons with vested interests hurl at them when their desired outcome of a decision is thwarted. In effect, a person who serves the people must pass the tests described in the eulogy that Senator William Pitt Fessenden of Maine delivered on the death of Senator Solomon Foot of Vermont in 1866:

> When, Mr. President, a man becomes a member of this body he cannot even dream of the ordeal to which he cannot fail to be exposed;
> of how much courage he must possess to resist the temptations which daily beset him;
> of that sensitive shrinking from undeserved censure which he must learn to control;
> of the ever-recurring contest between a natural desire for public approbation and a sense of public duty;
> of the load of injustice he must be content to bear, even from those who should be his friends;
> the imputations of his motives;
> the sneers and sarcasms of ignorance and malice;
> all the manifold injuries which partisan or private malignity, disappointed of its objects, may shower upon his unprotected head.
>
> All this, Mr. President, if he would retain his integrity, he must learn to bear unmoved, and walk steadily onward in the path of duty, sustained only by the reflection that time may do him justice, or if not, that after all his individual hopes and aspirations, and even his name among men, should be of little account to him when weighed in the balance against the welfare of a people of whose destiny he is a constituted guardian and defender.[26]

Such is the price of true social-environmental leadership—to be the keeper of everyone else's dignity by keeping one's own in the eternal, present moment—leading by example. (For an in-depth discussion of sustainable leadership, see *Decision Making for a Sustainable Environment: A Systemic Approach.*[27])

We, the human component of the world, must understand and accept that these Laws of Reciprocity are an interactive thread in the tapestry of the social-environmental world that must be accounted for—and honored—if society is to become a sustainable partner with its various environments. As such, these laws are an essential and unavoidable part of any human legal system based on the Rights of Nature, which includes such things as clean air, pure water, fertile soil, viable forests, and bountiful oceans as the birthright of every human being and life in general. Protecting the long-term biophysical viability of Nature is the foremost responsibility of every designated decision maker—regardless of the nomenclature by which they are known.

Endnotes

1. Russ Beaton and Chris Maser. *Economics and Ecology: United for a Sustainable World*. CRC Press, Boca Raton, FL. (2012) 191 pp.
2. Chris Maser. *Decision Making for a Sustainable Environment: A Systemic Approach*. CRC Press, Boca Raton, FL. (2013) 304 pp.
3. Johan Rockström, Will Steffen, Kevin Noone, and others. A safe operating space for humanity. *Nature*, 461 (2009):472–475.
4. Jan Zalasiewicz, Will Steffen, and Paul Crutzen. The New World of the Anthropocene. *Environmental Science and Technology*, 44 (2010):2228–2231.

5. (1) Dawn of the Anthropocene Epoch? Earth Has Entered New Age of Geological Time, Experts Say. http://www.sciencedaily.com/releases/2010/03/100326101117.htm (accessed February 4, 2011). And (2) Colin N. Waters, Jan Zalasiewicz, Colin Summerhayes, and others. The Anthropocene is functionally and stratigraphically distinct from the Holocene. *Science*, 351 (2016) http://www.sciencemag.org/content/351/6269/aad2622 (accessed January 8, 2016).
6. Russ Beaton and Chris Maser. *Economics and Ecology: United for a Sustainable World. op. cit.*
7. Donna Blankinship and Lisa Baumann. Northwest Storms Ease Drought Worries; Mudslide Risks Remain. ABC News, December 11, 2015. http://abcnews.go.com/US/wireStory/wild-northwest-weather-eases-drought-worries-35709582 (accessed December 11, 2015).
8. Russ Beaton and Chris Maser. *Economics and Ecology: United for a Sustainable World*. CRC Press, Boca Raton, FL. 2011.
9. (1) Yuan-Chug Cheng and Graham R. Fleming. Dynamics of light harvesting in photosynthesis. *Annual Review of Physical Chemistry* 60 (2009):241–262; (2) Paul May. Chlorophyll. http://www.chm.bris.ac.uk/3motm/chlorophyll/chlorophyll_h.htm (accessed January 5, 2009).
10. Chris Maser. *Earth in Our Care: Ecology, Economy, and Sustainability*. Rutgers University Press, Piscataway, New Jersey. (2009) 262 pp.
11. (1) Sid Perkins. As ozone hole heals, Antarctic could heat up. *Science News* (July 5, 2008):10; (2) S.-W. Son, L.M. Polvani, D.W. Waugh, and others. The impact of stratospheric ozone recovery on the Southern Hemisphere westerly jet. *Science*, 320 (2008):1486–1489; (3) J. Perlwitz, S. Pawson, R.L. Fogt, and others. Impact of stratospheric ozone hole recovery on Antarctic climate. *Geophysical Research Letters*, 35, L08714 (2008):1–5. doi:10.1029/2008GL033317; and (4) D.W. Waugh, L. Oman, S.R. Kawa, and others. Impacts of climate change on stratospheric ozone recovery. *Geophysical Research Letters*, 36, L03805 (2009):1–6. doi:10.1029/2008GL036223.
12. (1) Solomon M. Hsiang, Kyle C. Meng, and Mark A. Cane. Civil conflicts are associated with the global climate. *Nature*, 476 (2011):438–441 and (2) J.P. Reganold, D. Jackson-Smith, S.S. Batie, and others. Transforming U.S. agriculture. *Science*, 332 (2011):670–671.
13. Chris Maser. *Earth in Our Care: Ecology, Economy, and Sustainability. op. cit.*
14. Francis Bacon. http://Science.prodos.ORG (accessed January 2, 2009).
15. John J. Magnuson. Long-term ecological research and the invisible present. *BioScience*, 40 (1990):495–501.
16. Winston Churchill's speech to the British Parliament in 1935. *In*: T.A. Warren. Leaders Need Followers. *The Rotarian*, 1945 (October):10–12.
17. (1) Seth Borenstein. New Federal Map for What to Plant Reflects Warming. (accessed July 16, 2015) and (2) Bill Blakemore. (Nov 5, 2011) Shakespeare, Global Warming, Sunset, and You. http://abcnews.go.com/blogs/technology/2011/11/shakespeare-global-warming-sunset-and-you/(accessed July 16, 2015).
18. (1) V.M. Tiwari, J. Wahr, and S. Swenson. Dwindling groundwater resources in Northern India, from satellite gravity observations. *Geophysical Research Letters*. 36, L18401 (2009):1–5. doi:10.1029/2009GL039401 and (2) Devin Powell. Satellites show groundwater dropping globally. *Science News*, 181 (2011):5–6.
19. J.P. Reganold, D. Jackson-Smith, S.S. Batie, and others. Transforming U.S. agriculture. *Science*, 332 (2011):670–671.
20. Andrew Kemp, Benjamin P. Horton, Jeffrey P. Donnelly, and others. Climate related sea-level variations over the past two millennia. *Proceedings of the National Academy of Sciences*, 108 (2011):11017–11022.
21. B.M. Jones, C.D. Arp, M.T. Jorgenson, and others. Increase in the rate and uniformity of coastline erosion in Arctic Alaska. *Geophysical Research Letters*, 36, L03503, (2009):1–5. doi:10.1029/2008GL036205.
22. Western Environmental Law Center. In Advance of Paris Climate Talks, Washington Court Recognizes Constitutional and Public Trust Rights and Announces Agency's Legal Duty to Protect Atmosphere for Present and Future Generations (Press Release 11/20/15). *op. cit.*
23. Per Bak and Kan Chen. Self-organizing criticality. *Scientific American*, January (1991):46–53.

24. (1) Devin Powell and Alexandra Witze. Weather affects geologic activity. *Science News*, 180 (2011):8; (2) Daniel Brothers, Debi Kilb, Karen Luttrell, and others. Loading of the San Andreas Fault by Flood-Induced Rupture of Faults beneath the Salton Sea. *Nature Geoscience*, 4 (2011):486–492; (3) William L. Ellsworth. Injection-induced earthquakes. *Science*, 341 (2013):1225942-1–1225942-7. http://www.gwpc.org/sites/default/files/files/Earthquakes%20and%20fracking(2).pdf (accessed July 12, 2015); and (4) K.M. Keranen, M. Weingarten, G.A. Abers, and others. Sharp increase in central Oklahoma seismicity since 2008 induced by massive wastewater injection biophysical. *Science*, 345 (2914):448–451.
25. Chris Maser. *Interactions of Land, Ocean and Humans: A Global Perspective*. CRC Press, Boca Raton, FL. (2014) 308 pp.
26. John F. Kennedy. *Profiles in Courage*. Harper & Row, New York, NY. 1961.
27. Chris Maser. *Decision Making for a Sustainable Environment: A Systemic Approach. op. cit.*

2

Rights of Nature—The Emerging Legal Paradigm

Although the emerging Rights of Nature legal paradigm does not specifically highlight the individual principles per se, it is based on the Laws of Reciprocity. Now this paradigm must be nourished, strengthened, expanded, and lived to shift the short-term economic priorities of the wealthy minority to encompass biophysical necessities and rights of all peoples and generations for a sustainable and dignified life.

Ecuador Speaks: The Rights of Nature in the Constitution

It began with Ecuador. That South American country is the first nation in the world to place the "Rights of Nature" in its Constitution (Figure 2.1). In the process of revising its Constitution, Ecuador included Rights of Nature along with other sweeping rights. The genesis of these provisions was a potent combination of Ecuador's experience with environmental abuses from foreign corporations, traditional Ecuadorian indigenous philosophy, and a dialogue between Ecuadorian activists from the international Pachamama Alliance, Ecuadorian delegates from its Constitutional Assembly, and the American activist group Community Environmental Legal Defense Fund, founded in 1995.[1]

An additional spur for placing the Rights of Nature in its Constitution was Ecuador's extremely high level of biodiversity. It is considered one of the most biologically diverse countries in the world in both flora and fauna. However, roads into the Ecuadorian Amazon to access oil and minerals greatly threaten this biodiversity.[2] The citizenry ratified the new Constitution in 2008.[3]

Apart from the specific provisions for the Rights of Nature, there are many other revolutionary provisions in the Ecuadorian Constitution that lead to a changed worldview in which the Rights of Nature are a necessary and essential aspect of living fully in community. For example, water is "for use by the public, and it is unalienable, not subject to a statute of limitations, immune from seizure and essential for life."[4] Furthermore, the Constitution says that Ecuador "shall promote food sovereignty," because of the right to healthy, nutritious, and locally produced food.[5]

An entire section of the new Constitution is devoted to a healthy environment and begins with the inalienable right to live in an ecologically sustainable environment. This section leads into the State's obligation to promote environmentally clean technologies. However, the most radical and farsighted of all is the new, constitutional prohibition on development, ownership, marketing, and use of biological and nuclear weapons, organic pollutants, agrochemicals, and genetically modified organisms that are harmful to human health or that jeopardize ecosystem integrity.[6]

Chapter Four of the Constitution provides lists of protections for the rights of communities, peoples, and nations in Ecuador, focusing especially on indigenous communities

FIGURE 2.1
The plaza of San Francisco in old town Quito, Ecuador, the capital of the country. Photographer unknown (http://www.publicdomainpictures.net/view-image.php?image=85231&picture=quito-ecuador).

(Figure 2.2). This is critical, as there have been many, highly publicized, abuses of Ecuadorian resources and sovereignty by foreign—often U.S.-owned—corporate interests.[7]

The new protections stated in the Constitution face these abuses head on, requiring informed consultation with these communities before the extraction of a nonrenewable resource that could affect them culturally or environmentally. The intent is to protect the communities' rights to develop their ancestral traditions and societies, maintain ownership of their community lands, and keep the fruits of renewable resources located thereon. They also have the right to protect their collective knowledge (such as traditional medicines), ancestral wisdom, and holy places—as unequivocally state in the Constitution: "All forms of appropriation of their knowledge, innovations and practices are forbidden."[8]

Communities also have the right to remain on their ancestral lands. This right is essential because the erosion and ultimate extinction of indigenous forms of knowledge and languages through which they understand their world have been—and are—coupled with worldwide ecosystem degradation.[9] No one people can comprehend the subtleties and ecological rhythms of all parts of the Earth. For this reason alone, local knowledge and languages are not only an important but also a critical aspect of ecosystem functioning and human understanding. Every language develops its own symbolic understanding for the surrounding world, and each language provides crucial feedback about the world and about a people's cultural relationships to it, as well as being shaped by it:

> The extinction of languages is part of the larger picture of worldwide near total ecosystem collapse. Our research shows quite striking correlations between areas of biodiversity and areas of highest linguistic diversity, allowing us to talk about a common repository of what we will call "biolinguistic diversity": the rich spectrum of life encompassing all the Earth's species of plants and animals along with human cultures and their languages. The greatest biolinguistic diversity is found in areas inhabited

FIGURE 2.2
Ecuadorian market women at Zumbahua native market. Photographer unknown (http://www.publicdomainpictures.net/view-image.php?image=28458&picture=pretty-in-pink).

> by indigenous peoples, who represent around 4 percent of the world's population, but speak at least 60 percent of the world's languages. Despite increasing attention given to endangered species and the environment, there has been little awareness that peoples can also be endangered.[10]

Indigenous societies are often looked upon as guardians of Nature and/or of their specific ancestral lands, where their knowledge, language, rituals, and spiritual understanding run deep. From the protective provisions in the new Constitution toward such indigenous communities—who have suffered greatly in past eras from colonial exploitation—the new Rights of Nature provisions flow naturally. This is Title II, Chapter Seven of the new Constitution, which turns away from an anthropocentric vision of the world for the first time in modern legal systems.

The provisions begin with a blunt declaration, "Nature is subject to those rights given by this Constitution and Law." Article 71 then lays out the basic framework: "Nature, or Pacha Mama, where life is reproduced and occurs, has the right to integral respect for its existence and for the maintenance and regeneration of its life cycles, structure, functions and evolutionary processes."[11] Nature also has the right to be restored, and its integral restoration is held to be independent of "natural persons or legal entities."[12] However, in cases of severe impact resulting from resource extraction, the State is to establish the most efficient (= effective) means of restoration.

The State is also required by the new Constitution to prevent or restrict activities leading to species extinction, ecosystem destruction, and permanent alteration of natural cycles.[13] In a decisive turn away from indiscriminate resource extraction and ownership by multinational corporations, "environmental services shall not be subject to appropriation," but must, instead, be regulated by the State.[14] However, as history makes manifestly clear, the difficulty living up to these provisions will lie in their enforcement, which means that legal documents are only as good as the collective will to abide by Nature's Laws of Reciprocity—the *heart of the law* rather than the *letter of the law.*

Starting in 2009, China has lent Ecuador billions of dollars in exchange for shipments of oil. So it is that Ecuador, like many of the world's poorer countries, has become dependent on natural-resource extraction to obtain foreign money to support many basic services—a dependency that historically weakened the political will of many a national government. Thus, heavily dependent on Chinese money to finance infrastructure and programs against poverty, the Ecuadorian government has been willing to compromise the nation's environmental integrity.

For example, as of 2013, Ecuador began planning to lease vast tracts of pristine Amazonian forest, up to a third of the total, to China for oil exploration in an effort to help pay its gigantic foreign debt of more than $7 billion. That lease would amount to approximately 7,413,161 acres of Ecuador's more than 19,768,431 acres of pristine Amazon rainforest. Moreover, seven indigenous groups who inhabit the area were not consulted as required, and would not have given consent if they had been.[15]

Implementation of the Rights of Nature provisions is still in its infancy for these budgetary reasons, and simply because they are so new and unprecedented. They cut directly across the current power structure. Ecuador's government has not yet appointed an environmental ombudsperson, as directed by the new Constitution, nor has it approved laws creating a framework for management of the country under the Rights of Nature provisions. Furthermore, in 2012, Rafael Correa, the President of Ecuador, proposed large-scale, open-pit mining projects in both the Andean highlands and the Amazon rainforest.[16]

Continued mining, oil drilling, and oil exploration in the Amazon led several indigenous groups in 2011, under the umbrella of the Confederation of Indigenous Nationalities of The Ecuadorian Amazon, to sue both President Correa and several high-ranking ministers for "cultural and physical disappearance, which amounts to the crime of ethnocide or genocide."[17] It was the first time the indigenous group had filed such a lawsuit.

But even if the required laws and enforcement are so far lacking, the new Constitution, with its vitally important Rights of Nature provisions, shows a fundamental shift in consciousness, as well as a new template into which the Ecuadorian people *can* shift their economy and their future in the years ahead.

The First Rights of Nature Legal Case in the World

Floods began to inundate the farm—adjacent to the Vilcabamba River—of Americans Richard Wheeler and Eleanor Huddle, who had moved to Ecuador. Through their attorney, Carlos Eduardo Bravo Gonzalez, they asked the Provincial Justice Court of Loja for a Constitutional Injunction against the project. In 2011, the Provincial Court ruled in favor of the Vilcabamba River, using the precautionary principle as a basis for its ruling. The judges were Provincial Judge Dr. Luis Sempértegui Valdivieso, Interim Provincial Judge Dr. Galo Arrobo Rodas, and Associate Judge Dr. Galo Celi Astudillo.[18]

The Court stated, "(2) That, based on the precautionary principle, until it is objectively demonstrated that the probability of certain danger that a project undertaken in an established area does not produce contamination or lead to environmental damage, it is the responsibility of the constitutional judges to incline toward the immediate protection and the legal tutelage of the rights of nature, doing what is necessary to prevent contamination of call for remedy ... (3) The recognition of the importance of nature, raising the issue that damages to nature are generational damages, defined as such for their magnitude that impact not only the present generation but also future ones... "[19]

The Court required the Provincial Government to provide a remediation plan and to undertake immediate corrective actions, such as storing construction materials elsewhere than in the river. The ruling also set up a delegation to watchdog the cleanup and ensure proper follow-up took place. In addition, "The defendant must publicly apologize on one-fourth page in a local newspaper for beginning construction of a road without the necessary environmental license."[20]

As revolutionary as it is, enforcement has remained a problem in implementing this first Rights of Nature ruling. The provincial government of Loja was not able to complete the impact studies by 2012 or the remediation and rehabilitation plans to restore the river. The government had only erected a few signs and completed a partial and superficial restoration of the riverside with removal of a small amount of waste. These cursory and token efforts have not come close to restoring the river's vital cycles or functions as the Constitution requires.[21]

By 2013, it appeared the Ecuadorian legal process to enforce the restoration of the Vilcabamba River had completely stalled. Perhaps this was attributed to the Constitutional language, which in Ecuador often provides guidance rather than mandates, or perhaps the judges still had no precedent for integrating the Rights of Nature into decision making. Rapid change in the way things are done is difficult, despite the urgent need for action. Paradigms in human culture do not shift overnight, even once the law is changed, because entrenched behaviors retain tremendous power for a time, both in legal systems and outside.[22]

Bolivia Speaks: The Rights of Nature in the Law

At the same time as Ecuador was working on its new national Constitution, the nearby country of Bolivia was debating its own "Mother Earth Law."[23] Their new Framework Law on Mother Earth and Integral Development for Living Well was passed and finalized in October 2012 (Figure 2.3). The six principles guiding the statute, as given in Chapter 1, are:

> **Article 1. (SCOPE).** This Act is intended to recognize the rights of Mother Earth, and the obligations and duties of the Multinational State and society to ensure respect for these rights.
>
> **Article 2. (PRINCIPLES).** The binding principles that govern this law are:
>
> 1. **Harmony.** Human activities, within the framework of plurality and diversity, should achieve a dynamic balance with the cycles and processes inherent in Mother Earth.
> 2. **Collective good.** The interests of society, within the framework of the rights of Mother Earth, prevail in all human activities and any acquired right.
> 3. **Guarantee of the regeneration of Mother Earth.** The State, at its various levels, and society, in harmony with the common interest, must ensure the necessary conditions in order that the diverse living systems of Mother Earth may

FIGURE 2.3
The Gateway of the Sun in Bolivia from the ancient Tiwanaku civilization. Photograph by Mhwater (https://commons.wikimedia.org/wiki/File:Zonnepoort_tiwanaku.jpg).

absorb damage, adapt to shocks, and regenerate without significantly altering their structural and functional characteristics, recognizing that living systems are limited in their ability to regenerate, and that humans are limited in their ability to undo their actions.

4. **Respect and defend the rights of Mother Earth**. The State and any individual or collective person must respect, protect and guarantee the rights of Mother Earth for the well-being of current and future generations.
5. **No commercialism**. Neither living systems nor processes that sustain them may be commercialized, nor serve anyone's private property.
6. **Multiculturalism**. The exercise of the rights of Mother Earth requires the recognition, recovery, respect, protection, and dialogue of the diversity of feelings, values, knowledge, skills, practices, skills, transcendence, transformation, science, technology and standards, of all the cultures of the world who seek to live in harmony with Nature.[24]

Bolivia's law defines Mother Earth as "a dynamic living system comprising an indivisible community of all living systems and living organisms, interrelated, interdependent and complementary, which share a common destiny."[25]

And what of her rights? Chapter III, Article 7 enumerates them:

1. **To life**: The right to maintain the integrity of living systems and natural processes that sustain them, and capacities and conditions for regeneration.
2. **To the diversity of life**: The right to preservation of differentiation and variety of beings that make up Mother Earth, without being genetically altered or structurally modified in an artificial way, so that their existence, functioning or future potential would be threatened.

3. **To water**: The right to preserve the functionality of the water cycle, its existence in the quantity and quality needed to sustain living systems, and its protection from pollution for the reproduction of the life of Mother Earth and all its components.
4. **To clean air**: The right to preserve the quality and composition of air for sustaining living systems and its protection from pollution, for the reproduction of the life of Mother Earth and all its components.
5. **To equilibrium**: The right to maintenance or restoration of the interrelationship, interdependence, complementarity and functionality of the components of Mother Earth in a balanced way for the continuation of their cycles and reproduction of their vital processes.
6. **To restoration**: The right to timely and effective restoration of living systems affected by human activities directly or indirectly.
7. **To pollution-free living**: The right to the preservation of any of Mother Earth's components from contamination, as well as toxic and radioactive waste generated by human activities.[26]

The other side of the law is the obligation of the state and the people. The state is required to prevent human activities that cause extinctions and "the alteration of the cycles and processes that ensure life," including safeguarding the cultural systems that are part of Mother Earth. The state is also newly called upon to develop policies to protect Mother Earth from multinational exploitation by demanding "international recognition of environmental debt through the financing and transfer of clean technologies that are effective and compatible with the rights of Mother Earth…."[27] In addition, the people have the obligation to uphold and respect the rights of Mother Earth; promote earth-based harmony, including the sustainable use of natural resources; and report violations of the rights of Mother Earth.[28] This revolutionary statute concludes by establishing an "Office of Mother Earth" to ensure compliance with the rights of Mother Earth; its details are to be worked out in a subsequent statute.[29]

This astounding statute is the result of strong and deeply rooted indigenous and *campesino* (small-scale farmer) movements in Bolivia. The law takes its underlying vision from indigenous concepts that understand Nature ("Pachamama") as a being on whom we are completely dependent.

The indigenous and *campesino* movement has more than 3.5 million members, and its leaders were instrumental in drafting the law. It was based, in part, on their feelings of exclusion from the government of President Evo Morales Ayma, elected in 2005, though he is himself indigenous.[30]

However, the Pacto de Unidad, a coalition of Bolivia's largest social movements, a critical force behind the law, recognizes that law alone will not change environmental practices. As with Ecuador, a major obstacle to actually implementing "The Law of Mother Earth" is the fact that Bolivia is structurally dependent on extractive industries, and has been since the Spanish discovered silver in the 16th century, which tied Bolivia's history to ruthless exploitation of its people and its environmental resources. In fact, 70 percent of Bolivia's exports in 2010 were still in the form of minerals, gas, and oil. Moreover, there is opposition from such powerful sectors as mining, agro-industry, and energy producers to any environmental law that threaten their profits.[31]

The goal for Bolivia must therefore be to slowly develop an economy wherein the extractive industries have less and less presence and power. Since The Framework Law on Mother Earth and Integral Development for Living Well was passed in 2012, implementation has been slow. The Plurinational Authority for Mother Earth, required by the law, was set up in 2014, and presented the first national workshop on policies for dealing with climate change, which targeted government officials, academic, and private organizations.

The law is not written with quantifiable targets or goals, making it more difficult to implement or assess in the face of an economy based in part on exploitation of natural resources. New legislation is required to implement and enforce the Framework Law, but Bolivian officials report that environmental awareness is increasing since the law's passage.

In the meantime, other desperately necessary environmental initiatives are beginning. Bolivia's needs are many: 2 million of the country's 10.5 million population lack clean drinking water, and nearly 4 million have no access to sanitation. Bolivia has initiated concrete efforts outside the nascent Framework Law to deal with changing water-supply problems in the face of climate change, which causes increased glacial melting and winter flooding. Bolivian activists are also hoping to succeed in efforts to clean up the Pilcomayo River, contaminated by mining near its headwaters, and reforestation of mining regions in the Potosí area, home to centuries of silver mining.[32]

First Steps in the United States: A Beginning for the Rights of Nature

The United States has done little toward implementing a nationwide Rights of Nature legal system. The country's environmental protection statutes are based on meeting human necessities and merely trying to curb environmental harm that results from development projects. The environmental-protection framework is almost entirely regulation-based, requiring permits and governmental oversight. Consequently, it is a permissive system, in which human needs and desires are met, and environmental protections are secondary. Further, the American legal framework is multi-layered, with the broadest level of regulation promulgated in federal laws. Regulatory powers narrow and focus more locally at state and city, county and township levels. Thus, placing the Rights of Nature in the legal framework of towns or counties will not, by itself, be enough to guarantee a nationwide Rights of Nature structure. Nevertheless, a number of communities have taken the first step and placed the Rights of Nature in their legal codes.

Community Environmental Legal Defense Fund and the Rights of Nature

The Community Environmental Legal Defense Fund, which aided in drafting the Ecuadorian constitutional language, has led the charge for the Rights of Nature in the United States, beginning among Midwestern communities faced with the devastating environmental consequences of coal mining and hydraulic fracturing. As the Legal Defense Fund explains, "Existing environmental legal frameworks anchor the concept of nature as a commodity. …Today, communities in the United States … are setting course to redefine ecosystems as vibrant, rights-bearing entities. Through Rights of Nature laws and policies, they are recognizing that ecosystems have an independent and inalienable right to exist and flourish. Those rights can be enforced by people, government, and communities on behalf of nature."[33]

Thus far, the few scattered Rights of Nature ordinances in the United States have all been local, passed by city or township governing bodies, usually with guidance and expertise from the Community Environmental Legal Defense Fund. As is already clear from the Ecuadorian and Bolivian experiences, the only effective way of implementing the Rights of Nature is by placing them in the Constitution—that is, the governing blueprint under which all laws and ordinances are formulated. Otherwise, it will not be the central and overarching principle that shapes all other laws relating to human activity that affects the environment. But local ordinances provide a crucial beginning.

The first community to pass a Rights of Nature ordinance in the United States was Tamaqua Borough, deep in Pennsylvania's coal-mining country. The new ordinance, passed in 2006, allowed the residents of Tamaqua to bring suits to defend the Rights of Nature. Damages gained through legal action had to be used to restore the ecosystems and natural communities. More than a hundred communities in Pennsylvania's coal region have passed similar ordinances.[34]

A Watershed Tries to Protect Itself in Court

Inevitably, these local Rights of Nature ordinances are being tested in the courts. The first groundbreaking action took place in Grant Township, Indiana County, Pennsylvania. In June 2014, the elected officials of Grant Township passed a Community Bill of Rights Ordinance, joining scores of municipalities in Pennsylvania that have passed similar ordinances. The impetus for the Grant Township ordinance was protection of Little Mahoning Creek, which flows through the township. Because all 200 residents of the area depend on private wells for drinking water, concern in the community for protection of the groundwater and the watershed was high. Residents were troubled about the effects of underground injection of wastewater in the process of hydraulic fracturing.[35]

The ordinance set forth the rights of communities and Nature: "All residents of Grant Township, along with natural communities and ecosystems within the Township, possess the right to clean air, water, and soil, which shall include the right to be free from activities which may pose potential risks to clean air, water and soil within the Township, including the depositing of waste from oil and gas extraction."[36] Furthermore, and very importantly, "Natural communities and ecosystems within Grant Township, including but not limited to rivers, streams and aquifers, possess the right to exist, flourish and naturally evolve."[37] The ordinance declares all these rights, and the associated community rights, to be "inherent, fundamental and unalienable."[38]

The ordinance required that any action brought to defend the rights of ecosystems had to file suit in the name of the ecosystem itself. The language says, "Any action brought by either a resident of Grant Township or by the Township to enforce or defend the rights of ecosystems or natural communities secured by this Ordinance shall bring that action in the name of the ecosystem or natural community in a court possessing jurisdiction over activities occurring within the Township."[39] The Ordinance goes on to describe how damages to the ecosystem shall be measured: "by the cost of restoring the ecosystem or natural community to its state before the injury."[40]

In other words, the new ordinance gave the Little Mahoning Watershed a kind of personhood, able to defend its rights. Two months after the Bill of Rights was passed, Pennsylvania General Energy, an oil and gas company, filed a lawsuit in federal court against the ordinance. Pennsylvania General Energy argued communities have no rights to prohibit waste injection. The watershed, through the township's attorney, intervened in the lawsuit to protect its rights to flourish. It was the first time in United States history that a part of Nature—a watershed, in this instance—had participated in the legal system in its own name to protect its right to flourish.[41]

Nevertheless, the Environmental Protection Agency, in 2014, granted a permit to Pennsylvania General Energy to inject hydraulic fracturing wastewater in Little Mahoning Watershed. The Environmental Protection Agency did so despite its own declaration that it would not issue a permit if contaminated fluid would move into an underground source of drinking water. This action is evidence that injection wells allowed by the Environmental Protection Agency have contaminated groundwater. The Pennsylvania Department of Environmental Protection followed with a permit of its own. Residents have appealed these separately.

In October 2015, the U.S. District Court for the Western District of Pennsylvania invalidated parts of the Community and Nature's Rights ordinance, on grounds that municipalities must allow all legitimate uses, of which development of oil and gas wells is one. By seeking to regulate oil and gas wells, the Township exceeded its authority. In addition, the judge ruled there was no "clear and convincing evidence" that an ecosystem had standing under the law, since its interests were adequately represented by Grant Township. Essentially, the judge ducked the legal question of whether any part of Nature has a right to sue in its own name, and be a party entitled to damages.[42]

Inserting the Rights of Nature into the legal system at the local level is insufficient, given the hierarchical nature of the American legal system. Nevertheless, the Little Mahoning Watershed case has been, and continues to be, groundbreaking—the first Rights of Nature legal action in the United States, where the intervener was a watershed.

Highland Township and Other Communities Embrace the Rights of Nature

Ohio's Highland Township passed a Rights of Nature ordinance in January 2013. The new ordinance not only banned the injection of fracking wastewater but also created the rights of human and natural communities to a healthy environment, including ecosystems' right to exist and to flourish. Seneca Resources Corporation sued the Township in 2015 to overturn the Bill of Rights, claiming that, "the Bill of Rights violates the constitutional right of the corporation to inject frack wastewater in the Township." The Crystal Spring ecosystem, along with the Highland Township Municipal Authority, filed a motion to intervene in the case. The defending attorneys came from the Community Environmental Legal Defense Fund. The Crystal Spring ecosystem thus became the second in the nation seeking to defend itself in court, after the Little Mahoning.[43]

Hostile rulings are likely to be repeated against other community-based Rights of Nature ordinances when enacted at the local level alone. However, these local ordinances, even when ignored by the courts or overturned by adverse rulings, provide the first effort to build a Rights of Nature legal framework in the United States. The Community Environmental Legal Defense Fund continues to spearhead the movement, aiding communities and states in setting up Community Rights Networks. With the Fund's expertise, communities under environmental threat are increasingly passing Rights of Nature ordinances. More than 150 communities nationwide had passed such ordinances as of 2015.[44]

Several larger cities have also voted for Rights of Nature ordinances in an effort to protect their local environments, including Barrington, New Hampshire, and Athens, Ohio.[45] Moreover, in 2010, the Community Environmental Legal Defense Fund aided the city of Pittsburgh, Pennsylvania, the first major city in the United States to pass a community rights and Rights of Nature ordinance. Pittsburgh was seeking a means to prevent fracking in their community[46] (Figure 2.4).

FIGURE 2.4
Pittsburgh skyline 1908. Photographer unknown (https://upload.wikimedia.org/wikipedia/commons/4/40/PSM_V72_D446_Pittsburgh_skyline_1908.png).

Santa Monica, California, approved a Rights of Nature ordinance in 2013, the first West Coast city to do so. Santa Monica had already approved a Sustainable City Plan and sought greater authority to implement it and work toward a sustainable future via its new Sustainability Rights Ordinance. As Linda Sheehan, a principal advocate for the Sustainability Rights Ordinance, stated, "Current environmental statutes do not acknowledge the inherent rights of the natural world to be healthy, thrive and evolve. Accordingly, they only move slowly, rather than reverse, the trend of degradation. A rights-based approach is essential to begin steadily and consistently improving the health of the natural world and we who depend on it."[47]

Recognizing that changing a state's Constitution is the only way to secure a Rights of Nature base for the legal system, the Community Environmental Legal Defense Fund and Community Rights organizers in at least three states are working on drafting Rights of Nature language for their Constitutions: Colorado, New Hampshire, and Oregon. Discussion also continues among organizers in Ohio, Pennsylvania, Washington, and Illinois about the best strategies to pursue a Rights of Nature Constitutional amendment.[48]

The Rights of Nature Vision Takes Shape

All these efforts, both international and national, toward Rights of Nature governance—which usually includes human rights placed in a Rights of Nature framework—are the cusp of a deep paradigm shift that is desperately needed if the Earth's ecosystems are to flourish and support all life as they have done under Nature's Laws of Reciprocity. This new paradigm involves creation of more holistic economic frameworks, governance of communities and nations based on protection of the commons and Rights of Nature, and increasing efforts worldwide to rein in the corporate despoliation of the environment.

However, these steps will not suffice by themselves, even if the Rights of Nature is placed in state or federal Constitutions. In order to truly allow Nature to flourish, Nature's

Rights must be placed in legal systems, as the *first* among rights. It is insufficient to have Nature's Rights merely constitute one set of rights among others in our rights-based system: human rights, community rights, rights to equal protection of the laws, and so forth. All the other rights are important, but they must operate *within* the Rights of Nature framework. Otherwise, they become subject to balancing acts among the numerous rights, as courts and society seek to decide which rights should take precedence in situations of conflict. Without a resilient and healthy Nature, all human rights will be under duress—or worse. Society must make the leap of raising to primacy the natural system upon which all life, including human life, inseparably depends.

Thus, as Rights of Nature language is drafted and voted into the legal codes of towns, cities, counties, and state and federal Constitutions, it must be clearly stated that Nature's Rights have legal primacy over any other right. It is critical that Nature have full ability to flourish, be restored, and maintain its overall resiliency. Then, and only then, will human society be on track to creating a truly sustainable culture in which all other rights can likewise flourish. Difficult though some of the transitions may appear to be at the outset, this revamped priority system will also recreate the economy on a sustainable and holistic basis, so that humans do not use more of the Earth's resources than Nature can afford to give up and still remain fully resilient.

How "The Universal Rights of Mother Earth" Came into Being

Globally concerned citizens noted that governments worldwide, often allied to corporations masquerading as "persons" with rights, were failing massively to protect Earth's ecosystems. The United Nations delegates failed in a Bonn, Germany, meeting in June 2009 to create a means of enforcing the gigantic gap in emission reductions necessary to meet scientific targets deemed vital to averting the serious consequences of climate change.[49] Instead of focusing on ways to reduce emissions, the talks centered around discussions of new market mechanisms to transfer pollution opportunities, without considering the real-time climate effects of failing to reduce pollution and greenhouse gases.

Both at Bonn and subsequently, delegates from Bolivia led the discussion of the problems and the need for a complete change of framework in which to deal with the issues, calling for a consideration of Nature's Rights.[50] Bolivian President Evo Morales Ayma addressed the United Nations General Assembly in April 2009 expressing the hope that the 21st century would be known for addressing the Rights of Mother Earth (Figure 2.5). The Bolivian Alliance for the Peoples of Our America followed up that same year supporting the call for a Universal Declaration of the Rights of Mother Earth.[51]

The United Nations began to pay attention as the Rights of Nature movement burgeoned in Ecuador, Bolivia, and beyond. Under the leadership of Bolivia's President Morales Ayma, the General Assembly in 2009 designated April 22nd of each year—already observed as Earth Day in many countries—as International Mother Earth Day. The Resolution stated, "Acknowledging that the Earth and its ecosystems are our home, and convinced that in order to achieve a just balance among the economic, social and environmental needs of present and future generations, it is necessary to promote harmony with nature and the Earth." The Resolution invited "…all Member States…civil society, non-governmental organizations and relevant stakeholders to observe and raise awareness of International Mother Earth Day, as appropriate."[52]

FIGURE 2.5
Evo Morales Ayma, President of Bolivia, holds up a manual of the World People's Conference on Climate Change and the Rights of Mother Earth during a press conference at UN Headquarters on May 7, 2010. Photograph by Eskinder Debebe, United Nations. (Photograph # 435836 in the United Nation's photo archive.)

The General Assembly also passed a related Resolution inviting its Member States and United Nation's organizations to make use of International Mother Earth Day to "exchange opinions and views" on experiences and principles "for a life in harmony with nature." The Resolution requested the Secretary-General to submit a report on the theme of "Harmony with Nature," collating the testimonies, views and comments received.[53]

The Secretary-General submitted the initial, highly symbolic document in 2010: "The report provides an overview of how the lifestyle of the twenty-first century, through its consumption and production patterns, has severely affected the Earth's carrying capacity and how human behaviour has been the result of a fundamental failure to recognize that human beings are an inseparable part of nature and that we cannot damage it without severely damaging ourselves."[54]

The report summarizes many traditional, cultural understandings of Harmony with Nature, and focuses on "sustainable development: a holistic paradigm for Harmony with Nature in the twenty-first century." It also contains examples of sustainability initiatives from around the world and recommends furthering such projects, as well as education that focuses on sustainability and the philosophy of holism. In 2010, the United Nations set up a website on Harmony with Nature, which contains a timeline, all the pertinent resolutions and reports, and announcements of upcoming activities.[55]

Giving Mother Earth a Voice

Concerned about the Bonn talks' failure to grapple with climate change, President Morales Ayma and environmentally concerned citizens responded by convening a conference in Cochabamba, Bolivia, on April 22, 2010. It was entitled, "World People's Conference on Climate Change and the Rights of Mother Earth."[56] Organizations from around the world called for the conference, including groups from nations that have suffered greatly from corporate exploitation of their natural resources: countries such as India, Bangladesh, Belarus, Bolivia, Chile, Ecuador, Nepal, Philippines, Togo, and Uruguay; as well as many groups from Canada, Europe, and the United States.

Building on the success of Bolivia's fledgling Law of Mother Earth, the conference created many working groups ranging over such vast problem areas as:

1. Destruction of global forests
2. Protection of the "Rights of Nature"
3. The growing ramifications of climate change
4. Social-environmental equality for all indigenous peoples
5. Democratic sharing of technology
6. Dangers of carbon markets to achieve their stated objective
7. Climate change as one of the most serious threats to agriculture and food sovereignty

These working groups promulgated declarations and statements concisely stating the problems in these areas, frequently calling upon the United Nations to address them.[57]

Most important of all, the conference had a working group on Mother Earth Rights, led by Bolivian organizations dedicated to social change. The group promulgated the "Universal Declaration of the Rights of Mother Earth," modeled on Bolivia's "Mother Earth Law." This Universal Declaration was accompanied by a "People's Agreement," which laid out, in detail, the imperative for changing not only the consciousness of all peoples but also the legal framework that currently allows such massive transnational despoliation by corporations seeking only profit under a capitalist system. The People's Agreement states:

> It is imperative that we forge a new system that restores harmony with nature and among human beings. And in order for there to be balance with nature, there must first be equity among human beings. We propose to the peoples of the world the recovery, revalorization, and strengthening of the knowledge, wisdom, and ancestral practices of Indigenous Peoples, which are affirmed in the thought and practices of "Living Well," recognizing Mother Earth as a living [sentient] being with which we have an indivisible, interdependent, complementary and spiritual relationship. To face climate change, we must recognize Mother Earth as the source of life and forge a new system based on the principles of:
>
> - harmony and balance among all and with all things;
> - complementarity, solidarity, and equality;
> - collective well-being and the satisfaction of the basic necessities of all;
> - people in harmony with nature;
> - recognition of human beings for what they are, not what they own;
> - elimination of all forms of colonialism, imperialism and interventionism;
> - peace among the peoples and with Mother Earth.[58]

Based on the Agreement, the Universal Declaration of the Rights of Mother Earth was promulgated. It is a revolutionary document, changing the entire legal framework upon which world economies are currently based. It begins boldly with the statement in Article I, Section I, "Mother Earth is a living being."[59]

The Declaration then defines Mother Earth as "a unique, indivisible, self-regulating community of interrelated beings that sustains, contains and reproduces all beings." The heart of all the Principles is relationship, and the Declaration recognizes that relationship by stating that "each being is defined by its relationships as a part of Mother Earth."[60]

Reaching far beyond the ideas of "rights" in Western legal systems, the Declaration states that Mother Earth's rights are inalienable and inherent, "without distinction of any kind, such as may be made between organic and inorganic beings."[61] In other words, the Declaration refuses one of the most basic dichotomies in Western thought—that between living and non-living entities; instead, it holds that all natural entities are living and part of the indivisible whole, whether a granite boulder, a wetland, a scarlet flower, the river in the valley, or a snow leopard in her cave high in the mountains of Asia.

What are the rights of Mother Earth, and all beings? Beginning with the most basic, the first is "the right to life and to exist."[62] This might seem elementary, but in an era in which many species—from the humblest (such as frogs) to the most majestic (such as tigers)—are imperiled, dwindling, and in danger of extinction, this right is of burning importance on its own. But the rights continue, the second being equally basic: the right to be respected. All beings have the right to continue their own lifecycles free of human disruption, which leads naturally into rights to clean water, air, to be free of contamination, genetic modification, and torture—and the right to restoration when rights are violated.[63]

Since humans have, by all accounts, now created a vast stockpile of problems, disrupting ecosystems worldwide and causing many species to teeter on the brink of extinction, the Declaration enumerates the current responsibilities of human beings—but no rights over and above those of other living beings—and many additional responsibilities.[64] The first one requires all humans to respect and live in harmony with Mother Earth. The Declaration places detailed burdens on individuals, States, and "all public and private institutions" to implement the Declaration, encourage and promote harmonious living with Mother Earth, and pass laws for protection and conservation of Mother Earth's rights. They must also restore the integrity of vital ecological cycles and processes of Mother Earth, where these have been disrupted, compromised, or weakened.[65]

Perhaps toughest of all, the Declaration requires those responsible for violations of inherent rights of Mother Earth to "restore the integrity, of the vital ecological cycles, processes and balances of Mother Earth"[66]—surely among the most difficult of tasks, as the corporate-capitalist economic system can easily attest. They must also empower people and institutions to defend Mother Earth's rights, including curbing destructive activities and promoting harmonious economic systems. The Declaration also highlights one of the most critical requirements: to guarantee peace and eliminate nuclear, chemical and biological weapons."[67]

The International Tribunal: Addressing Abuse of Mother Earth

The difficulty with the Universal Declaration of the Rights of Mother Earth is, of course, implementation. Such is the case not only because its requirements cut directly into economic systems worldwide, which are based on procuring "natural resources" for profit, but also because those systems externalize the environmental costs onto the public.[68] In addition, the Declaration is completely outside the current, nation-based legal systems of the world. Though they vary in details, no contemporary legal system (with the exception of Ecuador and Bolivia) places Nature's Rights before the rights of human beings or calibrates human activities on whether they harm the inherent rights of other beings, including the Earth itself.

Moreover, the Universal Declaration of the Rights of Mother Earth is also outside the framework of international law. Though it was presented to the United Nations in 2010, formally considered at the United Nations' April 2011 Dialogue on Harmony with Nature, and prominently highlighted at the United Nations' June 2012 Conference on Sustainable Development, the United Nations did not formally endorse it.

So, how to begin implementing the Declaration? Earth lawyers and advocates from around the world who attended the 2010 conference in Cochabamba, and helped draft the Declaration, also formed the Global Alliance for the Rights of Nature in 2010.

The Alliance, having no other way to implement the Universal Declaration of the Rights of Mother Earth, created the "International Rights of Nature Tribunal," though neither international law nor the vast majority of nation-states recognize any such thing as "Rights of Nature." The International Tribunal sat for the first time in 2014 in Quito, Ecuador, despite having no legal authority whatsoever. But its ethical authority, as it brings forth the first fruits of a desperately needed new paradigm, is tremendous.[69]

The Tribunal and its founding document, the Universal Declaration of the Rights of Mother Earth, are evidence of a truly new paradigm that—after a long gestation—arose out of the existing paradigm as a revolutionary shift in consciousness. Although the new paradigm incorporates much of the existing one (albeit from a pristinely imagined, and completely changed, viewpoint), it nevertheless allows avant-garde thinking to emerge that heretofore had no outlet—tenets that could not even have been formed within the old paradigm.[70]

Already new concepts and ideas are flowing from it; the paradigm has catalyzed the new field of "earth jurisprudence" and provided a forum for judgments in an extensive and far-reaching field of what should become "earth criminal law": crimes against the Rights of Nature. Earth jurisprudence takes, as its starting point, the idea that humans are an inseparable part of the interconnected whole of the Earth. Even in the West, these are not new ideas—just not mainstream ideas that shape economies or legal systems. Many philosophers, jurists, sociologists, and artists have explored ideals of eco-centrism, the idea that humans and their needs are part of the Earth, rather than above it. To wit:

> What sets us apart from our fellow creatures is not some higher sense of spirituality or some nobler sense of purpose but our deeming ourselves wise in our own eyes. Therein lies the fallacy. We are no better than or worse than other kinds of animals; we are simply a different kind of animal—one among the many. We are thus an inseparable part of Nature, not a special case apart from Her.
>
> As a part of Nature, what we do is natural even though it often is destructive. This is not to say our actions are wise, or ethical, or moral, or desirable, or even socially acceptable and within the bounds of Nature's laws. As a part of Nature, we will of necessity change what we call the "natural world," and it is natural for us to do this, since people are an integral part of the total system we call the Universe. The degree to which we change the world, and the motives behind and the ways in which we make these changes, however, are what we may justifiably question. And it is the motive behind the creation—God's spirituality versus humanity's materialism—that is knocking at the door of our consciousness.[71]

The Tribunal gathered as an international panel of judges, lawyers, and ethics experts from around the world. Ramiro Avila, prosecutor for the Earth, declared, "We the people assume the authority to conduct and International Tribunal for the Rights of Nature. We will investigate cases of environmental destruction, which violate the Rights of Nature."[72]

At its first meeting, the Tribunal heard the case of the Great Barrier Reef of Australia, presented by the Australian Earth Laws Alliance. Advisory opinions were requested on the danger to life of the so-called "genetically modified organisms" and the problems of "Defenders of Nature" persecuted by the Ecuadorian government. The main case argued that Mother Earth's Rights are being violated because the very existence of the Great Barrier Reef is threatened (Figure 2.6). The threats are many, ranging from coal-port development to land developments causing ocean pollution, combined with the insidious effects of climate change. In 2012, the United Nations' Educational, Scientific, and Cultural Organization (UNESCO) issued a warning to Australia that, because the Reef was under threat, Australia might lose its World Heritage listing for the internationally renowned Reef.

At the end of the case, an attorney spoke on behalf of the Great Barrier Reef, as a member of the Earth community. She spoke from the point of view of the Reef, as best she understood it, arguing for its right to exist from its own point of view, again as best as she could conceive it. Only the new paradigm allowed this breakthrough; under the old paradigm one could not imagine an advocate speaking for the Earth's needs *from the point of view of the Earth's ecosystems*, as best as a human could understand them and speak for them.

This kind of advocacy would never be permitted in a court of law, and was revolutionary. It went far beyond a presentation of scientific data and legal analysis within a human framework. The advocate spoke for the Reef, as if the Reef was arguing not only for its

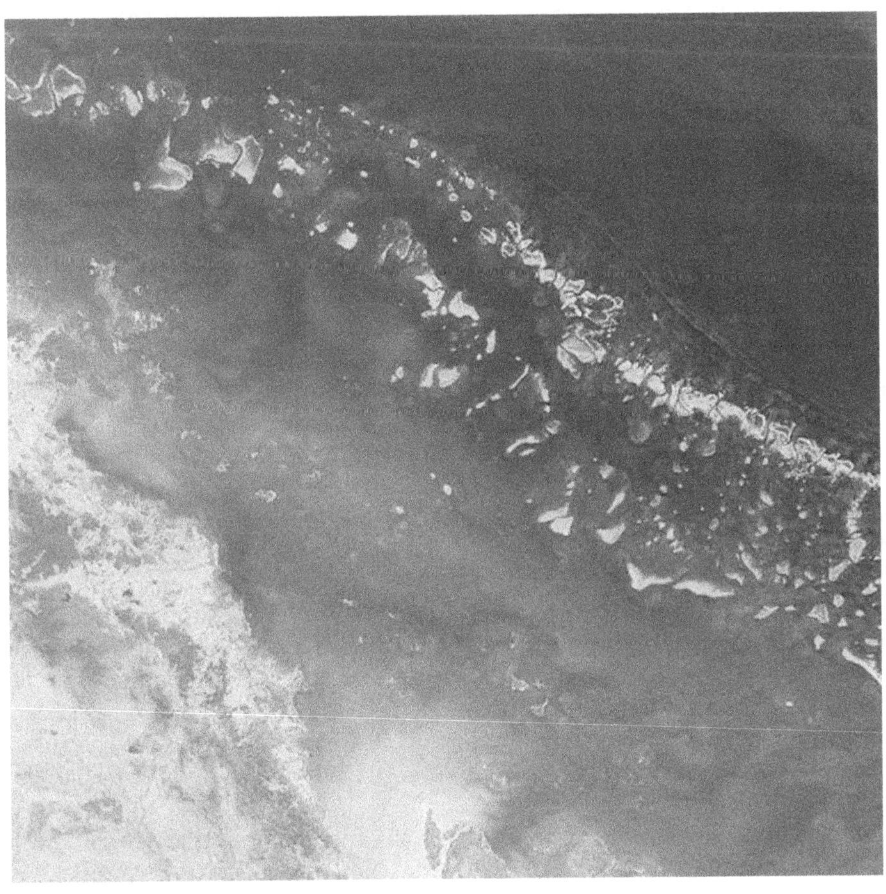

FIGURE 2.6
Satellite image of the Great Barrier Reef, Australia. Photograph by MISR, U.S. National Aeronautics and Space Administration (https://commons.wikimedia.org/wiki/Great_Barrier_Reef#/media/File:GreatBarrierReef-EO.JPG).

inherent right to exist but also for its *inherent right to be protected* from pollution or other human activity that would compromise its biophysical integrity.

After the International Tribunal's decision that the Great Barrier Reef's inherent rights were being violated, Australians convened that country's first "Rights of Nature Tribunal" in Brisbane in October 2014. This was a Regional Chamber of the International Tribunal; its purpose was to hear evidence from local witnesses concerning the Great Barrier Reef. As part of examining the Reef's plight, the Regional Chamber placed on trial the legal and economic systems that were legalizing the Reef's destruction. Its judges were leading Australian scientists and an indigenous leader.

The Regional Chamber's conclusions concerning the ongoing destruction of, and irreversible impacts to, the Reef were forwarded to the International Tribunal for its December 2014 meeting in Lima, Peru. The International Tribunal's official findings condemned Australia's government for violating the Great Barrier Reef's rights, demanded many solutions to reduce human interference and destruction of the Reef, and petitioned the government to comply with UNESCO's recommendations for the Reef's protection.

Even without any legal justification, the Regional Chamber's case in Brisbane had tremendous impact, catalyzing discussions and networking; creative, impassioned dialogues about ways to revamp legal systems toward an earth-oriented jurisprudence; critiquing existing legal assumptions; and giving birth to new Earth-focused campaigns.[73]

The Tribunal Continues Its Work

The International Tribunal has a staggering amount of work to do, sitting in judgment of, and speaking for, the inviolable Rights of Mother Earth, which are being violated worldwide in an astounding number of ways. To this end, a meeting in Paris on December 2015 was chosen to coincide with the United Nations Framework Convention on Climate Change. Once again, concerned citizens from around the world testified publicly against their own governments for allowing continual abuse of the Earth. The Tribunal's renowned lawyers and leaders again heard cases in such matters as hydraulic fracturing, a form of natural gas extraction that is devastating environments across the United States and Britain, endangering humanity and ecosystems as a whole.[74]

The Universal Declaration of the Rights of Mother Earth, which underpins the Tribunal's authority, is a key document for the movement. Proponents are in the process of designing potential amendments to the Rome Statute of the International Criminal Court to recognize, for the first time, the crime of "Ecocide."[75] The International Criminal Court was created by the United Nations' member states in 2003.

The International Criminal Court's purpose is to help end immunity for perpetrators of serious crimes against humanity. The International Criminal Court is a permanent, independent, treaty-based entity, headquartered in The Hague, Netherlands. Instead of the shifting, temporary international tribunals dealing with such acts as mass slaughter, ethnic cleansing, war crimes, and similar crimes against humanity, the International Criminal Court was created as a permanent court when the need for such a tribunal became overwhelming.[76]

How does the International Criminal Court work? Its Prosecutor can investigate based on a referral from any State Party, or from the United Nation's Security Council. A Prosecutor can also convene proceedings based on information about crimes from individuals or organizations. Since 2003, Uganda, the Congo Republic, the Central African Republic, and Mali have referred problems to the Court; the United Nation's Security Council has referred the problems in Darfur (Sudan) and Libya to the court. The Prosecutor analyzes information and subsequently conducts investigations. The court has the power to arrest and detain suspects in its cases, and then to judge guilt and imprison those found guilty. The court can also decide on reparations to victims and create agreements, trust funds, and other arrangements.[77]

It is clear from this discussion that adding a crime of "Ecocide" to the International Criminal Court would be a major step forward. The international tribunal, created by member states of the United Nations, has the power to hear cases, arrest and detain suspects, pronounce guilt, imprison perpetrators, and require reparation to victims of the crimes. Ecocide is often perpetrated by corporations, registered in one country but international in scope and organization, often outside the reach of nation-state governments. Just as frequently, one nation-state's laws may encourage corporations to commit ecological aggression and natural resource possession against a weaker state or people. The International Criminal Court would be an excellent tribunal in which the Universal Declaration of the Rights of Mother Earth could flourish and give yet more strength to the new paradigm.

In addition to this exciting frontier for the international Rights of Nature movement, the United Nations itself has continued to adopt Resolutions on the Harmony with Nature, publish annual Reports from the Secretary-General on the subject, and host an Interactive Dialogue of the General Assembly on Harmony with Nature in commemoration of International Mother Earth Day. Most auspiciously, the United Nations, in 2015, decided to begin a "virtual dialogue" with experts on Earth Jurisprudence worldwide, "in order to inspire citizens and societies to reconsider how they interact with the natural world in order to implement the Sustainable Development Goals in Harmony with Nature, noting that some countries recognize the rights of nature in the context of the promotion of sustainable development."[78]

Though these actions by the United Nations are non-binding and largely symbolic, they have tremendous prestige, provide conceptual approval from the international community, and create invaluable opportunities for the exchange of ideas, research, discussions on the harmony with Nature, and, increasingly, the Rights of Nature.

Endnotes

1. Peter Burdon. The rights of nature: Reconsidered. *Australian Humanities Review*, 49 (2010):69–89.
2. Melissa Arias. Conversation with Natalia Greene about the Rights of Nature in Ecuador. http://environment.yale.edu/envirocenter/post/conversation-with-natalia-greene-about-the-Rights of Nature-in-ecuador/ (accessed July 23, 2016).
3. The following discussion of the Ecuadorian Constitution is based on: [Ecuadorian] Constitution, full language: http://pdba.georgetown.edu/Constitutions/Ecuador/english08.html (accessed August 10, 2015).
4. *Ibid.* (Title II, Chapter 2, Sec. 1, Article 12).
5. *Ibid.* (Article 13).
6. *Ibid.* (Title II, Sec. 2, Art. 15).
7. (1) Erica Glaser. Chevron in Ecuador Representative of Multinationals' Continuing Abuse of Indigenous Peoples. http://www.truth-out.org/opinion/item/23717-chevron-in-ecuador-representative-of-multinationals-continuing-abuse-of-indigenous-peoples# (accessed August 11, 2015); (2) Ecuador's Foreign Minister: Force Multinational Corporations to Respect Human Rights and Environment. http://www.telesurtv.net/english/news

/Ecuadors-Foreign-Minister-Force-Multinational-Corporations-to-Respect-Human-Rights-and-Environment—20141025-0032.html (accessed August 11, 2015); and (3) Karen Alter. *The European Court's Political Power.* Oxford University Press, Oxford, UK. (2009) 364 pp.
8. Ecuadorian Constitution. *op. cit.* (Title II, Chapter 4, Article 57 (12)).
9. Chris Maser. *Global Imperative: Harmonizing Culture and Nature.* 1992. Stillpoint Publishing, Walpole, NH. (1992) 267 pp.
10. Daniel Nettle and Suzanne Romaine. *Vanishing Voices: The Extinction of the World's Languages.* Oxford University Press, New York. (2000) p. ix of 243 pp.
11. Ecuadorian Constitution. *op. cit.* (Title II, Chapter 7, Article 71).
12. *Ibid.* (Title II, Chapter 7, Article 72).
13. *Ibid.* (Title II, Chapter 7, Article 73).
14. *Ibid.* (Title II, Chapter 7, Article 74).
15. Jonathan Kaiman. Ecuador auctions off Amazon to Chinese oil firms. http://www.theguardian.com/world/2013/mar/26/ecuador-chinese-oil-bids-amazon (accessed August 10, 2015).
16. Burns H. Weston and David Bollier. *Green Governance: Ecological Survival, Human Rights and the Law of the Commons.* Cambridge University Press, New York. (2013) 363 pp.
17. Indigenous Ecuadorian group sues for 'genocide.' http://www.heraldsun.com.au/news/breaking-news/indigenous-group-sues-for-genocide/story-e6frf7k6-1226030863231 (accessed July 21, 2016).
18. Natalia Greene. The first successful case of the Rights of Nature implementation in Ecuador. http://therightsofnature.org/first-ron-case-ecuador/ (accessed July 13, 2016).
19. *Ibid.*
20. *Ibid.*
21. Gabriela León Cobo. Vilcabamba River case law: 1 year after. http://therightsofnature.org/Rights of Nature-laws/vilcabamba-river-1-year-after/ (accessed July 13, 2016).
22. Norie Huddle. World's First Successful 'Rights of Nature' Lawsuit. http://www.kosmosjournal.org/article/worlds-first-successful-Rights of Nature-lawsuit-2/ (accessed July 13, 2016).
23. Language of the Bolivian law (compete text): http://www.worldfuturefund.org/Projects/Indicators/motherearthbolivia.html (accessed August 12, 2015).
24. *Ibid.*
25. *Ibid.*
26. *Ibid.*
27. *Ibid.* (Chapter IV, Article 8).
28. *Ibid.* (Chapter IV, Article 9).
29. *Ibid.* (Chapter IV, Article 10).
30. Peter Neill. Law of Mother Earth: A Vision From Bolivia. *Huffington Post.* http://www.huffingtonpost.com/peter-neill/law-of-mother-earth-a-vis_b_6180446.html (accessed August 12, 2015).
31. (1) Nick Buxton. The Law of Mother Earth: Behind Bolivia's historic bill. https://therightsofnature.org/bolivia-law-of-mother-earth/ (accessed August 12, 2015); and (2) Eduardo Galeano. *Open Veins of Latin America: Five Centuries of the Pillage of a Continent.* Monthly Review Press, New York. (1997) 317 pp.
32. Franz Chávez. Bolivia's Mother Earth Law Hard to Implement. http://www.ipsnews.net/2014/05/bolivias-mother-earth-law-hard-implement/ (accessed July 13, 2016).
33. Community Environmental Legal Defense Fund. International Center for the Rights of Nature. http://celdf.org/rights/Rights of Nature/ (accessed July 24, 2016).
34. Kate Beale. Rights for Nature: In PA's Coal Region, a Radical Approach to Conservation Takes Root. http://www.huffingtonpost.com/kate-beale/rights-for-nature-in-pas_b_154842.html (accessed July 24, 2016).
35. Melissa Troutman. Pennsylvania Ecosystem Fights Corporation for Rights in Landmark Fracking Lawsuit. http://publicherald.org/grant-township-speaks-for-the-trees-in-landmark-fracking-lawsuit/ (accessed February 19, 2016).

36. Grant Township, Indiana County, Pennsylvania Community Bill of Rights Ordinance. https://assets.documentcloud.org/documents/1370022/grant-township-community-bill-of-rights-ordinance.pdf (accessed February 19, 2016). (Sec. 2(b)).
37. *Ibid.* (Sec. 2(c)).
38. *Ibid.* (Sec. 2(g)).
39. *Ibid.* (Sec. 4 (c)).
40. *Ibid.*
41. Melissa Troutman. Pennsylvania Ecosystem Fights Corporation for Rights in Landmark Fracking Lawsuit. *op. cit.*
42. (1) Ellen M. Gilmer. DISPOSAL: Pa. judge throws out ban, sidesteps 'rights of nature.' http://www.eenews.net/stories/1060026418 (accessed February 19, 2016) and (2) Nicole Jacobs. Federal Judge Throws Out CELDF "Rights of Nature" Fracking Ban. http://energyindepth.org/marcellus/federal-judge-throws-out-celdf-Rights of Nature-fracking-ban/ (accessed February 19, 2016).
43. Bob Downing. Ecosystem seeks to intervene in Pennsylvania drilling suit. http://www.ohio.com/blogs/drilling/ohio-utica-shale-1.291290/ecosystem-seeks-to-intervene-in-pennsylvania-drilling-suit-1.615966 (accessed July 20, 2016).
44. Simon Davis-Cohen. Why Are Fracking Hopefuls Suing a County in New Mexico? http://readthedirt.org/why-are-fracking-hopefuls-suing-a-county-in-new-mexico (accessed July 20, 2016).
45. The Community Environmental Legal Defense Fund. International Center for the Rights of Nature. *op. cit.*
46. The Community Environmental Legal Defense Fund. Pittsburgh Bans Natural Gas Drilling: Adopts first-in-the-nation ordinance—Elevates the right of the community to decide, not corporations. http://celdf.org/2010/11/press-release-pittsburgh-bans-natural-gas-drilling/ (accessed July 20, 2016).
47. Linda Sheehan. Santa Monica Passes West Coast's First Rights of Nature Ordinance. http://readthedirt.org/santa-monica-passes-west-coast's-first-Rights of Nature-ordinance (accessed July 20, 2016).
48. (1) The Community Environmental Legal Defense Fund. International Center for the Rights of Nature. *op. cit.*; and (2) Simon Davis-Cohen. Why Are Fracking Hopefuls Suing a County in New Mexico? *op. cit.*
49. World People's Conference on Climate Change and the Rights of Mother Earth. https://pwccc.wordpress.com/2010/04/24/proposal-universal-declaration-of-the-rights-of-mother-earth/ (accessed August 16, 2015).
50. (1) 2009 United Nations Climate Change Conference. https://en.wikipedia.org/wiki/2009_United_Nations_Climate_Change_Conference (accessed August 16, 2015); and (2) Ros Donald. The Carbon Brief. http://www.carbonbrief.org/blog/2014/06/whats-going-on-with-the-un-climate-talks-in-bonn/ (accessed August 16, 2015).
51. Peter Burdon. The rights of nature: Reconsidered. *op. cit.*
52. Ambitious human rights and socially just pathways to address climate change, disaster and risk reduction, and sustainable development. http://www.un.org/en/ga/search/view_doc.asp?symbol=A/RES/63/278 (accessed August 21, 2016).
53. Harmony with nature: Resolution/adopted by the General Assembly. http://repository.un.org/handle/11176/148458 (accessed August 21, 2016).
54. Harmony with Nature: Report of the Secretary-General. http://www.un.org/ga/search/view_doc.asp?symbol=A/65/314 (page 3). (accessed August 21, 2016).
55. Chronology of Harmony with Nature. http://www.harmonywithnatureun.org/chronology.html (accessed August 21, 2016).
56. World People's Conference on Climate Change and the Rights of Mother Earth. https://pwccc.wordpress.com/2010/04/24/proposal-universal-declaration-of-the-rights-of-mother-earth/ (accessed August 16, 2015).

57. World People's Conference on Climate Change and the Rights of Mother Earth: Working Groups. https://pwccc.wordpress.com/category/working-groups/ (accessed August 16, 2015).
58. World People's Conference on Climate Change and the Rights of Mother Earth: Peoples [sic.] Agreement. https://pwccc.wordpress.com/support/ (accessed August 16, 2015).
59. Universal Declaration of Rights of Mother Earth: Global Alliance for the Rights of Nature. https://therightsofnature.org/universal-declaration/ (accessed August 16, 2015).
60. *Ibid.* (Article 1(2)).
61. *Ibid.*
62. *Ibid.* (Article 2, sec. 1).
63. *Ibid.*
64. *Ibid.* (Article 3, sec. 1 and 2).
65. *Ibid.*
66. *Ibid.*
67. *Ibid.*
68. Erin Fitz-Henry. International Rights of Nature Tribunal, Lima, Peru—December 2014. http://therightsofnature.org/great-barrier-reef-case-lima/ (accessed August 17, 2015).
69. Michelle Maloney. Finally being heard: The Great Barrier Reef and the International Rights of Nature Tribunal. *Griffith Journal of Law & Human Dignity*, 3 (2015):1–58.
70. The following discussion is based on: (1) Erin Fitz-H. *op. cit.*; and (2) Michelle Maloney. *op. cit.*
71. Chris Maser. *Global Imperative: Harmonizing Culture and Nature*. Stillpoint Publishing, Walpole, NH. (1992) 267 pp.
72. International Rights of Nature Tribunal—Paris. http://therightsofnature.org/Rights of Nature-tribunal-paris/ (accessed 18 August, 2015).
73. The foregoing discussion of the Great Barrier Reef is based on: Michelle Maloney. Finally being heard: The Great Barrier Reef and the International Rights of Nature Tribunal. *op. cit.*
74. (1) International Rights of Nature Tribunal—Paris. http://therightsofnature.org/Rights of Nature-tribunal-paris/ (accessed 18 August, 2015); (2) United States Environmental Protection Agency. The Process of Hydraulic Fracturing. http://www2.epa.gov/hydraulicfracturing/process-hydraulic-fracturing (accessed July 12, 2015); (3) United States Environmental Protection Agency. Basic Information about Injection Wells. http://water.epa.gov/type/groundwater/uic/basicinformation.cfm (accessed July 12, 2015); and (4) Roger Harrabin. Fracking bids to be fast-tracked. http://www.bbc.com/news/science-environment-33894307 (accessed August 18, 2015).
75. Rome Statute of the International Criminal Court. http://www.un.org/law/icc/index.html (accessed August 18, 2015).
76. International Criminal Court. http://www.icc-cpi.int/en_menus/icc/about%20the%20court/Pages/about%20the%20court.aspx (accessed August 18, 2015).
77. *Ibid.*
78. Chronology of Harmony with Nature. *op. cit.*

3

Other International Voices for the Rights of Nature

New Zealand: A National Park, a River, and Indigenous Peoples

Independently of the Universal Declaration of the Rights of Mother Earth and the International Tribunal for the Rights of Nature, New Zealand created another facet of the emerging paradigm in two contrasting, but complimentary, ways.

In the Te Urewera Act of 2014,[1] New Zealand granted its 821-square mile Te Urewera National Park the status of a person, citing its intrinsic worth and the integrity of its values. Instead of giving direct ownership to the Maori indigenous group (Tuhoe) who claimed the land as expressing their language, customs, and identity, the government and the Maori agreed to give the land itself legal personhood. This is the first national park in the world to receive such protection.[2]

This revolutionary statute states, "Tuhoe and the Crown share the view that Te Urewera should have legal recognition in its own right, with the responsibilities for its care and conservation set out in the law of New Zealand."[3] The law seeks to assuage the grief of the Maori, who are most closely bound to Te Urewera, by giving the park a powerful means of protecting itself from future threats. As the statute describes it, Te Urewera is prized by all New Zealanders for its "outstanding national value and intrinsic worth; it is treasured by all for the distinctive natural values of its vast and rugged primeval forest, and for the integrity of those values; for its indigenous ecological systems and biodiversity, its historical and cultural heritage, its scientific importance, and as a place for outdoor recreation and spiritual reflection."[4]

The law then states, "Te Urewera is a legal entity, and has all the rights, powers, duties, and liabilities of a legal person."[5] These rights, powers, and duties are to be exercised by the Te Urewera Board. The land may not be alienated, mortgaged, or disposed of, except for such minor reasons as boundary adjustments if a parcel of land does not have the values to justify its further inclusion, or for concessions consistent with the Te Urewera management plan.[6]

New Zealand, like Ecuador and Bolivia, is showing how the new paradigm can protect ecosystems more powerfully than regulatory environmental laws. The Te Urewera Act has caught the attention of Rights of Nature activists worldwide studying how to apply its principles, such as in American communities faced with hydraulic fracturing.[7]

In India, the National Ganga Rights Movement advocates protection of the highly polluted Ganges under a rights-based system. The High Court of the Indian state of Uttarakhand in 2017 issued a ground-breaking ruling granting the Ganges River and its major tributary the Yamuna the rights of personhood. Then, in an expanded ruling responding to a citizen petition, the Court granted the entire Ganges ecosystem "the status of a legal person, with all corresponding rights, duties and liabilities of a living person, in order to preserve and conserve them." The ruling describes the streams, springs, lakes, air, jungles, meadows and forests on which the Ganges' health depends. The Court thus conveys its fundamental point that it would be impossible to protect one part of the ecosystem without protecting the whole. Now, polluting the Ganges or harming its ecosystem is equal to harming a person—a dramatic paradigm change. How these provisions will operate in practice, such as who represents the Ganges ecosystem in which situations of environmental harm, will take time to determine. But the legal framework of respect for Nature's right to flourish is in place.[8]

A similar, long-standing, unresolved situation in New Zealand faces the country's third-longest river, the Whanganui. Negotiators are hammering out a potentially similar solution to Te Urewera, based on a creative and novel fusion of indigenous Maori concepts and Western legal frameworks. The framework centers around "rangatiratanga," a Maori term that is exceedingly difficult to translate into English concepts. Perhaps it is best defined as Maori sovereignty asserted through a collective exercise of responsibilities to protect, conserve, and enhance places or processes for the security of future generations.

The Whanganui River of New Zealand—one of the country's most iconic and revered waterways—will now be the beneficiary of a partnership between the Maori and the New Zealand government to protect it (Figure 3.1). As of an Agreement, finalized in 2012,

FIGURE 3.1
Whanganui River, New Zealand (1913). Photograph by Josiah Martin (https://commons.wikimedia.org/wiki/File:Wanganui_river_-_page_138.jpg).

the Whanganui will become an entity, the first time in New Zealand—and possibly the world—in which a river has had a legal identity bestowed on it. The river will be defined and the activities will be governed by the Maori view of the river.

The Whanganui Iwi, the indigenous group that possesses rangatiratanga over the Whanganui River, is the local group whose entire history is tied to the River. Their definition has been adopted into the Agreement. The river will be considered "Te Awa Tupua," a living and integral whole whose life is inseparable from the life of the Whanganui Iwi people whose lives, in turn, are so defined by the river.

The new Agreement, entitled "Tu-tohu Whakatupua," essentially protects the rangatiratanga through a non-Maori guardianship model set up through a Western legal system construct in such a way that it allows the river to be considered a legal entity. It will be recognized as a person under the law, which will give it legally protected rights and interests. The Agreement was signed by the Crown, and by the Whanganui River Maori Trust, representing Maori groups for whom the river is not only home but also spiritual lifeblood. Guardians will be appointed to protect the river; one Guardian appointed by the Crown and one by the Whanganui Iwi. The guardians must focus not only on the river's physical and ecological rights but also on assuring its spiritual and cultural rights.

This Agreement is not a court settlement with enforceable powers behind it, nor does it carry independent binding authority. But it has the moral force of negotiations over Maori rights that began between the Maori and the British Crown in 1840, via the Treaty of Waitangi. Courts in New Zealand have held that, although the specific duties under the Treaty change as times and needs shift, there is nevertheless a requirement of partnership that continues, one that imposes a duty of good faith and reasonable conduct between the Maori and the government. Among other things, this binds the government to protect Maori property rights and redress past wrongs that violated the spirit of the Treaty. This process requires the Maori to file a claim with the Waitangi Tribunal, which then leads to settlement talks and compensation for proven failures to live up the Treaty's provisions or principles that have endured beyond specific, enumerated obligations.

Whanganui Iwi have struggled through the New Zealand legal system to have their rights and needs recognized by the Crown since 1873, and the Whanganui River agreement is a fruit of recent negotiations. The Tribunal issued a report on the Whanganui River in 1999. Talks between Whanganui Iwi and the government focusing on the River's needs started in 2002, began again in 2009, and reached the stage of framing further negotiations in 2011. The on-the-ground details of the Agreement, and Whanganui Iwi claims under the Treaty of Waitangi, specific to the Whanganui River, have yet to be filed. Legislation to implement the Agreement, and to grant legal personhood to the River, as was done for Te Urewera National Park, became law in March 2017.

Tu-tohu Whakatupua holds tremendous promise for fusing indigenous concepts and Western legal concepts in order to protect the natural resources all humans depend on. According to the Agreement, the River will not be owned in any Western legal sense, but will be governed by the Whanganui Iwi's understanding of the river as a living being, integrated and whole, whose survival means the survival of the group's culture as well. Since all peoples depend on the local environment for survival, this model is one that could be widely adapted to provide, and given binding authority and powers of enforcement in other contexts far from New Zealand and the beautiful Whanganui River.[9]

Pope Francis Speaks for the Earth: The Encyclical *Laudato Si*

On May 24, 2015, Pope Francis, head of the Roman Catholic Church, issued his encyclical *Laudato Si*—Latin for "Praise Be to You," a line from the *Canticle of the Sun*, a famous hymn praising God for the handiwork of Nature, written by St. Francis of Assisi, who lived in 13th-century Italy.[10]

In the ancient Catholic Church, an "encyclical" referred to a letter sent to all the churches in a particular area about a given subject. But in more modern times, an encyclical is primarily the provenance of the Pope, the supreme pontiff of the Roman Catholic Church, whose seat is the Vatican in Rome. A Papal encyclical is a teaching document in the form of a letter sent out by the Pope to all Catholic bishops, often including such Catholic groups as clergy, monastic religious, or laypeople, and sometimes to a wider audience that includes "all people of goodwill." Encyclicals are the result of intense prayer, study, thought, reflection, and often consultation with others, over long periods. Consequently, they are solemn documents for the Church and its members.

Laudato Si is different from other encyclicals because Pope Francis addressed it to every person on the planet, not just the 1.25 billion Catholics who might be expected to take the encyclical most seriously. Although encyclicals are primarily focused on Catholic religious doctrine, *Laudato Si* is unique. Though it contains many elements of Catholic teaching, the encyclical devotes much time to summarizing the world's environmental problems from a scientific point of view (Figure 3.2).

Pope Francis invites all people to begin a dialogue about the world we share. Catholics, however, have an additional responsibility toward this encyclical. Despite the unique emphasis on scientific consensus concerning environmental dangers, the Pope is also speaking authoritatively on the faith and morals centered in Catholic doctrine. Catholics must consider these teachings carefully, as they are obligated to discern and contemplate them as part of their faith.

Laudato Si ranges widely across environmental problems in six chapters, beginning with a summary of current problems, exploring a Gospel of Creation and the human roots of the crisis at hand. Pope Francis then discusses "integral ecology," by which he means any ecological concept will be incomplete that leaves humankind out of the equation (Chapter 4). His point is that such an omission produces failures in any social-environmental vision because humans and nature are an inseparable part of one another.[11]

Our human inseparability from Nature is based on the concept of "ecology." In Greek, both "ecology" and "economy" have the same root, *oikos*, meaning "house." *Ecology* is the knowledge or understanding of the house, and *economy* is the management of that house, and it is the same house—one that cannot be divided against itself and remain standing. That is to say, Nature's environmental processes provide the energy that drives our economically competitive society—a fact misunderstood, forgotten, overlooked, or deliberately ignored in the way people treat the sustainability of such things as water catchments and the streams they provide.

With the understanding of what "ecology" means, we can expand the concept to "ecosystem," which adds the interactions between the living and non-living parts of our environment. Grasping these interactions will help us take care of our "house" in a way that protects the sustainability of its long-term productive capacity for our benefit, that of our children, their children, and the generations yet unborn. These concepts are the foundation of Pope Francis's vision.

FIGURE 3.2
Pope Francis, Pontiff of the Roman Catholic Church. Photograph by Casa Rosada (Argentina Presidency of the Nation) (https://commons.wikimedia.org/wiki/File:Franciscus_in_2015.jpg).

Though the encyclical is based on ecologically oriented pronouncements of prior popes, especially Benedict XVI and St. John Paul II, Pope Francis ranges much more widely than his predecessors. He makes it clear that the basis of the created world is relationship:

> The divine Persons are subsistent relations, and the world, created according to the divine model, is a web of relationships. Creatures tend towards God, and in turn it is proper to every living being to tend towards other things, so that throughout the universe we can find any number of constant and secretly interwoven relationships. This leads us not only to marvel at the manifold connections existing among

> creatures, but also to discover a key to our own fulfillment. The human person grows more, matures more and is sanctified more to the extent he or she enters into relationships, going out from themselves to live in communion with God, with others and with all creatures.[12]

This spiritual understanding leads the Pope to an analysis of current unsustainable consumerism, fed by feelings of instability, uncertainty, self-centeredness, and greed. He calls for a universal awareness that can metamorphose into more sensible and sustainable habits, linked to an environmental education that makes the "leap towards the transcendent, which gives environmental ethics its deepest meaning."[13]

The Pope's clarion call is "For a new dialogue about how we are shaping the future of our planet. We need a conversation which includes everyone…"[14] He also chastises those who fail to care about this deeply troubling set of problems out of indifference, self-interested opposition, quietist resignation or—a common problem in technological societies—blind faith in technological innovation.

The first part of *Laudato Si* explores the many deep-rooted troubles affecting the planet, basing its summaries on the best available scientific consensus of the day. He begins with pollution and climate change, calling the latter, "a global problem with grave implications."[15] From there, he moves on to water, the most important natural resource for all peoples, but whose quality is frequently compromised so severely as to endanger life. He then discusses the loss of biodiversity. Boldly, the encyclical states:

> It is not enough, however, to think of different species merely as potential "resources" to be exploited, while overlooking the fact that they have value in themselves. Each year sees the disappearance of thousands of plant and animal species which we will never know…. The great majority become extinct for reasons related to human activity. Because of us, thousands of species will no longer give glory to God by their very existence, nor convey their message to us. We have no such right.[16]

True to his intention of ensuring that human necessities—and the breakdown of human society as part of the ecological decline—are an integral part of his vision, the Pope next discusses the decline in the quality of human life, as part of the overarching carelessness toward both environment and society. This includes a discussion on global inequality, as the human and natural environments deteriorate in tandem; one cannot be separated from the other. Always, Pope Francis brings his vision back to the added burden placed on the poor of the world, who disproportionately bear the weight of environmental degradation.

Integrated into the Pope's enumeration of current crises is a deeply religious and catholic (in the widest sense of the word) understanding of the problems. "Respect must also be shown for the various cultural riches of different peoples, their art and poetry, their interior life and spirituality. If we are truly concerned to develop an ecology capable of remedying the damage we have done, no branch of the sciences and no form of wisdom can be left out, and that includes religion and the language particular to it."[17]

The encyclical's discussion of faith gives the Pope the opportunity to decisively reject the narrow interpretation of Genesis, in which the act of creation has so often been deemed to give humans "dominion" over the earth:

> …nowadays we must forcefully reject the notion that our being created in God's image and given dominion over the earth justifies absolute dominion over other creatures. The biblical text[s]…implies a relationship of mutual responsibility between human beings and nature…The responsibility for God's earth means that human beings, endowed

> with intelligence, must respect the laws of nature and the delicate equilibria existing between the creatures of this world...[18]

The encyclical quotes the bishops of many regions of the world, ranging from Japan to Canada to Brazil, on the value of Nature, reaffirming that no creature is excluded from being a manifestation of God; that Nature's myriad relationships reveal the infinite glory of God; and that Nature is a locus of God's presence that continuously, unceasingly, calls humans into relationship with God. The Pope calls for human hearts to be open to universal communion with the rest of Nature.

The encyclical takes aim at the current plethora of perils that, in some ways, are themselves unique: technology's creativity and power, and the problem of globalizing the technocratic paradigm—essentially, working toward the further development of technology in a one-dimensional, narrowly focused way. This robs people and societies of the ability to develop in their own richness outside of the straitjacket of technological thinking. Ironically, technocentrism has, in the Pope's view, led to an even more dangerous anthropocentrism, in which the messages from individuals and society calling for change go unheeded. Instead, every new technological innovation is embraced uncritically.

Pope Francis devotes an entire chapter to integral ecology—the placing of human beings in the framework of Nature, in a wide-ranging discussion of human ecology that includes the uniqueness of each place's cultural ecology, currently threatened by mass consumerist ideology, leaving both despair and degraded physical environments behind. The encyclical asks us to show special care for indigenous communities and their traditions because their traditional lands, sacred to their cultures and the resting place of their ancestors, are often under the most intense pressure of development worldwide.

In addition, he discusses the problems of poor living spaces, common areas, degraded housing, public transport, and the degradation of social-environmental sustainability through our human refusal to acknowledge a moral law, which is "inscribed in our nature."[19]

In discussing the principle of the common good, the Pope pinpoints its requirements of respect for the human person and a call for solidarity across cultural boundaries within our global society. But, more than this, the common good also extends across generations.[20] "Doomsday predictions can no longer be met with irony or disdain. We may well be leaving to coming generations debris, desolation and filth.... Our difficulty in taking up this challenge seriously has much to do with an ethical and cultural decline which has accompanied the deterioration of the environment."[21]

Calling for many discussions between religion and science and between peoples of many cultures within the international community, Pope Francis culminates *Laudato Si* with a call for ecological conversion of those who, whether from pragmatism or opposition, refuse to recognize the magnitude of the problems confronting us. What is ecological conversion? It consists of gratitude, of recognizing that the world is God's gift, and of loving awareness that humans are connected to other creatures in universal communion—along with the responsibility humans have by reason of their unique capacities. Uncompromisingly and boldly, the Pope states: "Care for nature is part of a lifestyle which includes the capacity for living together and communion.... Love, overflowing with small gestures of mutual care, is also civic and political, and makes itself felt in every action that seeks to build a better world."[22]

Laudato Si is so vast, and incorporates such a profoundly new vision of Catholic theology and traditional concerns, that it will no doubt take years for its effects to be fully felt throughout the international communities and the global environment. Several Catholic organizations have already begun the work of endorsing its principles and collaborating with Catholics and non-Catholics to widen the Catholic environmental message. Prominent among these is the Catholic Climate Covenant, which works to raise awareness, build a network of Catholics, make the moral case (especially the effects of climate change on the world's poor) for limiting the carbon footprint, host campus and community events on these topics, and collaborate with parishes, monastic communities, and other Catholic organizations to take practical steps toward realizing Catholic values in the climate-change crisis and related environmental issues.[23]

As is often the case with a breathtaking shift in an old paradigm, the new ideas burst forth in a seemingly sudden, bewildering, and many-faceted complexity. The cascade of ideas and new relationships embrace all the narrow vistas of the old paradigm and ultimately obliterate them. Nevertheless, the new paradigm also grows out of the old, builds on it, enlarges its horizons, and makes novel approaches possible through a broader foundation than the preceding one.

Some religious traditions, such as Catholicism, have in their histories one or more ugly periods of exploitation and victimization. For example, Christopher Columbus's 1492 encounter with the indigenous peoples of what was to become the Americas set in motion efforts within the Catholic Church, Spain, and later the Holy Roman Empire to define the rights of Spain and Portugal to the newly "discovered" lands.

Within the Catholic Church, these efforts (usually in the form of papal bulls and encyclicals) directly or indirectly defined the rights of the indigenous peoples, as a matter of what might today be considered international law. A papal bull is a formal proclamation issued by the pope and sealed with a "bulla," which was originally a circular plate or boss of metal, so called because it resembled in a bubble floating on water.[24] Though originally addressing the imperial claims of Spain and Portugal, these papal bulls also influenced the legal ethics applied to the claims of other European imperial powers and their relationships with the indigenous peoples of the Americas.

Political goals and pressures heavily influenced these papal pronouncements. Pope Alexander VI's Bull, *Inter Caetera Divinae* (issued May 4, 1493), wasted no time in addressing the subject of Spanish imperial control in the Americas. It allocated the newly discovered lands to the king and queen of Castile and Leon (Spain) along with the responsibility of converting the indigenous peoples to the Catholic faith:

> [W]e, of our own accord ... and out of the fullness of our apostolic power..., which we hold on earth, do by tenor of these presents ... give, grant, and assign to you and your heirs and successors, kings of Castile and Leon, forever, together with all their dominions, cities, camps, places, and villages, and all rights, jurisdictions, and appurtenance, all islands and mainlands found and to be found, discovered and to be discovered towards the west and south, no matter whether the said mainlands and islands are found and to be found in the direction of India or towards any other quarter, the said line to be distant one hundred leagues towards the west and south from any of the islands commonly known as the Azores and Cape Verde.[25]

By its terms, *Inter Caetera* purported to grant all the newly "discovered" lands to the Spanish Crown. Beginning in 1503, it granted them the right of *encomienda* (the right to demand tribute and forced labor from the indigenous inhabitants of an area), whereby the grantee (the *encomiendero*) was supposedly entrusted with the protection of a specified number of Natives, along with the responsibility to teach them Spanish and convert them to Catholicism. In exchange, the *encomiendero* had the right to use the Natives as slave labor. The *encomienda* was an adaptation of the European feudal system of tribute employed by the Spanish Crown after the conquest of Granada. It granted the invaders, conquistadors, and colonial officials trusteeship over the very people they had conquered.

After the conquest of Mexico (1519–1521), the Spanish co-opted the administrative structures of the indigenous peoples, which allowed the king of Spain and the new colonial government to usurp the traditional revenues of the region. The directive from the king of Spain to Cortés in 1523 stated explicitly that the indigenous peoples must pay monetary tribute to Spain, to wit: "They will give and pay us for each year as much revenue and tribute as were given and paid until now to their priests and lords."[26] In order to accomplish this in terms of organization, the lower echelons of the indigenous administration had to be maintained and integrated into the structures of colonial government.

According to the wishes of the king, personal *encomiendas* were to be reserved only for the first generation of conquistadors, yet, following the lead of Cortés, these privileges were repeatedly expanded with the support of the colonial administration.

Although *encomienda* did not legally confer the rights of private property for land, the holders of the privilege received most of the earliest land grants and had the advantage of virtual slave labor during the seasons of planting and harvesting. As the system of *encomienda* was gradually reformed and phased out, farming, which included raising livestock, became the major source of income for the increasing number of rural Spanish squatters. By 1525, ownership of public lands was overwhelmingly given to officials, the military, and the colonial elite.[27]

However, on May 24, 2015 (522 years after Pope Alexander VI's Bull of May 4, 1493), Pope Francis issued his *Laudato Si* in which he outlined opportunities to speak for the Earth, represent its needs, and its rights to life.[28] It is our obligation to encourage and nurture this newly emerging paradigm that it can replace the extant one of such narrowed vision that it is responsible for many social-environmental crises, the resultant suffering (both human and nonhuman), and the growing impoverishment of all future generations.

Islamic Declaration on Global Climate Change

A few months after *Laudato Si* and after more than a year's work of refining drafts, a coalition of Muslim scholars, environmentalists, academics, and specialists unveiled the Islamic Declaration on Global Climate Change at a symposium held in Istanbul in August 2015. The driving force behind the Declaration was the United Kingdom-based Islamic Foundation for Ecology and Environmental Sciences, founded in 1994. They have concentrated on developing a knowledge base and producing educational material, training programs, and on-the-ground projects, as well as reaching out and networking with other nonprofit organizations and governments via seminars and conferences.[29]

The Islamic Declaration begins by noting that, "The pace of Global climate change today is of a different order of magnitude from the gradual changes that previously occurred. … Moreover, it is human-induced: we have now become a force dominating nature."[30] The Declaration goes on to note that, since the Industrial Revolution, humans have used massive amounts of the Earth's resources: "We are driven to conclude from these warnings that there are serious flaws in the way we have used natural resources—the sources of life on Earth. An urgent and radical reappraisal is called for."[31]

The Declaration affirms that Allah created each thing for truth and for right, and encompasses all of creation,[32] as well as the corruption humans have caused by unceasing economic growth and consumption, the consequences of which range from global climate change to contamination of land, air, water, and oceans, which says nothing of the destruction of habitats, impairment of ecosystems, introduction of alien species and genetically modified organisms, and severe impacts to human health.[33] In Section 2.6, the Declaration states, "We are but one of the multitude of living beings with whom we share the earth. We have no right to oppress the rest of creation or cause it harm."[34] Muslims are thus reminded: "We recognize that we are accountable for all our actions."[35] Furthermore, they have a responsibility to act according to the Prophet Muhammad's example of frugal living, and his actions to protect the rights of all living beings.[36]

The Declaration ends with a call to the parties to the "well-off nations and oil-producing states" to exercise leadership in drastically reducing greenhouse gas emissions and consumption, and "re-focus their concerns from unethical profit from the environment to that of preserving it and elevating the condition of the world's poor," as well as provide financial and technical assistance to less wealthy nations.[37] The Declaration requests the leaders of other nations to commit to renewable energy, and to "realize that to chase after unlimited economic growth in a planet that is finite and already overloaded is not viable."[38] Finally, the Declaration calls upon all Muslims to tackle the root causes of climate change, environmental degradation, and loss of biodiversity, by following the example of Prophet Muhammad.[39]

Endnotes

1. Te Urewera Act 2014. http://www.legislation.govt.nz/act/public/2014/0051/latest/DLM6183601.html#DLM6183888 (accessed August 23, 2016).
2. Devon O'Neil. Parks Are People Too. http://www.outsideonline.com/2102536/parks-are-people-too-legally-speaking (accessed August 23, 2016).
3. *Ibid.* (Part 1, Subpart 1).
4. *Ibid.*
5. *Ibid.* (Part 1, Subpart 3).
6. Te Urewera Act 2014. *op. cit.*
7. Devon O'Neil. Parks Are People Too. *op. cit.*
8. (1) The Ganga Rights Act: Recognizing the Rights of the Ganga River Basin. http://www.gangarights.org/ganga-right-act/ (accessed August 23, 2016); and (2) Indian court awards legal rights of a person to entire ecosystem. http://www.climatechangenews.com/2017/04/03/indian-court-awards-legal-rights-person-nature/ (accessed April 4, 2017).
9. The foregoing discussion of the Whanganui River in New Zealand is based on: (1) Kate Shuttleworth. Agreement entitles Whanganui River to legal identity. http://www.nzherald.co.nz/nz/news/article.cfm?c_id=1&objectid=10830586 (accessed August 18, 2015); (2) Brendan Kennedy. I Am the River and the River is Me: The Implications of a River Receiving Personhood Status. http://www.culturalsurvival.org/publications/cultural-survival-quarterly/i-am-river-and-river-me-implications-river-receiving (accessed August 18, 2015); and (3) Stephen Messenger. New Zealand Grants a River the Rights of Personhood. http://www.treehugger.com/environmental-policy/river-new-zealand-granted-legal-rights-person.html (accessed November 28, 2016).
10. Eric Doyle. *St. Francis and the Song of Brotherhood and Sisterhood*. Franciscan Institute Publications, St. Bonaventure University, St. Bonaventure, NY. (1996) 244 pp.

11. The foregoing discussion of Pope Francis and his *Laudato Si* is based on: Encyclical Letter Laudato Si of The Holy Father Francis on Care for Our Common Home [in English]. http://w2.vatican.va/content/francesco/en/encyclicals/documents/papa-francesco_20150524_enciclica-laudato-si.html (accessed August 19, 2015).
12. *Ibid.* (Chapter 6, Section VII, subsection 240).
13. *Ibid.* (Chapter 6, Section II, subsection 210).
14. *Ibid.* (Introduction, subsection 14).
15. *Ibid.* (Chapter 1, Section 1, subsection 25).
16. *Ibid.* (Chapter 1, Section III, subsection 33).
17. *Ibid.* (Chapter 2, Section I, subsection 63).
18. *Ibid.* (Chapter 2, Section II, subsections 67–68).
19. *Ibid.* (Chapter 4, Section II, subsection 155).
20. *Ibid.* (Chapter 4, Sections IV–V).
21. *Ibid.* (Subsections 161–162).
22. *Ibid.* (Chapter 4, Sections IV–V, subsections 228 and 231).
23. Catholic Climate Covenant. http://www.catholicclimatecovenant.org/ (accessed August 20, 2015).
24. Kevin Knight. Catholic Encyclopedia: Bulls and Briefs. *New Advent*, http://www.newadvent.org/cathen/03052b.htm (accessed August 20, 2015).
25. The Bull *Inter Caetera* (Alexander VI), May 4, 1493 [in English]. http://www.nativeweb.org/pages/legal/indig-inter-caetera.html (accessed August 19, 2015).
26. *Ibid.*
27. The preceding three paragraphs are based on: Hanns J. Prem. Spanish colonization and Indian property in Central Mexico, 1521–1620. *Annals of the Association of American Geographers*, 82 (1992):444–459.
28. Encyclical Letter Laudato Si of The Holy Father Francis on Care for Our Common Home [in English]. *op. cit.*
29. ifees. ecoislam. http://www.ifees.org.uk/about/our-story/ (accessed July 15, 2016).
30. Islamic Declaration on Global Climate Change. http://www.ifees.org.uk/wp-content/uploads/2016/06/Climate_Declaration_final_edit_web.pdf (Preamble, Section 1.5) (accessed July 15, 2016).
31. *Ibid.* (Preamble, Section 1.3).
32. *Ibid.* (Section 2: 2.1, 2.2).
33. *Ibid.* (Section 2.5).
34. *Ibid.*
35. *Ibid.*
36. *Ibid.* (Sections 2.7 and 2.8).
37. *Ibid.* (Section 3.2).
38. *Ibid.* (Section 3.3).
39. *Ibid.* (Section 3.6).

4

Relationship: The Basis for Rights of Nature

Nature's primary Law of Reciprocity—the underlying basis and foundation for all other laws—is *relationship.* It is also the foundation of any Rights of Nature legal system, which is designed to expand the law of relationship beyond human beings. It is easy to announce this, but much more difficult to grasp in practice, because relationships can be, and usually are, complex, intricate, and even disguised. They are, by definition, reciprocal in that every relationship is an amalgam of all the relationships that comprise Nature, such as those between and among the physical elements and the biological components (= biophysical) that comprise all life. But relationships can be of many kinds—some emphasizing giving and caring, others self-centered taking.

The kind of behavior best suiting the ability of individuals to continue their relationships depends on each party's psychological nature, in addition to such external circumstances as the environment in which they live. In the end, however, all relationships depend on personal responsibility, which may simply be inherent in the relationship, though not formally stated, culturally explicit, or consciously observed.

Nevertheless, each party knows when responsibility has been breached, because there no longer is confidence in the relationship due to lack of trust. The discontinuance of a relationship may be caused by external factors, such as the death of one party, or may stem from the irresponsible behavior of one party or both parties.

Although we do not usually think of it this way, relationship is an aspect of sustainability, because it keeps the web of interaction intact and thriving. However, despite the inviolable laws of reciprocity that govern all interrelationships, including human societies, the Western industrial paradigm—beginning with the East India Company in 1600—is counter to the ability of the Earth to maintain its biophysical integrity, resilience, and productivity, which is the foundation of human survival.

In essence, a human society that disregards the reciprocity of its relationships with its natural environment is raising red flags about its ability to live sustainably in the world it depends on. Western culture, for example, in the exaltation of its capitalistic economic system, has most frequently denied all sense of responsibility for and relationship with the environment, whether soil, water, animals, the ocean, birds, future human generations, or any other aspect of the ecosystem in which we are embedded.

However, there are models of doing things a different way, and these models can be expanded and retooled to fit human–environmental interactions. Here, we discuss two very different aspects of relationship, historically separated in time, place, and focus: The Feudal Chain of Obligation and Indigenous System of Sustainability. But both of them can be used as beginning models for building sustainability into the legal system via a Rights of Nature legal framework that expands human relationships and the attendant responsibilities.

The Feudal Chain of Obligation

The feudal system in medieval Europe (roughly, 1000 CE–1300 CE) was complex and variable. It was also a more highly integrated system of relationships than any other in Europe before or since. As such, it deserves an examination for its governing rules, because it provides a model for a relationship that can be designed to fit current problems stemming from a lack of relationship between humans and the global ecosystem. Because it created a system of richly interacting relationships and responsibilities, it provides the starting point for distilling its rules into frameworks more workable for modern problems. Although the society it produced was rigidly controlled and highly unequal, it needs to be examined to show the perils of arranging responsibilities too narrowly or without adequate loyalty to the foundations of society, where they can flourish. Though feudal ties of obligation bound only human beings, they nonetheless allow current theorists to explore how these rules of reciprocity could be expanded to encompass relationships among humans, non-human species, and their collective biophysical environment.

The feudal system of continental Western Europe arose out of the ashes of strongly state-based societies that preceded it, most notably the Roman Empire, which slowly disintegrated in the face of Germanic invasions. The subsequent Carolingian Empire, also in its decay, left a social vacuum that needed to be filled. The Carolingian Empire was a confederation of Germanic tribes during Late Antiquity (about 235–284 CE) and the Early Middle Ages (about 1001–1300 CE).[1] The allegiances between and among men, for which feudal society is so noted, arose because blood kinship was insufficiently broad-based to create binding and overarching ties that could hold an entire society together.[2]

Servile homage carried over from one generation to the next, allowing no choice of independence, and gradually evolved into serfdom. Bound by their dependent status to working the land, serfs were considered "commended men" during the Carolingian age, because they were bought and sold with the land. In contrast, the ties of vassalage began to differentiate shortly before the Carolingian age; such a tie ended the moment one of the two men died. Not being hereditary, vassalage bore no stigma of constraining the vassal's liberty. A "vassal" was someone obligated to show loyalty and homage (respect) to a feudal lord, while occupying the lord's land and receiving his protection.

The Feudal Chain of Obligation, which created a system of richly interacting relationships and responsibilities, provides the starting point for distilling its rules into frameworks more workable for modern problems. Fundamentally, feudal society was strictly hierarchical in that it required the submission of many dependent people to a few powerful lords. There were many powerful ties of vassalage in the feudal regions of Europe:

> In feudal society the characteristic human bond was the subordinate's link with a nearby chief. From [one] level to another the ties thus formed—like so many chains branching out indefinitely—joined the smallest to the greatest.... For amongst the highest classes, distinguished by the honorable profession of arms, relationships of dependence had assumed, from the outset, the form of contracts freely entered into between two living men confronting one another. From this necessary personal contact derived the best part of its moral value.[3]

The medieval theory about a feudal chain of legal obligation and authority—acts of homage—running from serf to monarch through the manorial system, as well as among the free knights and nobles, was in reality much more intricate and elaborate than it might seem today. With time, the feudal chain of obligation became less of a direct "chain," not only as military service due the crown changed to payments but also as the obligations of vassals multiplied, from those holding a small manor to vassals above them holding large manors and scattered estates.

The central basis of the feudal system was the manor—the large, rural estate owned by a family of the nobility—on whom both the villages and agricultural serfs on the estate depended. The "manorial system" was the economic and social basis of medieval Europe. All legal and economic power belonged to the lord, who was economically supported by his land and by contributions from the peasant population under his authority. Land was valued above all, as the basis of wealth. It thus provided the foundation for the principal tenets of European feudalism, which included:

- Serfs, most frequently bound to the land and without mobility
- Use of land and services as payment, rather than a salary
- Binding ties of obedience and protection extending down the feudal chain, beginning with the king and creating obligations man to man
- Increasing fragmentation of authority, caused by such factors as family loyalty, inheritance, and the rules of family succession, which eroded feudal obligations
- Increasing obligations of multiple loyalties pulling at vassals, especially of the warrior class[4]

England was somewhat different, because a partial feudalism and a fragmented manorial system existed at the time of the Norman Conquest in 1066. But William the Conqueror united the two and created a full-fledged, manor-based feudalism across the country. He did this by making himself landlord of England and then placing large land grants in the hands of trusted followers, as tenants-in-chief, to hold and manage the land for the Crown.[5]

It is essential to remember that, in feudalism, vassalage was always a relationship based on responsibility, from the King down the entire chain. "Vassal homage was a genuine contract and a bilateral one. If the lord failed to fulfill his engagements he lost his rights.... The originality of the ... system consisted in the emphasis it placed on the idea of an agreement capable of binding the rulers...."[6]

The heart of the obligation was a man-to-man act of homage. It was a simple and striking rite, depicted in many illustrations and pieces of artwork from the era:

> Imagine two men face to face; one wishing to serve, the other willing or anxious to be served. The former puts his hands together and places them, thus joined, between the hands of the other man—a plain symbol of submission ... sometimes further emphasized by a kneeling posture. At the same time, the person proffering his hands utters a few words—a very short declaration—by which he acknowledges himself to be the 'man' of the person facing him. Then chief and subordinate kiss...[7] (Figure 4.1).

The legal details of the deeply symbolic Act of Homage were contained in the oath of fealty, which arose during the Carolingian period. The new vassal, swearing on a Bible or sacred relics, vowed to be true to his master. Homage bound the whole man at once, and was therefore never renewed. The oath of fealty, on the other hand, could be taken many times in different circumstances. "Fealty" is usually thought of as an oath of loyalty sworn

FIGURE 4.1
A vassal paying homage to a medieval feudal lord in a public ceremony. Painter unknown (https://en.wikipedia.org/wiki/Homage_(feudal)#/media/File:Hommage_au_Moyen_Age_-_miniature.jpg).

to a feudal lord by a vassal or tenant but, in the broad sense, it is loyalty or allegiance shown to anyone. For example, royal officials took it as part of accepting their duties; higher clergy demanded it of lesser clergy, and lords of the manor even required it sometimes of their peasants. At times, an oath of fealty was essentially a detailed, legal contract in circumstances like those above.

The Act of Homage, by which a vassal simply and starkly agreed to submission in aid of his lord, eventually broadened from purely military service to include such commitments and duties as might be required at court in times of peace. Though technically vassalage focused on, and grew out of, the need for military protection, especially between noble classes and the royal house, these ties of dependence were both used by and linked to all levels of society during the feudal era.

Although vassalage and a bond of fealty did not necessarily entail the granting of an estate, it often did. Moreover, the "fief" soon began to pass as a bequest from one generation to another in a hereditary manner. (A "fief," in the Middle Ages, was a piece of land granted to someone by a feudal lord, rather than money, in return for services rendered.) The lord had little means of preventing this impromptu act of inheritability, save to require homage of the new heir. To refuse to give homage under these circumstances became a principal means by which a family's fief might be lost during the feudal era.[8]

Over time, the circumstances surrounding the inheritability of a vassal's land led to increasingly difficult situations, in which a vassal would become enfeoffed to several lords in order to gain yet another fief—and also a heavy roster of vassalage duties, which could, upon occasion, conflict deeply with one another. (Under the European feudal system of the Middle Ages, "feoffment" or "enfeoffment" was the deed by which a person was given land in exchange for a pledge of service.) The practice of paying vassals for their services by granting a fief added to the problem.[9] "The diversity of ties was embarrassing at any time. In moments of crisis the problems it engendered became so urgent that both the theory and the practice of feudalism were obliged to look for a solution. When two of his lords were at war with each other, where did the duty of the good vassal lie?"[10]

As feudalism developed in complexity, a lord might give or sell his retainers' loyalty along with his woods, fields, and castles, but the men of the fief were still bound to their duties. Ties of dependence between lord and tenants were not usually formal acts of homage, but regulated by "the custom of the manor." They were communal, and constantly changing, but very detailed. In the ninth-century Frankish state, these customary obligations were enforced by royal courts, but later manorial lords often usurped the judicial powers and enforced customs beneficial to themselves.[11] The Franks were a Germanic tribe who invaded the ancient territory of Gaul (largely encompassing modern-day France and parts of modern-day Germany), where they established and ruled the Frankish Empire, known as Francia, from the 5th through the 10th centuries following the fall of the Roman Empire, usually cited as 476 CE.[12]

Manorial custom was dazzling in its local complexity:

> Nothing varied more from manor to manor according to locality, nothing exhibited more diversity, than the burden of tenancy in the feudal age. On certain days the tenant brings the lord's steward perhaps a few small silver coins or, more often sheaves of corn harvested on his fields, chickens from his farmyard, cakes of wax from his beehives.... At other times he works on the arable or the meadows of the demesne. Or else we find him carting casks of wine ... on behalf of his master to distant residences. His is the labour which repairs the walls or the moats of the castle.... When the hunting season comes round, he feeds the pack[13] (Figure 4.2).

The picture in feudal England was similar. In the mid-14th century, Henry de Bray, a petty knight and lord of Harlestone in Northamptonshire, owned a small manor of 500 acres. His 24 tenants had an annual obligation of "cash rents, a pound of pepper, and eight fowls, and performing harvest services."[14] But the earls or counts above him might have had manors with multifarious obligations stretching across half of England.

At the bottom of the system, as previously discussed, was the serf. The status of serfs varied by region in England, as opposed to the Continent, but fundamentally, his status revolved around whether he was free or bound, and how much land or wealth he owned. Bound serfs were, however, not exactly slaves; instead, bondage consisted of a cluster of restrictions, such as inability to inherit or sell land. His status was defined by the feudal system of obligation: "The unfreedom of the villein or serf was never a generalized condition, like slavery, but always consisted of specific disabilities: he owed the lord substantial labor services; he was subject to a number of fines or fees, in cash or in kind; and he was under the jurisdiction of the lord's courts."[15] In Frederic Maitland's words, the serf, or villein, "remained a free man in relation to all men other than his lord."[16] Thus, while the legal subtleties of vassalage could become astoundingly complex, the serf's economic status was usually obvious and unmistakable. He owned or farmed a given number of acres and possessed a given number of cattle or sheep.[17]

FIGURE 4.2
Medieval peasants harvesting grain with threshing flails. Painter unknown (https://commons.wikimedia.org/wiki/File:Battage_à_fléau_(original).jpg).

The Challenge of Feudalism

Feudalism was a system of legal obligations based on the manor, the rural estate owned by nobles or knights, thus linking the highest nobility to the serf, who was tied to the land. As such, it encapsulated a truth that remains important today: wealth is based on access to land and water. There are also essential obligations between a landowner and his or her laborers that have remained intact since feudal times.

Surely a central rule of feudalism—the larger the manor, the more extensive the obligations, both above and below—could be retooled for the current ecological crisis. As a starting point: the largest landholders are currently multinational corporations and, in some cases, the state. It is imperative that they have the highest obligations for the protection, creation, and maintenance of social-environmental sustainability. They must also carry the greatest responsibility for the reciprocal relationships on which all humanity depends for clean air, fertile lands, and pure waters. Meeting that goal will require removing the fictitious "legal personhood" from corporations so that each individual's personal obligations (rather than an abstract desire for profit) rule the reciprocity of his or her relationships—an inescapable requirement of social-environmental sustainability and justice. The abstract entities involved in the relationship will be units of government, which are public, and made up of the members of the public who are residents in that nation or state.

The ideals of feudalism were based on relationship. These ideals of reciprocal obligation, including spiritual obligation, can be extended fairly easily to non-human species and their habitats, local landscapes, and oceans, as well as human habitats, such as communities, cities, states, and nations. It is precisely this widening of relationship that we contemplate by placing Rights of Nature at the center of the legal system. A legal framework of this kind is the modern equivalent of the homage, feudal oaths, and vassalage that knitted medieval society together.

Modern society is even more complex than feudal society. But in the feudal era, the ties of vassalage and homage became more and more complex until they finally became unworkable, as more members of society sought unencumbered access to the ultimate source of wealth—land. In adapting and expanding the feudal ideals of responsibility, modern people can avoid creating a system too unwieldy to work well, by placing Nature's right to flourish first, extending responsibilities to all in society, and limiting the abilities of all to plunder Nature solely for their own benefit.

Indigenous Systems of Sustainability

Indigenous communities worldwide and in the United States not only differ substantially in many ways but also, as often noted, have such things in common as community elders instructing younger generations in traditional values and practices of spiritual sensitivity to plants, animals, and natural processes (Figure 4.3). This legacy often shaped, and still shapes, these cultures more deeply than has generally been, or is, true in Western society. We have picked a few recently studied examples to such enhanced reciprocity to illustrate the point that culture can be based on social-environmental reciprocity between the human and non-human species inhabiting our planet. When successful, it enhances and strengthens all the interdependent parties in their overall relationship.

Such a relationship must be community to community. That is, the relationship must be based on ideals of respect. This can be designed to include situations uncommonly thought of in terms of relationship, such as agriculture or forestry. For example, in a recent article focusing on marine cultivation by Northwest coast indigenous people, the authors define cultivation in this context as, "*any conscious effort to create specific conditions for advantageous engagement or relations with another being*; as such, cultivation includes a range of processes that often enhance resource output, even if their proximate goals may accentuate comparatively immaterial objectives, such as demonstrating interspecific respect between human communities and communities of prey species."[18]

Such a complex relationship has many goals for the human community, as the authors point out—not only an increase of practical knowledge necessary to succeed with cultivation, but also the result of making the cultivated species more "accessible, resilient, and predictable." When human communities succeed in creating and sustaining such relationships, they achieve a surprising status: "humans, as cultivators of keystone species, may have become keystone species themselves in terms of their contribution to ecosystem functioning and stability."[19]

In other words, if humans place themselves in relationship with other communities of beings, the result is both a strengthening of the other community and human importance

FIGURE 4.3
Elder native Alaskan woman. Photograph by Jo Keller, U.S. Fish and Wildlife Service (https://commons.wikimedia.org/wiki/File:Elder_native_Alaskan_woman_1979_FWS.jpg).

in enriching that particular ecosystem's resilience, which helps maintain its sustainability to the benefit of all concerned. Simply entering an ecosystem and beginning to "manage" it for human consumptive needs is not the same as entering into a mutually respectful relationship, in which species important to humans are observed sensitively and related to in ways that aid both the human community and the species being cultivated.

For example, this kind of respectful and reciprocal relationship existed between the Klamath and Modoc peoples and several vegetative species in their traditional territory, such as the huckleberry. The return to Huckleberry Mountain, west of Crater Lake, followed a regular annual rhythm, as Native families packed their belongings and arrived at their ancestral campsites on the mountain. Family members picked huckleberries at the same traditional sites annually, and processed them nearby. But the Klamath also worked to increase huckleberry production, by setting fires at the end of the season every other year to enhance future harvests.[20]

But this was only one means of encouraging an increase in the berries; equally important was their respectful spiritual ceremony: "[t]he Klamath and possibly other groups appear to have enhanced berry productivity at Iwamkani [Huckleberry Mountain] through first food ceremonies and other rituals conducted as part of the harvest. Some interviewees suggest that people had to 'feed the earth' there and if this respect was shown properly and regularly, 'you would always find lots of berries.' A small number of elderly women in the tribe recalled that the central focus of this ritual effort was the 'first huckleberry ceremony.' One tribal elder recalled: *The old Indians used to tell us that you had to take the first berries that you gathered up and throw them to the ground, and give them back to 'Mother Earth.' It was an offering.*"[21]

The Klamath people cultivated a similar reciprocal relationship with the yellow pond lily, found in the shallow, slow-moving waters of the Klamath marshes and associated lakes. The Klamaths removed plants that competed with the water lily, using fires and weeding as tools in order to increase the lily's presence near Klamath settlements. As with huckleberries, there appears to have been a "first foods ceremony": "[w]omen may have gathered up the first harvest and scattered it back to the ground in close proximity to their seasonal villages… as a kind of offering: 'returning to the earth I guess, returning it to their spirits.'"[22]

In this and other traditional Klamath relationships with plants—including lodgepole pine and Ponderosa pine, whose cambium was used for food and medicine—the central concern was always reciprocal obligations between the human and the plant communities, which was the foundation for the active enhancing of the resource: "In part, these values were rooted in oral traditions that described trees and other organisms as sentient beings that willingly accepted their functions within human societies; by these traditions, humans had obligations to care for and honor these organisms as part of a reciprocal obligation that binds humans to particular plant and animal communities."[23]

A second example of this kind of reciprocal relationship is the so-called "clam gardens" of Northwest coast peoples. The indigenous peoples used various techniques to enhance production of intertidal clam beds, which at the same time gave the clams opportunity to expand into new parts of the tidal zone and thus strengthened these marine invertebrates' resiliency. Techniques included cleaning the gardens of shells and refuse, litter or other waste, and rolling larger rocks out of the intertidal area to clear more ground, as well as stirring the soil with digging sticks to loosen it, and making it more able to breathe. These activities were part of the clam and cockle harvest of the Broughton Archipelago region's indigenous peoples, of what today is British Columbia, Canada.[24]

Such activity both created nearby stable clam beds, and increased clam habitat by removing boulders and thus extending the sandy beach regions preferred by local species of clams. In addition, the clams were harvested with an eye to maintaining their reproductive capacity, and giving the colony further room to expand by thinning the gardens when they became overcrowded. Clams were only harvested at certain periods of the year, when they were felt to be at their best. A garden's maintenance was a recognized clan resource: "The *loxiwey's* [clam garden's] defining characteristic is that they were actively monitored and maintained over generations. Their ownership by a clan or lineage was recognized; not only the harvesting and distribution of the clams, but also the traditional maintenance of the *loxiwey* itself was under the authority of the Ugwamay, or Clan Chiefs and their designates…"[25]

Such a clam garden also attracted other animals to scavenge, feed, or live in proximity to the enhanced clam beds, such as barnacles and sea cucumbers. Thus, the human communities benefited from a stable and plentiful food source and a growing oral tradition of techniques needed to enhance ecosystem resilience as part of the clam garden activities. The intertidal marine zone gained expanded clam habitat that benefited many other species, as well as the clams themselves.[26]

A third example concerns traditional cultivation of salmon on the Northwest coast of North America before the European settlement (Figure 4.4). Western concepts of "management" and conservation are narrower than the corresponding indigenous terms that cultivate a respectful and mutual relationship with an animal and its habitat—rather than "management" of a resource simply as a commodity for monetary gain. Respect, like relationship, requires reciprocal action by both parties, and thus requires a spiritual element—completely lost in mere "resource management."[27]

The authors refer to this kind of model as an "ecological-moral paradigm." The entire framework might often include communication with the spirits of non-human beings in specific ritual protocol to avoid offense and maintain respect:

> These paradigms can be highly developed, including stories and observations from the past, personal negotiations with non-human species, and a range of practical techniques for effective interaction development from experiment and trial and error, which, in turn, are transmitted intergenerationally ... the emphasis remains on cultivating and regulating respectful relationships with local species towards a goal of "keeping it living," rather than curtailing or restricting resource use.[28]

FIGURE 4.4
Atlantic salmon. Painting by Timothy Knepp, U.S. Fish and Wildlife Service (https://commons.wikimedia.org/wiki/File:Atlantic_salmon_Atlantic_fish.jpg).

Only part of the paradigm included "management" techniques, such as adjusting weirs to allow adequate escapement to continue a salmon run, curtailing net fishing, "streamscaping," predator control, or enhancing habitats by creating more pools or removing beaver dams. The goal was to "keep it living" and be respectful toward salmon, by (in part) enhancing the habitat so that salmon would desire to, and be able to, return to their natal stream.[29]

Transplanting of salmon or salmon eggs from one stream to another was also practiced, though always in the same context of respect, skilled trusteeship, and use of knowledge in ways that would not offend the Salmon-people. Many Northwest coast peoples had a so-called "salmon chief" "who oversaw human–fish relations both materially and ceremonially—monitoring harvests, enforcing proscriptions on over harvest, and leading the first salmon ceremony for his counterpart, the chief of the salmon."[30] Thus, it appears that robust, cultivated salmon runs may be said to have co-evolved with the indigenous communities wherein the people practiced the kind of reciprocity that ultimately led to enhanced resilience and stability for both.

The final example from the Northwest coast concerns traditional herring cultivation by the Tlingit, Haida, and Tsimshian peoples of Alaska. It was a process that combined both traditional knowledge and techniques to maintain and strengthen an important species for marine food webs across the entire North Pacific. The underlying goal was to deepen the indigenous relationships with the herring population. Spiritual practice was an integral part of this cultivation, as the fish were considered sentient beings.

The cultivation appears to have resulted from a mutual relationship and a desire to care for an important, local species. Nevertheless, cultivation techniques in the nearshore area were carried out sensitively to avoid disrespecting herring in their critical periods of spawning, during which the Haida and Tlingit practiced both spiritual and physical limitations in their activities. They enforced prohibitions on disturbing spawning areas, and implemented sanctions against individuals who treated the resource disrespectfully. But more than that, the herring cultivation included increasing the resilience and productivity of the herring fishery by expanding the spawning habitat, creating formal protocols to encourage herring to come inshore, spreading eggs into areas most suitable for producing high survival rates, and transplanting herring to new regions. These techniques not only enhanced the productivity for the people but also reinforced the ecosystem in which herring were, and are, a culturally important species. In addition, the Natives dried and smoked the fish so that they would last and could be used sparingly. Thus, herring were honored ecologically, practically, ceremonially, and spiritually.[31]

The Challenge of Indigenous Cultivation

The challenge is to look at such paradigms as the Klamath–Modoc plant relationships and the three Northwest coast mariculture examples, and work with them to fit the current situation of worldwide depletion of resources as human-influenced biophysical systems lose resilience in the wake of excessive and insensitive human use. ("Mariculture" is the cultivation of marine plants and animals in their native habitats, generally for human use.) There is no single way forward, but many. Communities vary, as do the habitats in which they are embedded. Nevertheless, there are some similar considerations.

Respect has always been central to traditional indigenous cultivation. What that term might mean outside its original indigenous context is hard to gauge. However, at the very least, we submit that it clearly means the cessation of industrial/commercial "harvest" of species far beyond any sustainable baseline. Such ventures always seek to suppress any notion of material respect—to say nothing of spiritual respect—for the other species involved. The only way a species can be used by humans is within the context of a reciprocal relationship between the human and the species itself, its habitat, and the other species that also depend on it. This requires conscious caretaking, not the illusion of control through "management."[32]

Such caretaking includes our human responsibility to the non-human species, through which we become the underpinning of sustainability by working within Nature's Laws of Reciprocity to enhance the overall well-being of the species on which we and all many others depend. A Rights of Nature legal system must be formulated to incorporate some form of these ideals and practices that allow Nature not only to flourish but also to increase in resiliency.

On the other hand, it is essential to recognize that the worldwide pressure on the natural environment is based in part on the necessities of a large, and growing, human population. How best to meet these requirements without destroying the various systems that produce the goods and services we require is surely one of the largest challenges. Small-scale cultivation of resources, such as those practiced by Northwest coast people, will not solve the problems, but they do provide pointers in terms of respect, and integrating spiritual understanding with techniques of trusteeship. The critical necessity of protecting viable populations, coupled with the integrity of their habitat, as governed by Nature's Laws of Reciprocity, provides the key to understand how people in widely varying circumstances can retool their reciprocal relationship with the ecosystems of the world toward social-environmental sustainability.

The United States, as other modern nation states, has multiple population groups and varied environments, ranging from forests to grasslands, and wetlands to alpine peaks. Clearly, a simple transfer of these indigenous ideals to modern legal systems would be infeasible, because indigenous-cultivation systems are deeply and intimately local. However, the principle of localism is critically important to all Rights of Nature systems. Thus, while the overall framework is broad, the responsibilities, restoration activities, and enforcement of Nature's Rights must happen place by place and community by community, in accordance with the federal and state laws—setting all human activity in a Rights of Nature legal system. Though the spiritual aspects of indigenous cultivation cannot be duplicated outside indigenous societies, the creation of a legal structure that requires people to think of Nature's Rights first, and reshape their activities not only to maintain Nature's Rights but also to *allow Nature to flourish* and to live free of pollution, creates a form of spiritual responsibility of the sort the Western legal system has not so far envisioned. Clearly, the opportunities for creative and wide-ranging collaboration are many.

Endnotes

1. (1) Written by the Editors of *Encyclopædia Britannica*. Carolingian dynasty. *Encyclopædia Britannica*. http://www.britannica.com/topic/Carolingian-dynasty (accessed September 9, 2015); and (2) The Saylor Foundation. The Fall of the Carolingian Empire. http://www.saylor.org/site/wp-content/uploads/2012/10/HIST201-1.1.3-FallofCarolingianEmpire-FINAL1.pdf (accessed September 9, 2015).
2. Discussion of the feudal system is based on: Marc Bloch. *Feudal Society, Vol. 2: Social Classes and Political Organization*. Translated by. L.A. Manyon. University of Chicago Press, Chicago. (1961) 229 pp.

3. *Ibid.* (p. 445).
4. *Ibid.*
5. Frances and Joseph Gies [correctly cited from the book cover]. *Life in a Medieval Village.* Harper Collins Publishers, New York. (1990) 257 pp.
6. Marc Bloch. *Feudal Society, Vol. 2: Social Classes and Political Organization. op. cit.* (pp. 541–452).
7. Marc Bloch. *Feudal Society, Vol. 1: The Growth of Ties of Dependence. op. cit.* (pp. 145–146).
8. The preceding four paragraphs are based on: Marc Bloch. *Feudal Society, Vol. 1: The Growth of Ties of Dependence. op. cit.*
9. *Ibid.*
10. Marc Bloch. *Feudal Society, Vol. 1: The Growth of Ties of Dependence. op. cit.* (p. 213).
11. Marc Bloch. *Feudal Society, Vol. 1: The Growth of Ties of Dependence. op. cit.*
12. (1) The Saylor Foundation. 2012. The Frankish Empire. http://www.saylor.org/site/wp-content/uploads/2012/10/HIST201-1.1.1-FrankishEmpire-FINAL1.pdf (accessed October 6, 2015); and (2) 6f. The Fall of the Roman Empire. http://www.ushistory.org/civ/6f.asp (accessed October 6, 2015).
13. Marc Bloch. *Feudal Society, Vol. 1: The Growth of Ties of Dependence. op. cit.* (pp. 249–250).
14. Frances and Joseph Gies [correctly cited from the book cover]. *Life in a Medieval Village.* (p. 44).
15. Frederic William Maitland. Doomsday Book & Beyond: Three Essays in the Early History of England. Cambridge University Press, New York. (p. 68).
16. *Ibid.*
17. Frances and Joseph Gies. *Life in a Medieval Village. op. cit.*
18. Thomas F. Thornton and Douglas Deur. Introduction to the special section on marine cultivation among indigenous peoples of the Northwest Coast. *Human Ecology*, 43 (2015):187 (emphasis in original).
19. *Ibid.*
20. Douglas Deur. "A caretaker responsibility": Revisiting Klamath and Modoc traditions of plant community management. *Journal of Ethnobotany*, 29(2009):296–322.
21. *Ibid.* (p. 302) [emphasis in original].
22. *Ibid.* (p. 305).
23. *Ibid.* (p. 309).
24. Douglas Deur, Adam Dick, Kim Recalma-Clutesi, and Nancy J. Turner. Kwakwaka'wakw "clam gardens": Motive and agency in traditional Northwest Coast mariculture. *Human Ecology*, 43 (2015):201–212.
25. *Ibid.* (p. 206).
26. *Ibid.*
27. Thomas Thornton, Douglas Deur, and Herman Kitka Sr. Cultivation of salmon and other marine resources on the Northwest Coast of North America. *Human Ecology*, 43 (2015):189–199.
28. *Ibid.* (p. 191).
29. *Ibid.*
30. *Ibid.* (p. 193).
31. The preceding two paragraphs are based on: Thomas F. Thornton. The ideology and practice of Pacific herring cultivation among the Tlingit and Haida. *Human Ecology*, 43 (2015):213–223.
32. Chris Maser. 2005. *Our Forest Legacy: Today's Decisions, Tomorrow's Consequences.* Maisonneuve Press, Washington, DC. (2005) 255 pp.

Section II

Building Blocks for a Rights of Nature System

It is difficult to free fools from the chains they revere.

François-Marie Arouet (Pen name: Voltaire)[1]

Endnote

1. Voltaire. http://www.brainyquote.com/quotes/authors/v/voltaire_2.html (accessed October 23, 2015).

5

Precursors to Rights of Nature

The Misunderstood Privilege of Land and Resource "Ownership"

While I (Chris) may, in a legal sense, "own" (with title and deed) the land in which my garden resides, I can only borrow it in an ethical sense. I am therefore both a temporary custodian and a trustee of my garden for those who must someday live where I now dwell.[1]

Private ownership of land is a historically recent concept when one considers the length of time humans have been on Earth. For instance, the pygmies of central Africa—the most ancient of all forest dwellers—hold no enforceable claims to the forests they have inhabited for at least 40,000 years.[2] Generally speaking, indigenous peoples on every continent find the notion of private ownership of land to be ludicrous and contrary to their ethical and communal framework. This sense of ethics and community includes the societies indigenous to Europe when they were still tribally organized.

How can an individual human being own something that he or she has not created and therefore cannot control? How can a person own something that has been around for millennia before he or she was born and will continue for millennia after he or she has passed on?[3]

Although fervently divided on this issue, society must come to terms with it for the sake of all generations—born and unborn. Before a system of Nature's Rights can be put into place in the American legal system, the ownership of land and the unlimited rights of private property must be modified. Land ownership in the United States gives owners a nearly unlimited legal right to do whatever they want on their land, subject to limitations based on curbing at least some harm to neighbors or the larger community. From these concerns have sprung land-use regulations, as well as air and water quality laws, among others. But these regulations have not shaken the fundamental ideal of private land ownership or required landowners to bear all the costs of their activities, even when supremely destructive.

Edmund Burke, a British economist and philosopher, captured the limits of land ownership and the unlimited rights of private property as a sustainable ideal proposition when he wrote:

> Men are qualified for civil liberty in exact proportion to their disposition to put moral chains upon their own appetites.... Society cannot exist unless a controlling power upon will and appetite be placed somewhere, and the less of it there is within, the more there must be without. It is ordained in the eternal constitution of things that men of intemperate minds cannot be free. Their passions forge their fetters.[4]

The limits to the use of private property that underpin Western society are becoming apparent: they do not make the owners of the land pay for or live with all the environmental damages accrued by some activities. Instead, surrounding landowners—and usually the public at large—are forced to bear nearly all of the damages. Classic examples include mountaintop-removal mining,[5] clearcut logging,[6] and agriculture that depletes the groundwater used by all people in an area.[7] Since the landowners do not face the true costs, they are more easily able to propose and undertake destructive activities that fragile ecosystems and future generations can ill afford.

This is even true of damaging activities taking place on federal public lands, where the environmental costs of such things as clearcut logging or oil and gas drilling are more starkly unacceptable, since the lands are owned by all Americans. Yet land-owning agencies often make the same shortsighted calculus as do private landowners, even when weighing public values.

If such is the case with public and private property, how much the more so with land or waters deemed to be the "commons," or to use the Latin term, *res nullius* (meaning "nobody's property," a term derived from Roman law). Traditionally, the concept of *res nullius* was applied to the high seas in international law: it was outside anyone's jurisdiction and thus could be used freely by all. It was also commonly applied to the resources of the sea, such as its fisheries and plants.

The Commons

The commons is that part of the world's environment that is every person's birthright. There are two kinds of commons. Some are available to all, such as clean air, pure water, the ocean, uncontaminated soil, a rainbow, the northern lights, a beautiful sunset, or a tree growing in the middle of a village; others are the collective product of human creativity, such as the town well from which everyone draws water.

Origin of the Commons

The commons is the "hidden economy, everywhere present but rarely noticed."[8] It provides the basic social-environmental support systems of life and well-being. It is the vast realm of our shared heritage, which we typically use free of toll or price. The commons has an intrinsic quality of just being there without formal rules of conduct. People are free to breathe the air, drink the water, or enjoy a beautiful landscape without asking permission or having to pay a toll.[9]

All humans jointly inherit the commons, which is more basic to our lives and well-being than either the market or the state.[10] We are "temporary possessors and life renters," wrote Burke, and we "should not think it amongst [our] rights [to] commit waste on the inheritance."[11]

The Tragedy of the Commons

Despite the wisdom of Burke's admonishment, the commons is today almost everywhere assaulted, abused, and degraded in the name of economic development, as corporations are increasingly privatizing both Nature's services and every creature's birthright to those services. For example, the greatest threats to the world's oceans are increasing temperature, destructive fishing in deep waters, and point-source pollution, some of which is simply human sewage.[12] Pollution also despoils the air, defiles the soil, and poisons the water. City lights hide the stars by night. Urban sprawl; the disintegration of community; and attempts to control, engineer, and patent the very substance of life itself are all part of the economic raid on the commons for private monetary gain.[13]

To elucidate such irresponsible human behavior, biologist Garrett Hardin coined the phrase, "tragedy of the commons,"[14] which he based on a pamphlet published in 1833. In that leaflet, British economist William Forster Lloyd described the problems of cattle overgrazing the commons of English villages. In essence, Lloyd detailed how each herder's decision to compete even slightly for more grass for his cattle led to overgrazing, because the whole group shared the damage to the commons, yet individually received greater benefits for his own livestock. By this means, the commons could be destroyed.[15] (Lloyd's actual passage is found at endnote 16.)

And so it is that today, the global commons is being so manipulated for economic gain that its productivity and resilience are being rapidly destroyed. "Corporations," says author David Korten, "are pushing hard to establish property rights over ever more of the commons for their own exclusive ends, often claiming the right to pollute or destroy the regenerative systems of the Earth for quick gain, shrinking the resource base available for ordinary people to use in their pursuit of livelihoods, and limiting the prospects of future generations."[17]

This is not to say that all corporations are bad. But it is to say that both corporations and the market must have boundaries to keep them within limits. Only a legal system organized around the Rights of Nature is able to place human activity in a truly sustainable framework. Otherwise, the corporatized worldwide economic system, combined with increasing human needs at every level, will continue to plunder the commons and leave only minimal ecological functioning with greatly weakened resilience to combat future depredations, whether caused by human activities or natural events. "The supporters of the market economy do it the worst service by not observing its limits," says author Jonathan Rowe.[18]

For the sake of future generations, we must understand, acknowledge, and remember that we, in the biosphere, live sandwiched between the atmosphere (air) and the litho-hydrosphere (rock and water). Our lives depend on two great oceans: one of air and one of water, both of which have currents that circumnavigate the globe. Because each sphere is inexorably integrated with the others, if we degrade one, we degrade all three.

The oceans are a representative example of the commons, perhaps the quintessential one: beyond any nation's control, and available to all for travel, fishing, and other uses. Over-exploitation of the ocean's resources is a classic case of the tragedy of the commons. We discuss two examples of the tragedy affecting oceans worldwide to highlight the essential need for a Rights of Nature system—with the appropriate legal framework to encompass the global commons.

The Northern Atlantic Cod Fishery

The collapse of the cod fishery is a textbook example of how overfishing—combined with an inaccurate, unresponsive regulatory strategy—caused the collapse of the Northern Atlantic cod fishery on the Grand Banks off the eastern seaboard of Canada (Figure 5.1). Overfished for several centuries, but especially in the mid-20th century as industrial-scale offshore fishing trawlers took gigantic catches for a few years, the cod fishery collapsed by 1992, when the Canadian government issued a complete moratorium on commercial cod fishing.

FIGURE 5.1
Four-foot long cod are slung across the backs of cod fishermen, February 1915. Photograph by A. B. Wiltse, *National Geographic* (https://commons.wikimedia.org/wiki/File:Four-foot_long_cod_are_slung_across_the_backs_of_cod_fishermen.jpg).

After limited reopening of the inshore fisheries 10 years later, at what turned out to be unsustainable levels, the depleted stocks were further devastated. In 2003, Canada once again placed a total moratorium on the commercial fishing for cod, which is considered commercially extinct, and might be headed for biological extinction. The situation is unclear, however, because, as of 2004, some inshore cod populations seemed to be recovering, while offshore stocks remain severely depleted. As of 2015, research showed cautious optimism that some of the stocks seem to be rebounding, though not yet fully recovered. All signs indicate that the moratorium was essential in the recovery, and continues to be so.[19]

There were several interrelated problems leading to this tragedy of the commons. It seems to have been less a problem of overfishing by smaller inshore Newfoundland cod fisheries than by industrial corporations using technology in factory trawlers on the Outer Banks to catch northern cod on a gigantic scale that the inshore fisheries never reached. In addition, scientific assessment of cod populations, used in setting allowable catches, were deeply flawed. Among other things, the science failed to take into account the different structures of inshore and offshore cod populations, as well as other subpopulation dynamics. Data from trawlers used to predict stock size were also inaccurate. Beyond that, it may be that the overfishing changed cod's behavior or lowered the population below levels that allowed critical reproductive behaviors to take place.[20]

Most important of all in averting a further tragedy is to allow complete recovery of the cod, and then place a legal framework on the fishery that focuses on Nature's Rights, following Nature's Laws of Reciprocity that guarantee the ecosystem's right to life, rather than mining it to satisfy human wants. This is essential because many of the factors leading to the cod's decline and recovery are, and will always be, poorly understood.[21]

Plastic in the Oceans

Another example of the tragedy of the unregulated commons is marine debris. The problem of plastics in the ocean (a great deal of which is then deposited back onto the world's shorelines by ocean currents) has received much publicity; there are "garbage patches" in the Pacific Ocean where plastic articles of many sizes gyrate in the currents at various depths of the water column. The most well known is the Great North Pacific Garbage Patch, brought to public attention in 1997 (Figure 5.2).

It is not so universally understood that plastics do break down in the ocean into smaller components of toxic material, especially in warmer areas of the ocean. Becoming ever smaller, "plastic confetti" disintegrates into toxic chemicals, such as Bisphenol-A and PS oligomer. This alarming discovery was unveiled by Japanese chemist Katsuhiko Saido at

FIGURE 5.2
Pieces of discarded plastic that have collected on a black-rock beach on the island of Hawaii. As the plastic breaks down, the smaller pieces are often mistaken for food and ingested by marine birds, sea turtles, and fish and can kill the unsuspecting wildlife. Photograph by Eric Johnson, U.S. National Oceanic and Atmospheric Administration (https://commons.wikimedia.org/wiki/File:Fish1968_-_Flickr_-_NOAA_Photo_Library.jpg).

the 238th National Meeting of the American Chemical Society in August 2010.[22] Plastic-derived chemicals are easily ingested by small, marine animals of various kinds and move up the food chain, becoming more concentrated in the bodies of higher-order predators. This has made of the ocean a toxic soup, likely to be very damaging to marine wildlife, and ultimately to human beings as well.[23]

According to Charles Moore of the Algalita Marine Research Foundation, "Plastic debris, most of it smaller than five millimeters in size, is now dispersed over millions of square miles of ocean, and miles deep in the water column. The plastic soup we've made of the ocean is pretty universal—it is just a matter of degree. All these effects we're worried about are happening through the ocean as a unity."[24]

To examine the effects of microplastic polystyrene particles in environmentally relevant concentrations, the ingestion of 90-micron-sized (0.003 543 307 086 6-inch-sized) particles on larval, freshwater European perch was studied. The microplastic inhibited hatching, decreased growth rates, and altered feeding preferences, as the larvae favored the particles over more natural foods. Moreover, individuals exposed to microplastics did not respond to cues of olfactory threats, the consequence of which would potentially be increased predator-induced mortality rates. In essence, the ingested microplastic particles operated both chemically and physically on the development of the larval fish and the performance of their innate behaviors.[25]

Marine plastic is a stark example of the tragedy of the commons. Nations regulate their own resource use, but frequently dispose of waste in the sea. Belonging to no one, the sea has been considered a good place for trash, because the country placing it there had no more need to regulate it. Any costs associated with the trash would be borne collectively by the peoples of the world, all of whom depend on the oceans for everything from food to climate—including every drop of water on Earth.

As there is no effective international legal framework to force nations to stop polluting the global ocean, the tragedy continues. Nations have every reason to continue dumping plastics and other debris into the oceans, because there are no penalties, the costs are borne by others, and it removes plastic waste from the nation's boundaries. It is a classic example of failed communal responsibility. Moreover, the effect is magnified by the fact that, even if plastic pollution were ended tomorrow, the plastics already in the ocean will continue polluting it in damaging ways far into the future—likely centuries. Global use of plastics was 260 million tons in 2008. In 2012 global plastics use was projected to reach 297.5 million tons by 2015. But in fact, the most recent figures show actual plastics production was 342 million tons in 2014, projected to double again in the next twenty years. Currently, nearly nine million tons of plastic are estimated to enter the oceans each year, and the amount is estimated to double by 2030 if no action is taken. Only 14 percent of plastics at most are recycled. Clearly, the likelihood is minuscule that polluting the world's oceans will end anytime soon.[26] Once again, the only way to end this devastation of the world's oceans is to adopt a Rights of Nature system that gives the global commons the right to life, to flourish, and to continue the cycles of life unimpeded. To protect the world's oceans, it must be placed in an international framework binding on all nations.

The Global Commons and Rights of Nature

Treating the global commons as *res nullius* (nobody's property) has led to numerous cases of severely depleted natural resources and peril to the ecosystem's functioning, such as the two highlighted above. Recognizing the limitations of the traditional concept, international law has posited a more communal way of considering resources of the commons: *res communis*. "*Res*" is Latin for "property" and *communis* means "common" or "shared property," a term derived from Roman law pointing to things owned by no one and subject to use by all, such as light, air, the sea, running water, and so on. This idea proposes that such resources are the property of everyone, rather than no one; thus, their conservation is the responsibility of all, as is the quality of their use. The principle of *res communis* was extended in the 20th century to areas that technology has recently brought within human reach, such as outer space, minerals on and in the deep seabed, and the atmosphere.[27]

But even a framework implementing the more expansive principle of *res communis*, had it been fully extended to the global commons and enforced by international legal mechanisms, would not go far enough in this era of massive overappropriation of communal resources by many nations, leading to a deep violation of Nature's right to life by negatively effecting Nature's biological resilience. In essence, over-exploitation of communal resources harms ecosystem-scale processes on which all humans depend: "Further technological progress … as well as the realization that the riches of this planet are not infinite and that therefore all the elements needed for the survival of mankind must be considered resources have led to the conviction that a simple regime of non-appropriation and common use of those resources—which could be freely appropriated—was not sufficient."[28]

Endnotes

1. Chris Maser (with meditations by Zane Maser). *The World is in My Garden: A Journey of Consciousness*. White Cloud Press, Ashland, OR. (2005) 303 pp. (p. 81).
2. Kirk Talbott. *Central Africa's Forests: The Second Greatest Forest System on Earth*. World Resources Institute, Washington, DC (January 1993).
3. Chris Maser (with meditations by Zane Maser). *The World is in My Garden: A Journey of Consciousness. op. cit.*
4. Edmund Burke. http://www.goodreads.com/quotes/293641-men-are-qualified-for-civil-liberty-in-exact-proportion-to (accessed December 3, 2015).
5. Plundering Appalachia. http://www.plunderingappalachia.org/theissue.htm (accessed December 3, 2015).
6. Chris Maser and Walter Smith. *Forest Certification in Sustainable Development: Healing the Landscape*. Lewis Publishers, Boca Raton, FL. (2000) 235 pp.
7. Groundwater depletion. http://water.usgs.gov/edu/gwdepletion.html (accessed December 3, 2015).
8. Jonathan Rowe. The hidden commons. *Yes! A Journal of Positive Futures*, (Summer 2001):12–17.
9. Chris Maser. *Earth in Our Care: Ecology, Economy, and Sustainability*. Rutgers University Press, New Brunswick, NJ. (2009) 304 pp.
10. Karen A. Kidd, Paul J. Blanchfield, Kenneth H. Mills, and others, Collapse of a fish population after exposure to a synthetic estrogen. *Proceedings of the National Academy of Sciences*, 104 (2007):8897–8901.
11. Constitutional Law Foundation. Intergenerational Justice in the United States Constitution, The Stewardship Doctrine: II. The Intergenerational Philosophy of the Founders and Their Contemporaries. http://www.conlaw.org/Intergenerational-II-2.htm (accessed December 3, 2015).
12. Chris Maser. *Interactions of Land, Ocean and Humans: A Global Perspective*. CRC Press, Boca Raton, FL. (2014) 308 pp.
13. Chris Maser. *The Perpetual Consequences of Fear and Violence: Rethinking the Future*. Maisonneuve Press, Washington, DC. (2004) 373 pp.

14. Garrett Hardin. The tragedy of the commons. *Science*, 162 (1968):1243–1248.
15. William Forster Lloyd. *Two Lectures on the Checks to Population*. Oxford Univ. Press, Oxford, England. (1833).
16. William Forster Lloyd. Wikipedia, the free encyclopedia (under Influential lectures). https://en.wikipedia.org/wiki/William_Forster_Lloyd (accessed December 4, 2015).
17. David Korten. What to do when corporations rule the world. *Yes! A Journal of Positive Futures*, (Summer 2001):148–151.
18. Jonathan Rowe, The hidden commons. *op. cit.*
19. George A. Rose and Sherrylynn Rowe. Northern cod comeback. *Canadian Journal of Fisheries and Aquatic Sciences*, 72 (2015):1789–1798.
20. Bonnie J. McCay and Alan Christopher Finlayson. The Political Ecology of Crisis and Institutional Change: The Case of the Northern Cod. Presented to the Annual Meetings of the American Anthropological Association, Washington, DC. November 15–19, 1995. http://arcticcircle.uconn.edu/NatResources/cod/mckay.html (accessed December 1, 2015).
21. Douglas P. Swain and Robert K. Mohn. Forage fish and the factors governing recovery of Atlantic cod (*Gadus morhua*) on the eastern Scotian Shelf. *Canadian Journal of Fisheries and Aquatic Sciences*, 69 (2012):997–1001.
22. Hard plastics decompose in oceans, releasing endocrine disruptor BPA. http://www.acs.org/content/acs/en/pressroom/newsreleases/2010/march/hard-plastics-decompose-in-oceans-releasing-erine-disruptor-bpa.html (accessed December 1, 2015).
23. (1) Claire Le Guern Lytle. When The Mermaids Cry: The Great Plastic Tide. http://plastic-pollution.org/ (accessed December 1, 2015); (2) Chris Maser. *Interactions of Land, Ocean and Humans: A Global Perspective. op. cit.*
24. (1) Carolyn Barry. Plastic breaks down in ocean, after all—and fast. *National Geographic News*, August 20, 2009. http://news.nationalgeographic.com/news/2009/08/090820-plastic-decomposes-oceans-seas.html (accessed December 1, 2015); and (2) Karen Matthews. NYC [New York City] Waters Are Teeming With Plastic Particles, Study Finds. http://abcnews.go.com/Technology/wireStory/nyc-waters-teeming-plastic-particles-study-finds-36914574 (access February 13, 2016).
25. (1) Oona M. Lönnstedt and Peter Eklöv. Environmentally relevant concentrations of microplastic particles influence larval fish ecology. *Science*, 352 (2016):1213–1216; and (2) U.S. Fish and Wildlife Service. European Perch (*Perca fluviatilis*). https://www.fws.gov/injuriouswildlife/pdf_files/Perca_fluviatilis_WEB_9-15-14.pdf (accessed June 5, 2016).
26. (1) Claire Le Guern Lytle. When The Mermaids Cry: The Great Plastic Tide. *op. cit.*; (2) Chris Maser. Oceans in crisis—Human garbage. *Journal of Aquaculture & Marine Biology*, 2 (2015):1–6. DOI: 10.15406/jamb.2015.02.00030; and (3) World Economic Forum. The New Plastics Economy: Rethinking the future of plastics (January 2016). https://www.weforum.org/reports/the-new-plastics-economy-rethinking-the-future-of-plastics.
27. Alexandre Kiss. The common heritage of mankind: Utopia or reality? *International Journal*, 40 No. 3, *Law in the International Community* (1985):423–441.
28. *Ibid.* (p. 424).

6

International Steps toward Rights of Nature

The United Nations World Charter for Nature

In 1982, the international community first voiced its collective concern over the unimpeded despoliation of Nature. The General Assembly passed the United Nations World Charter for Nature that year.[1] It was the very first international document setting forth human responsibilities toward Nature. Though strictly aspirational, its moral and ethical impact was immense, and set the tone for future international activity to protect Nature from excessive exploitation through various types of human-oriented "developments."

In the Preamble, the Charter states, "Mankind is a part of nature," and affirms that the United Nations, "is convinced that every form of life is unique, warranting respect regardless of its worth to man.... [M]an can alter nature and exhaust natural resources by his action or its consequences and, therefore, must fully recognize the urgency of maintaining the stability and quality of nature and of conserving natural resources."[2]

The first of the General Principles is "Nature shall be respected and its essential processes shall not be impaired."[3] This is followed by prohibitions on compromising genetic viability, providing protections for unique areas, and managing all resources used by humans "to achieve and maintain optimum sustainable productivity, but not in such a way as to endanger the integrity of other ecosystems or species with which they coexist."[4]

Among the many Functions laid out in the Charter, requirements for planning in allocating resources for human use stand out; three Functions specifically call for planning as essential prerequisites to good decision-making on social and economic improvement schemes. Echoing later Rights of Nature language, the Charter states, "Living resources shall not be utilized in excess of their natural capacity for regeneration."[5] Other Functions require maintaining soil productivity, water reuse, restraint in use of non-renewable resources, and use of best available technology to minimize risks.

Significantly, the Charter enunciates an early version of the Precautionary Principle: "Activities which are likely to pose a significant risk to nature shall be preceded by an exhaustive examination; their proponents shall demonstrate that expected benefits outweigh potential damage to nature, and where potential adverse effects are not fully understood, the activities should not proceed."[6]

The Charter was drafted to provide moral and ethical guidance to nations. Thus, its principles were to be "reflected" in the laws and practices of each nation, as well as at the international level. The list of implementing guidance is long, including such things as:

1. Establishing standards for products and manufacturing processes that would otherwise adversely affect Nature
2. Monitoring to provide early detection of threats to an ecosystem
3. Providing environmental education
4. Ensuring that one nation's activities do not cause damage to natural systems in other states or outside national jurisdiction
5. Increasing knowledge of Nature by scientific research
6. Placing conservation of Nature at the heart of planning activities, including inventorying ecosystems
7. Assessing the effects of activities on natural systems[7]

The United Nations promulgated the World Charter for Nature on the initiative of Zaire. President Mobutu Sese Seko first proposed the idea to the International Union for Conservation of Nature and Natural Resources in 1975. On his suggestion, the United Nations created a multinational task force to draft the Charter; an initial draft was ready in 1979. The following year, Zaire presented the draft to the Secretary General of the United Nations, and took the initiative to steer it through two revisions. Zaire then formally introduced it to the General Assembly. More than 50 nations commented on the draft, expressing various concerns, and 15 commented on the revised draft. Zaire submitted the Charter for adoption to the UN General Assembly in 1982.

At this late juncture, the United States requested a delay to substantially revise the Charter—although the United States had not commented on any of the drafts. Other member states defeated the U.S. proposal, despite their own reservations about some of the Charter's language. The General Assembly adopted the Charter in 1982, by a vote of 111 to 1. The sole dissenting nation? The United States.

The World Charter was an initiative of the "non-developed" (= non-industrialized) nations and was meant to be aspirational from the beginning, though it contains a lot of mandatory language. But, though not legally binding on nations, it was a carefully written proclamation to create guiding principles that carry political and moral force. The Charter's purpose was to encourage voluntary enforcement, despite having aspirational language that is nonenforceable in the strict legal sense; it does not, for example, specify what kinds of environmental protections nations should put into place. But it does focus on the need for environmental study of proposed development projects, which led to much discussion among nations about what kinds of assessment might be valuable.

It was then, and remains, a primary issue in international forums whether environmental protection clashes with economic development. The World Charter ignited that discussion again at the time of its adoption, and such discussions need to continue even more urgently as time progresses.[8] At the time, "...the Amazonian countries reiterated the concern that environmental protection would hinder development. But several developing nations countered that the "protection of nature" intended in the Charter merely recognizes the limited capacity of environmental systems to support development, and this goal does not conflict with sustainable development."[9]

Now that environmental limits have been passed in many ways, it is essential to investigate how sustainability can be designed and maintained worldwide against all seekers of plunder. In the exploration for legal building blocks to a Rights of Nature system, the World Charter was an initial visionary document in which nations, for the first time, grappled with the problems of overuse of the commons—both the global commons and those inside the boundaries of nation states.

The Common Heritage of Mankind Principle

Recognizing the inadequacy of voluntary, cooperative restrictions to protect the global commons, Arvid Pardo of Malta enunciated the "Common Heritage of Mankind Principle" in the 1960s, and initially focused on the need for protection of the deep seabed.[10] Then, in December 1970, before the Parliamentary Assembly of the Council of Europe, he stated: "In ocean space, however, the time has come to recognize as a basic principle of international law the overriding common interest of mankind in the preservation of the quality of marine environment [sic] and in the rational and equitable development of its resources lying beyond national jurisdiction … in the longterm [sic] these interests can be protected only within the framework of a stable international regime of close co-operation between states."[11]

The Common Heritage Principle is deeper, broader, and more protective than the older and more diffuse idea of the commons. It posits that certain interests of all humankind must be safeguarded, which can be done only by specially tailored legal frameworks. Understandably, this idea has not been welcomed by many states or corporations.[12] In legal theory, the question remains whether the Common Heritage Principle actually creates communal ownership, or merely joint management of global resources held to be a common heritage outside national boundaries. Which definition applies has myriad consequences in terms of the subsequent legal framework.[13]

Nevertheless, elements of the Common Heritage Principle are becoming embedded in international law. What does it mean in practice? There are many definitions, but a clear one would be: "This means that the states suspend or do not assert rights or claims to territorial jurisdiction … for the benefit of the whole human community, without any immediate return, and conserve and if necessary manage areas in conformity with the common interest for the benefit of all mankind."[14] The Common Heritage Principle has been applied particularly to areas beyond the limits of national jurisdiction, such as the Antarctic, the deep seabed, and the moon.

The Race for the Antarctic

The earliest example is probably the Antarctic Treaty, signed by the United States in December 1959 (Figure 6.1). It was clear that such a treaty was desperately needed because Argentina, Australia, Chile, France, New Zealand, Norway, and the United Kingdom had already made sovereign claims over sections of Antarctica. It became clear to both the United States and the Soviet Union during the Cold War period that the continent had military uses and exploitable resources.[15]

The Antarctic Treaty applies to the Antarctic and the oceans around it. The Preamble states, "it is in the interest of all mankind that Antarctica shall continue forever to be used

FIGURE 6.1
The ceremonial marker at the South Pole is surrounded by flags of the original signatory nations of the Antarctic Treaty: Argentina, Australia, Belgium, Chile, the French Republic, Japan, New Zealand, Norway, the Union of South Africa, the Union of Soviet Socialist Republics, the United Kingdom of Great Britain and Northern Ireland, and the United States of America. The treaty was signed on December 1, 1959, in Washington, D.C. Photograph by Josh Landis, National Science Foundation (https://commons.wikimedia.org/wiki/File:Ceremonial_South_Pole.jpg).

exclusively for peaceful purposes and shall not become the scene or object of international discord."[16] The 12 original signatories did not renounce territorial claims, but did stipulate all use thereof must be peaceful, and they agreed to freedom of scientific exploration. The initial treaty was subsequently expanded in 1972, 1980, and later to limit human impact on the fragile Antarctic environment, protect wildlife, and put in place a regime of international cooperation and management of the living resources.

As a working legal concept, the Common Heritage Principle may be best embodied thus far in the Antarctic legal framework. However, the Antarctic Treaty was drawn up before the Common Heritage Principle was articulated, so it is not an overt organizing doctrine in the treaty. In addition, the Antarctic Treaty gives management responsibilities to those countries that negotiated it, those that agreed to it, and those that fulfilled the principal requirements in Article IX(2). These countries are the so-called "Antarctic Treaty Consultative Parties." Thus, the Treaty does not grapple with questions of equitable management or worldwide responsibility. These exclusivity provisions led to international dissatisfaction; in 1985, the United Nations made information regarding the treatment of Antarctica available to all[17] (Figure 6.2).

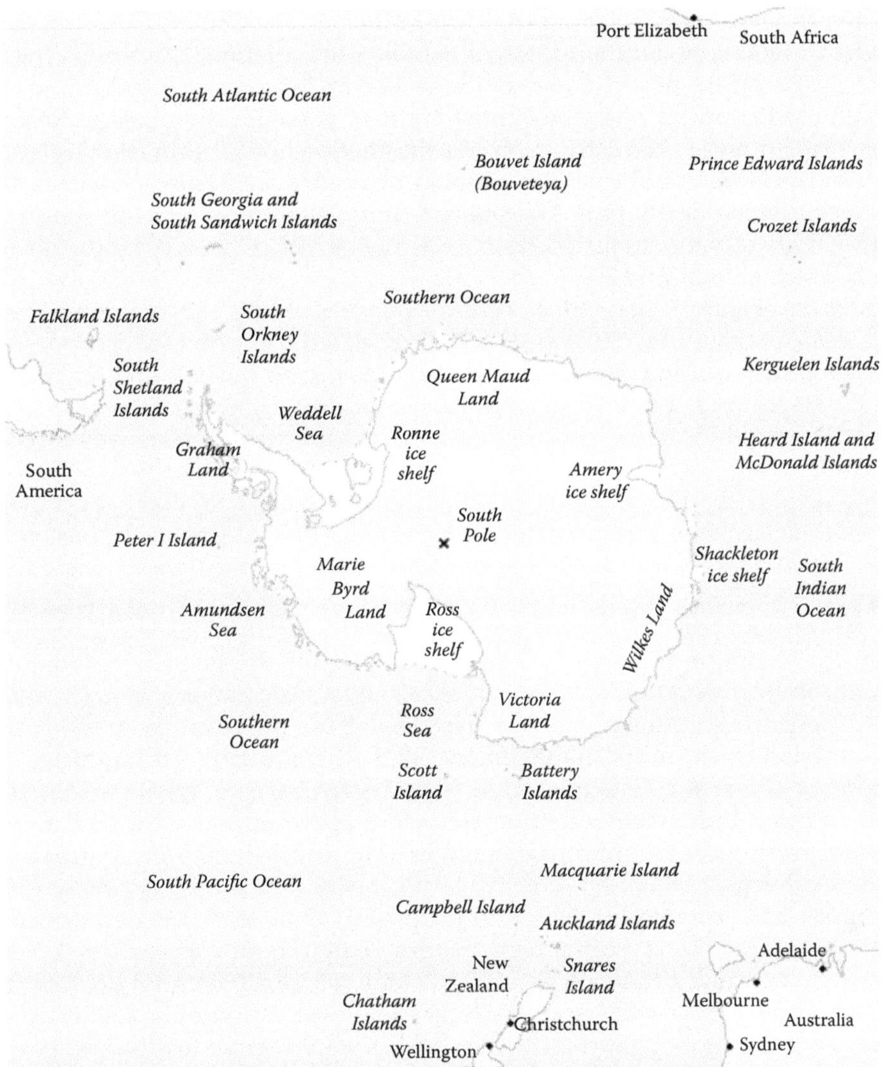

FIGURE 6.2
Map of Antarctica (https://commons.wikimedia.org/wiki/File:Antarctica_Map.png).

The Deep Seabed Controversy

The real genesis of the Common Heritage of Mankind Principle, and its most explicit wording, sprang from concerns in the 1970s about the riches of the deep seabed. The discovery of metallic nodules on the seafloor in the late 19th century, as well as the technological developments by the late 1960s, made commercial-mining ventures increasingly likely. The possibility of deep-sea mining became a major cause of concern, because the deep seabed lay outside the jurisdiction of all nations.

The Common Heritage of Mankind Principle was developed explicitly to establish a legal framework to regulate mining or other deep-seabed activity. The United Nations Secretary General outlined, in 1967, the need to establish a legal framework for natural resource extraction and other human activities on the deep seabed. The General Assembly followed the framework by forming the "Committee on the Peaceful Uses of the Seabed and the Ocean Floor beyond the Limits of National Jurisdiction," known as the "Seabed Committee," to draft the peaceful uses of the seabed and the ocean floor beyond the limits of national jurisdiction. In 1969, the United Nations passed a Moratorium Resolution to halt any efforts to mine the seabed until an international legal framework could be put in place. The Declaration of Principles adopted by the United Nations General Assembly in 1970 states unequivocally that "the sea-bed and ocean floor, and the subsoil thereof, beyond the limits of national jurisdiction ... as well as the resources of the area, are the common heritage of mankind."[18]

Essentially, developing (= non-industrialized) countries used the "Moratorium Resolution" to restrict seabed activity by developed (= industrialized) countries, which possessed the technology for such mining. However, industrialized countries felt that any moratorium blocking the development of technologies for seabed mining was contrary to a Common Heritage of Mankind Principle, because, under a moratorium, no state would receive any benefits at all.

Although deep-seabed mining has not materialized, other potential uses could equally damage seafloor ecology, including tourism, military tracking devices, bottom fishing, drilling for oil and gas, as well as some scientific investigations. All these would negatively affect hydrothermal vents and other fragile seafloor areas, if pursued in a non-regulatory manner.

Initially, the Common Heritage Principle, as applied to the seabed, included four elements: (a) prohibiting countries from claiming sovereignty over any part of the deep seabed, (b) requiring countries to use the seabed for peaceful purposes, (c) requiring all countries to share its management, and (d) requiring all countries to share the benefits of the seabed's exploitation. But disagreements between industrialized countries and non-industrialized countries were apparent early on, and the various ways of interpreting the Common Heritage of Mankind Principle have never been satisfactorily resolved.

For example, what constitutes "peaceful purposes?" This term has never been defined in the Law of the Sea treaty; subsequent treaties using the phrase, such as the Antarctic Treaty amendments and the Moon Treaty, have only narrowly focused definitions. There has been no sharing of benefits, and so no case-law exploration of factual dilemmas has developed. Perhaps, most importantly, the international regime legitimizes power sharing in ways that are exceedingly distant from a Common Heritage of Mankind Principle, which emphasizes mutual collaboration and a sharing of the benefits.[19]

Reaching the Moon

In 1959, the United Nations General Assembly established the UN Committee on the "Peaceful Uses of Outer Space." The committee drafted various resolutions in the General Assembly that, in 1969, became codified in the *Treaty on Principles Governing the Activities of States in the Exploration and Use of Outer Space, Including the Moon and Other Celestial Bodies.* This treaty established a basic framework for the international government in outer space[20] (Figure 6.3).

In 1971, Russia and Argentina proposed an agreement for the moon, which the Committee on the Peaceful Uses of Outer Space finalized in 1979. This agreement subsequently became the "Moon Treaty," the objectives of which are "the safe development and rational management of lunar resources and the equitable sharing of the benefits derived from those resources."[21]

FIGURE 6.3
Astronaut James B. Irwin, lunar module pilot, gives a military salute while standing beside the deployed U.S. flag during the Apollo 15 lunar surface extravehicular activity at the Hadley-Apennine landing site. The Lunar Module "Falcon" is partially visible on the right. The base of the mountain, Hadley Delta, which is about 3 statute miles in the background, rises approximately 13,124 feet above the plain. Photograph by Astronaut David R. Scott, Apollo 15 commander (https://commons.wikimedia.org/wiki/File:Apollo_15_flag,_rover,_LM,_Irwin.jpg).

Article 11 of the Moon Treaty provides that "the moon and its natural resources are the common heritage of mankind."[22] Article 11 further specifies that the moon cannot be used through force or without permission and that an international government is to be established when exploitation becomes feasible. Article 11(7)(d) requires the equitable distribution of benefits derived from potential resources, and, when distribution occurs, special consideration is to be given to non-industrialized countries, as well as those countries that contributed, directly or indirectly, to the moon's exploration.[23]

The current technological and economic limits preventing the search for mineral resources on the moon further reduce the likelihood that a workable definition for the Principle will be formulated in international law. As a result, the Common Heritage of Mankind Principle remains tentative in the context of outer space. This does not, however, preclude the operation of the Common Heritage of Mankind Principle in the future. If mining or other systematic activities were to commence, rapid agreement on the legal limits might result. But this has not yet occurred.

Applying the Common Heritage Principle in Other Ways

Can the Common Heritage principle be expanded beyond the exotic edge of technology's reach? Environmental philosophers and commentators have proposed extending its framework to non-common resources of widely shared importance to humans, such as rain forests, genetic resources, seed banks, cultural heritage, and other primary resources—even when located within national boundaries.

Exploring other ways of applying the Common Heritage Principle began with the Stockholm Conference on the Human Environment, convened by the United Nations General Assembly of 1972. This conference was the first to tackle global environmental problems in a comprehensive manner. Most importantly, the participants adopted the Stockholm Declaration, which contains a series of principles to guide nations in activities that influenced the environment.[24]

A follow-up gathering, the United Nations Conference on Environment and Development, was convened in Rio de Janeiro in 1992 (Figure 6.4). This conference was the turning point in making the United Nations Framework Convention on Climate Change and the Convention on Biological Diversity available for signatories. The Rio Conference also produced a non-binding, but authoritative, Statement of Principles for management, conservation, and the sustainable development of forests. Similar to its predecessor, the Conference produced the Rio Declaration: "Both the Stockholm Declaration and the Rio Declaration contained similar broad-based principles. The international community has generally accepted these

FIGURE 6.4
World leaders meeting during the Summit Segment of the United Nations Conference on Environment and Development—or Earth Summit—in Rio de Janeiro, Brazil, on June 3, 1992. The Conference's two-day Summit Segment convened a record 103 Heads of State and Governments. Photograph by Michos Tzovaras, United Nations (Photo # 281645 in the United Nations Photo Gallery).

principles, despite the fact that they contain undefined terms, and impose obligations upon both developing and developed states."[25]

Predictably, such calls for Common Heritage expansion have found few advocates among countries, corporate interests, or international bodies owing to the infringement on national sovereignty and the notions of private property, despite acquiescence in international forums. These ideas presuppose that humankind, as a whole, would be responsible for the administration of such Common Heritage resources. Definitions are neither firm nor agreed upon, but frequently discussed elements include:

1. A prohibition on the acquisition of, or exercise of, sovereignty over the area or resources in question
2. The granting of rights to the resources in question to humankind as a whole
3. Reservation of the area in question for peaceful purposes
4. Protection of the natural environment
5. An equitable sharing of benefits associated with the exploitation of the resources; in question, paying particular attention to the interests and needs of non-industrialized nations
6. Governance via a common management regime[26]

Problems with the Common Heritage Principle

But problems become immediately apparent. What would "protection of the natural environment" be—by whose definition, for how long, under whose jurisdiction, and for what purposes? Would it include "development?" If so, must it be sustainable, and how might that be defined? Other questions arise concerning "equitable sharing of benefits." For instance, should preferential treatment be granted non-industrialized nations, whose technological capacities for getting their portion of the goods is less sophisticated? Most troubling of all is how to create and police the common administration for any such resource, given that ecosystems, living beings, and Nature's Laws of Reciprocity themselves are outside of human control.

One of the most extensive experiments of how this might be achieved is found in Part XI of the Law of the Sea Convention signed in 1982. Uses of the seabed and seafloor were defined according to the Common Heritage Principle. They are to be organized and controlled by an international authority consisting of an assembly, a council, and a secretariat. Revenues are to be shared in the international community, with a focus on benefiting non-industrialized countries.

However, the problem is that even here, despite an international entity as the arbiter of development, the focus is still on human desires and the hoped-for revenues. Though definitions vary, it is clear that the Common Heritage Principle involves some form of living trust—the main goals of which are good administration and transmission of the trust's assets, intact and undamaged, to future generations. Clearly, for this to happen, all parties to a treaty or agreement must accept the underlying principles of the trust.

If one state refuses to abide by the Common Heritage Principle, what is the recourse? Currently, the only option involves international sanctions, or perhaps, the punishment that comes out of reciprocity: if one nation refuses to abide, then its interests are curbed in other international arenas, as other nations refuse to cooperate with its needs and requests.[27] Clearly, if one nation breaches the Principle, the other nations cannot respond similarly, because that would destroy the entire agreement. In effect, there is no national or international framework that has the power and enforcement authority over any resource that would be declared a Common Heritage of Mankind by the international community.

The Common Heritage Principle continues to evolve in international law, with more interest in the 1960s and 1970s during controversies over deep-seabed exploitation, and less in subsequent decades. It is recognized as an important principle in the realm of political or ethical precepts, non-binding "soft law." But it does not *strengthen* the international legal system's ability to halt the degradation of the world's resources.[28]

Only Nature's Rights Can Protect the Global Commons

A major concern is whether use of the Common Heritage Principle can stop, or at least slow, the fragmentation and degradation of the global commons' resources that continue to suffer from privatization and exploitation. Perhaps not, unless it is taken seriously in the laws and court decisions of individual nations, and defined more concisely as an aspect of the Rights of Nature—with mechanisms of enforcement in international law. Thus far, trends have been in the opposite direction: the Common Heritage Principle was rejected in the 1992 United Nations Convention on Biological Diversity, especially by non-industrialized nations that feared it would erode their sovereign rights to benefit from biological resources within their national boundaries.

Since the Common Heritage Principle has not been fully defined by existing treaties, it can be shaped to meet new exigencies. A core element, somewhat expanded from the original focus, but often left inchoate, is to protect the natural environment by incorporating some form of Nature's Rights. This is the requirement that the Common Heritage of Mankind "shall be transmitted to future generations in substantially unimpaired condition (protection of ecological integrity and inter-generational equity between present and future generations of humans)."[29]

However, thus far, even this most expansive doctrine for protecting the commons suffers from being anthropocentric, as well as outside the current power structure both in national and international law. The fault lines of the Western legal system have—and are—likely to continue to emasculate the principle, except as a lofty statement of noble intent that has little power to protect the resources.

Perhaps the current crises will rekindle interest in the Common Heritage of Mankind Principle, and the fundamental expansion necessary to make it an arm of Nature's right to life. It would then be not the Common Heritage of Mankind Principle, but the "Common Heritage of the Earth Principle," in which humans play a part, but not a role that dominates the fundamental rights of any other living being.

Placing the Rights of Nature at the center of the international legal system, and organizing them around Nature's Laws of Reciprocity, would solve the problem of continual marginalization of the Common Heritage Principle. It would become the centerpiece. The new Rights of Nature framework would create a "global environmental commons," of which humans are a member, along with all the others:

> In the ecological commons [global environmental commons] a multitude of different individuals and diverse species stand in various relationships to one another—competition and cooperation, partnership and predatorship, productivity and

> destruction. All those relations, however, follow one higher law: over the long run only behavior that allows for productivity of the whole ecosystem and that does not interrupt its self-production is amplified. The individual is able to realize itself only if the whole can realize itself. Ecological freedom obeys this form of necessity. The deeper the connections in the system become, the more creative niches it will afford for its individual members.[30]

"The commons" would not be an uninhabited area claimed by no nation or a resource common to humankind. Instead, it becomes an essential part of the biosphere, with a life and rights of its own, in its whole and all of its parts, that has rights in the legal system prior to those of humans. Rights of Nature must come before human needs for resources in order to ensure that ecosystems continue to function at their own level and remain resilient, rather than being brought down to barebones ecological function to meet maximum human use. Nature's Laws of Reciprocity ensure that humans receive feedback about the results of their actions and can regulate their activity accordingly.

If an expanded Common Heritage Principle is not embedded in an international Rights of Nature framework, then the best the legal system can create is probably a form of common ownership. But that is subject to exclusionary use and maximum exploitation by all parties with whatever minimal environmental restraints are in place, as well as all the other problems that plague the law of the commons, including how to adequately enforce current and future restraints placed on the resource.

Endnotes

1. World Charter for Nature, A/RES/37/7, United Nations General Assembly. http://www.un.org/documents/ga/res/37/a37r007.htm (accessed July 29, 2016).
2. *Ibid.*
3. *Ibid.* (Section I, General Principles 1).
4. *Ibid.* (Section I, General Principles 4).
5. *Ibid.* (Section II, Functions (10)(a)).
6. *Ibid.* (Section II, Functions (11)(b)).
7. *Ibid.* (Section III, Implementation).
8. The preceding four paragraphs are based on: Harold W. Wood, Jr. The United Nations world charter for nature: The developing nations' initiative to establish protections for the environment. *Ecology Law Quarterly*, 12 (1985):978–979.
9. *Ibid.* (p. 987).
10. Alexandre Kiss. The common heritage of mankind: Utopia or reality? *International Journal*, 40 No. 3, *Law in the International Community*, (1985):423–441.
11. *Ibid.* (p. 424).
12. *Ibid.*
13. Edwin Egede. *Africa and the Deep Seabed Regime: Politics and International Law of the Common Heritage of Mankind.* Springer, London. (2011) 271 pp.
14. Alexandre Kiss. The common heritage of mankind: Utopia or reality? *op. cit.* (p. 428).
15. Edward Guntrip. The common heritage of mankind: An adequate regime for managing the deep seabed? *Melbourne Journal of International Law*, 4 (2003) http://www.law.unimelb.edu.au/files/dmfile/downloadaf021.pdf (accessed December 5, 2015).
16. Alexandre Kiss. The common heritage of mankind: Utopia or reality? *op. cit.* (p. 429).
17. Alexandre Kiss. The common heritage of mankind: Utopia or reality? *op. cit.*

18. World Charter for Nature, A/RES/37/7, United Nations General Assembly. *op. cit.* (p. 1, paragraph 5).
19. The forgoing discussion of the deep sea and deep-sea mining are based on: Edward Guntrip. The common heritage of mankind: An adequate regime for managing the deep seabed? *op. cit.*
20. Edward Guntrip. The common heritage of mankind: An adequate regime for managing the deep seabed? *op. cit.*
21. *Ibid.*
22. *Ibid.* (Article 11(7)(d)).
23. *Ibid.*
24. Edward Guntrip. The common heritage of mankind: An adequate regime for managing the deep seabed? *op. cit.*
25. *Ibid.* (p. 25).
26. John E. Noyes. The common heritage of mankind: Past, present, and future. *Denver Journal of International Law & Policy*, 40 (2011):447–471.
27. Alexandre Kiss. The common heritage of mankind: Utopia or reality? *op. cit.*
28. John E. Noyes. The common heritage of mankind: Past, present, and future. *op. cit.*
29. The preceding two paragraphs are based on: Prue Taylor. The common heritage of mankind: A bold doctrine kept within strict boundaries. pp. 64–69 (emphasis within the quote) *In*: Common Heritage of Mankind Principle. (Klaus Bosselmann, Daniel Fogel, and J. B. Ruhl, Eds.) *The Encyclopedia of Sustainability, Vol. 3: The Law and Politics of Sustainability.* Berkshire Publishing, Great Barrington, MA. 2011.
30. Andreas Weber. The economy of wastefulness: The biology of the commons. (no page numbers given) *In*: *The Wealth of the Commons.* (David Bollier and Silke Helfrick, Eds.) The Commons Strategy Group, 2012. http://www.wealthofthecommons.org/essay/economy-wastefulness-biology-commons (accessed December 5, 2015).

7

Rights of Nature Concepts and Issues

Legal systems of today vary widely, based on their cultural histories and religious perspectives that, in turn, characterize their worldviews, including how they treat the environment. The Rights of Nature paradigm is so new that the legal theorizing necessary to create a supple legal system with clear and workable standards has scarcely begun. However, most—if not all—of the ideas, doctrines, balancing tests, and restructurings already exist in one place or another. Here, we identify some of the subjects that will be considered and hammered into shape to change the American legal system to one in which the Rights of Nature have priority. Other topics will arise as the language is debated, votes are taken to prioritize Nature's Rights in the legal system, and courts issue rulings.

Design of the Legal System: Regulatory or Market-Based?

The legal system in the United States, as it deals with environmental and natural resources, is primarily regulatory with some market-based aspects. The vast majority of efforts to curb human overexploitation of natural resources are based on a regulatory framework in which uses, even very harmful uses, are allowed in most instances—but with conditions that ostensibly protect the environment from degradation. This system is failing, however, as the current environmental crisis starkly demonstrates.

An often-explored alternative to the regulatory system is a market-based approach, such as a "pollution right" sold to a company in return for an offset that will theoretically increase ecological well-being elsewhere. A good illustration is the so-called "emission-trading programs," in which allowable greenhouse-gas emissions are capped, while market forces are allowed to allocate emission opportunities among the regulated industries. The goal, as in all market approaches, is to permit industrial dynamics to drive commercial desires for unlimited growth and profitability.

A common emission-trading program uses "carbon credits," a term for a certificate providing the right to trade a ton of carbon dioxide or other greenhouse gas for a tract of forest or tree plantation where the trees sequester a certain amount of carbon dioxide to offset that which a carbon credit allows the holder to produce.[1] The underlying idea is to create a price for greenhouse gases that, in turn, creates incentives for producers and consumers to invest in less climate-impacting technologies.

An international system of carbon credits was formalized in the Kyoto Protocol, which the participants agreed to in the later Marrakesh Accords. This was necessary to ensure that the environmental principles within the Kyoto Protocol remained intact, namely, to require the use of sound science and, most importantly, consistent methodologies in calculating carbon credits. The Marrakesh Accords also highlighted the obligation to revise carbon-credit calculations to account for—and subtract from the credit system—the release of greenhouse gases from natural events, such as forest fires, permafrost release of methane as a result of global warming, or human-caused environmental degradation.[2]

Fortunately, all countries in the European Union have adopted the Kyoto Protocol for carbon trading. In this way, projects to reduce greenhouse gas can be established in other areas, to the credit of the sponsoring country or industry.

Despite the seemingly noble intensions of the emission-trading program, the problem with such market-based mechanisms, apart from their incalculable complexity, is that they remain within the unworkable framework of market economics, which has led to the massive degradation the world now faces. A heavily polluting project can go ahead, as long as the polluter pays enough money for a distant, qualified endeavor, such as sponsoring a forest plantation in South America, to "offset" the pollution or other environmental damage in the United States. Clearly, however, such a distant venture has few, if any, benefits to the ecosystem being harmed by the project in question. Instead, it simply provides a financial rationalization to allow many massively destructive environmental projects to persist.

These market-based solutions do not place human activity in a new framework, such as providing for and prioritizing Nature's integrity and right to life. They are merely ways of exporting the worst excesses of the industrial complex and hiding them in distant ecosystems. Market-based maneuvers, such as emissions trading, simply create "sacrifice zones," where all kinds of pollution and degradation are deported to distant lands from where the supposed greenhouse-reduction projects take place. In other words, many ecosystems are still sacrificed under this market-based idea.

The answer to the original question, whether the legal system should be regulatory or market-based, is that it cannot be either. It must be revamped in a Rights of Nature framework, guaranteeing Nature's integrity and freedom to flourish. That framework must be designed, created, supported, and enforced at all levels of government.

Existing Law Shows the Way

The good news is that all the pieces are in place to create a legal system based on the primacy of Nature's Rights. The legal discussions have begun or, in some cases, progressed extensively. Most of the relevant legal doctrines exist in the United States or other countries. Court holdings have already carved out a generous understanding of Nature's role in human activities, and the need to safeguard it. There are no insurmountable problems to redesigning the legal system.

The threshold legal question of whether Nature has, or can have, rights seems important from the vantage point of theorizing, but can be rather easily answered in the affirmative. More important in the process of building a Nature's Rights legal system is what kind of rights Nature has, how these are articulated, how they affect other rights and responsibilities, and what the remedies are.

The question of whether Nature has rights was first debated in the United States in the now-famous dissent of Justice William O. Douglas in the U.S. Supreme Court case *Sierra Club v. Morton*, 405 U.S. 727 (1972).[3] The Forest Service had issued a permit for Walt Disney Enterprises to construct a complex of hotels, restaurants, and recreation facilities in Mineral King Valley in the Sierra Nevada (Figure 7.1). The Sierra Club sued for an injunction to stop the project, but the Appeals court held the Sierra Club Legal Defense Fund had no standing to bring the suit—that is, the organization would not be aggrieved or adversely affected if the project were built. This holding caught the attention of Christopher Stone, a professor at the University of Southern California School of Law, who authored a law

FIGURE 7.1
Mineral King Valley, now part of Sequoia National Park, California. Photograph courtesy of the U.S. National Park Service.

review article arguing that natural areas and objects should have legal rights. After all, the injury to the Sierra Club might be tenuous if the project were constructed, but the injury to Mineral King Valley would be drastic and perhaps permanent.[4]

Justice Douglas, dissenting from the majority upholding the Appeals court decision, embraced the idea of Nature having standing to bring suit for its injuries. He wrote, "Contemporary public concern for protecting nature's ecological equilibrium should lead to the conferral of standing upon environmental objects to sue for their own preservation."[5]

How Does Nature Have Rights?

Without going into the technical legal details, one might begin an overview of the question of Nature's Rights by asking what it means to be a legal rights-holder. Though there are no set standards, Stone proposes that to have rights, an entity must (1) institute legal rights at its behest, (2) a court must take injury to that entity into account, and (3) any relief the court gives must go to the benefit of the rights-holder.[6] It is of course no answer to say that Nature, or any being of Nature, should be denied rights because it cannot speak. The American legal system gives rights to many entities that cannot speak: cities, states, and the federal government are the most common examples. This could be multiplied to include water districts; religious organizations; social service organizations, such as the United Way; trusts and estates; joint venture partnerships; and many others. All these entities hire attorneys to speak for them and argue their cases in court.

Nature speaks all the time, through its Laws of Reciprocity, providing feedback on whether human activities are helpful or harmful. As Stone wrote, "...natural objects can communicate their wants (needs) to us, and in ways that are not terribly ambiguous. I am sure I can judge with more certainty and meaningfulness whether and when my lawn wants (needs) water than the Attorney General can judge whether and when the United States wants (needs) to take an appeal from an adverse judgment by a lower court. The lawn tells me that it wants water by a certain dryness in the blades and soil—immediately obvious to the touch—the appearance of bald spots, yellowing, and a lack of springiness after being walked on; how does "the United States" communicate to the Attorney General?"[7]

There are many ways Rights of Nature could enter the legal system. One way would be through a system of guardianship: an appointed representative who advocates for, and can sue on behalf of, an endangered tree, ecosystem, or other aspect of Nature. The American legal system uses guardianship now in many arenas, where an incompetent person needs to have a representative to advocate for his or her well-being: the elderly and infirm, the mentally compromised, infants, and many others. The questions to be answered in the case of a surrogate who advocates for Nature include the following: Who could be qualified? How they would be selected? What obligations would they have? And what powers granted by whom? Most importantly, to whom would a Nature's Rights guardian be accountable, and under what standards? But these are issues of an ordinary sort that have been answered in legal systems already; they are hardly insurmountable. They only need to be discussed in the new Rights of Nature forum.[8]

Such a guardianship could be created at either national or international levels, or both, to represent Nature in bilateral or multilateral venues. Stone proposes that such a system of national or international "commons guardians" might be financed via levies to nations on uses of the commons areas themselves—that is, charging a small percentage of the value of fair use of commons' resources, whether fish in the ocean or clean air and stable climate free of excess carbon. The money in the Global Commons Trust Fund would pay the Guardians for their advocacy on behalf of the commons. This model could be tailored to meet global, national, or local needs.[9]

A guardian (or ombudsman, to use a term for a governmentally appointed advocate, usually with limited powers) could be a government agency, but might prudently be expanded to include qualified environmental organizations that could act as additional, independent, guardians with rights to advocate for species or ecosystems in planning, and also through litigation. This is a system already in use in Germany.[10] Guardians could be given additional roles in protecting resources of the global commons, such as oceans or the air: monitoring the health of the commons, and having powers to enforce treaties and national laws affecting the commons. The guardian would have the power to appear before international rule-making and legislative bodies on behalf of the aspect of Nature the guardian represents, and also have powers to bring suit if the commons is endangered by human activity.

The Question of Legal Standing

In order for any guardianship or ombudsman to enforce Nature's Rights, the legal system must provide for Nature, or its surrogate, to have "standing" to fight for their rights and to defend against harms done to them. The astounding paradigm change is that even without guardianship, courts have been liberalizing standing requirements for decades since 1972. This has allowed groups to sue on the basis of recreational, aesthetic, or ecological losses, essentially "for" the place or being that will be harmed.[11]

But this expanded "standing" does not really reach the question of Nature's Rights, and courts have not directly ruled on whether Nature or its species have standing as entities, except in the new cases arising from local Rights of Nature ordinances. In those instances, there is a conflict owing to the hierarchical nature of the U.S. legal system, and courts have not reached the heart of the question. However, and crucial for changing the American legal system, "Nothing in the requirement of a 'case or controversy' should be read to forbid Congress from treating animals as owners of legal rights.... To be sure, the framers anticipated that plaintiffs would ordinarily be human beings. But nothing in the Constitution limits Congress' power to give standing to others."[12]

In some ways, Nature has already been given rights—or guardians—who provide legal representation and advocacy to ensure Nature's ability to flourish. The governmental agencies charged with protection of species, such as the U.S. National Oceanic and Atmospheric Administration or the U.S. Fish and Wildlife Service, negotiate on behalf of the species they protect for more and better habitat, and changes in public land management laws to benefit the species. Agencies can, and do, sue to stop harmful practices that affect whales, polar bears, salmon, or other species under their protection.[13]

As Stone points out, giving Nature standing changes the calculus a court must make in issuing an injunction or determining damages. It shifts the basis of argument from a balancing of hardships on human beings to the loss and detriment suffered by the environment itself.[14] A legal system where Nature has rights entirely avoids the problematic tests of balancing cost–benefit, tests of reasonableness, and similar efforts to weigh the relative seriousness of injuries. Rights can neither be calculated nor compromised by a cost–benefit analysis.

Courts are unlikely to be the ultimate leaders in expanding and systematizing Nature's Rights because, under the "political question doctrine," they will look to the legislative and executive branches of government to create the underlying legal framework.[15] This is appropriate, since only there will the citizenry have the opportunity to debate various provisions and work with legislators to craft a wide-ranging Rights of Nature system that will revamp both the law and the economy in which we live. But as in any rights-based system, the problem of balancing the rights that conflict with one another will come up. We propose that Nature must be the *primary* rights-holder in the legal system, and this must be clearly stated in both Constitutions and laws. Only in this way can the legal and economic paradigms truly become, and remain, sustainable without being compromised by human needs and desires that harm Nature's right to flourish—but are permitted by legal doctrine that places human desires over Nature's Rights.

Nature's Remedies When It Is Harmed

What will Nature's remedies be when conflicts arise and Nature is harmed? To begin with, how would the law account for Nature's "own" injury, especially if it is an ecosystem, or a riverine system whose injuries are very tough to even understand, much less calculate? In guardianship cases, the guardian speaks for the ward's "best interests," as provided by the relevant statutes. The law already provides a marker for this problem with respect to Nature: it gives endangered species—protected under the Endangered Species Act—what amounts to *property rights*. It protects the species from trespass on its habitat, despite human desires and needs. The Marine Mammal Protection Act also provides rights to marine mammals for the protection of both the animals themselves and their habitats from destructive human activities.[16] Thus, the framework already exists to pinpoint Nature's injuries and speak on behalf of the injured. It needs only to be expanded into a comprehensive Constitutionally-mandated system of Nature's Rights.

These and other statutes already provide an acceptable means to assess damages and provide recovery *for the benefit of* the injured species or ecosystem. Under the Comprehensive Environmental Response, Compensation and Liability Act (known as the Superfund Law), the polluters are assessed damages that may only be used to restore, replace, or create the equivalent of the damaged environment. For example, the natural resources settlement with the federal government and the state of Alaska from the disastrous 1989 *Exxon Valdez* oil spill in Prince William Sound amounted to $1.15 billion, earmarked for restoration of the devastated ecosystem.

In addition, we may look to civil recovery actions for guidance in assessing damages: "Units of the federal or state governments are authorized to sue polluters as trustees for the environment, to recover and apply the costs of restorations, *even if those costs exceed fair market value.* For example, when a mismanaged oil tanker ravaged a Puerto Rican mangrove swamp, the operators had to pay what was liberally estimated as the cost necessary 'to make the swamp whole.' "[17]

The building blocks are already in place to write Nature into the legal system, as the primary rights-holder, and create a system of Nature-based governance. As part of having rights, Nature can easily be granted standing in courts, a means of assessing injury and providing for damages to its own benefit to make it whole. The basic framework of statutory language and court holdings is already in place, ranging from local Rights of Nature statutes to court opinions in environmental damage and restoration cases. The only piece lacking is language in the state and federal Constitutions to draw the circle closed and galvanize a comprehensive legal framework creating Nature as a principal rights-holder, thereby beginning the process of environmental repair and living sustainably.

Nature Tribunals to Advocate for Nature's Rights

The exemplar for an official Nature's advocate already exists in the International Rights of Nature Tribunal, which has begun hearing cases of crime against the Rights of Nature. The advocate speaks on behalf of the abused, arguing for its *inherent* right to exist and flourish free of human dominance and abuse. The revised legal system needs to include such tribunals at both the state and federal levels, with each having jurisdiction over the abuse of Nature at their level, much as the American legal system's hierarchy operates today. As in other arenas of law, the federal decisions on Rights of Nature would be binding on all other units of government, including states. Cases would be buttressed by research, using the Precautionary Principle and the standards required to maintain the biophysical living trust to show the ways in which a given action does indeed harm the ecosystem, and must therefore either be ended or be denied all necessary permits.

Nature tribunals must have the power not only to deny permits for a project deemed damaging to Nature's inherent right to life but also to ensure that the project does not remain "on the books" of a government's planning processes that would allow it to emerge again in the years ahead. This is a common side effect of long-range planning, which unfortunately ensures that damaging or dangerous projects retain some shelf life, even when otherwise denied by the authorities.

A partially useful template might be Oregon's Land Use Board of Appeals, created in 1979 as part of Oregon's comprehensive land use system.[18] Its three judges are appointed by the Governor and confirmed by the Senate. The Land Use Board of Appeals is the first step into the judicial system for land use disputes and hears only cases relating to matters concerning the use of land. As a result, its judges have gained tremendous expertise in that field. Its holdings are binding on state agencies and local governments. The statutes provide for an expedited review timeline so that cases do not become hopelessly slowed down. The Land Use Board of Appeals' decisions can be appealed into the regular court system, but the Board's expertise in its rather specialized field is greatly respected.

We are mindful that Rights of Nature creates a deep paradigm shift in Western culture and will cause a complete reorganization of the American legal and social systems.

Methods of governance may well need to change from those currently used, especially the hierarchical bureaucracies and court structures that are common in current American legal systems. Similarly to ecosystems, local knowledge and conditions are important, even critical, because trusteeship of an ecosystem must be locally designed and its capabilities must be respected. When dealt with locally, the governance and enforcement of Nature's Rights can benefit greatly by turning away from a rigid "one-size-fits-all" design of the legal structure. This change allows human cultures to flourish in consonance with their local ecosystems, which creates a much deeper, interconnected state among them.

For example, "… human communities can evolve into higher, more complex forms of organization without the directive control or a central sovereign or bureaucracy … human communities have inborn capacities to create stable order. … effective governance need not be imposed through a comprehensive grid of uniform general rules embodied in formal State Law and administered by centralized legislatures, regulators and courts. Complexity science demonstrates that governance can be a highly distributed, evolving form of social practice and traditions … rooted in communities of decentralized responding to particular local circumstances."[19]

Nevertheless, since many opportunities for gain and corruption exist in the use of Nature's bounty (and undoubtedly will continue to do so), hybrid forms may develop that have elements of self-governing principles combined with traditional bureaucratic hierarchies to provide enforcement, whether at local levels or those of a state or nation.

The Precautionary Principle Underlies Nature's Rights

Composition defines a system's structure, which in turn governs how the system functions. It is therefore by its function that we must learn to characterize a system, including a legal system. As Aldo Leopold said, "To keep every cog and wheel is the first precaution of intelligent tinkering."[20] Put a little differently, "The first precaution of intelligent tinkering is to save all the pieces."[21] This philosophy, which underlies the so-called "Precautionary Principle," is an urgently needed change in Western, and American, thinking. What does it mean to have a mindset of saving all the pieces, of being cautious? Consider the current trend in biodiversity—the vast net of interacting species that makes up life on earth—as a perfect example of failing to honor this principle.

As a result of careless use of the environment, biodiversity is dropping below safe levels for the well-being of human societies. Humans depend on ecosystem services for crop pollination, waste recycling, and many other necessities. Ecosystem productivity, and the services humans depend on, requires biological diversity to function. However, a recent scientific model estimates that losing more than 10% of the local biodiversity places the ecosystem at risk. A recent report found that more than half of the world's land (58%) now has approximately 85% of its original biodiversity. That is, the biodiversity has dropped below safe levels in many regions of the Earth, where nearly three quarters of the world population resides.[22]

Ecosystems and biodiversity are exceedingly complex, and calculating the importance of the loss to ecosystem function is a difficult process; it will also vary locally. How much does it matter to drop below the 90% biodiversity level? Co-author Prof. Andy Purvis, from Imperial College London and the Natural History Museum, explained, "Once we're on the wrong side of the boundary it doesn't mean everything goes wrong immediately, but there is a markedly higher risk that things will go badly wrong."[23]

In other words, loss of biodiversity weakens Nature's resilience in the face of large-scale changes, such as climate change. This loss also threatens goals of creating long-term, sustainable development, and so has greater effects than first appear to be the case.[24]

The Precautionary Principle was first articulated in Germany, when private landowners noticed that their forests were dying. Appealing to the government for action, they incited an all-out effort to reduce acid deposition, most commonly thought of as "acid rain," from coal-fired power plants.

The German government translated the old adage "an ounce of prevention is worth a pound of cure," into law as the *Vorsorgeprinzip.*[25] which literally means "precautionary principle," or stated a little more accurately, *"caution in advance* principle."

The Precautionary Principle states that, "in cases of serious or irreversible threats to the health of humans or ecosystems, acknowledged scientific uncertainty should not be used as a reason to postpone preventive measures. The Principle originated as a tool to bridge uncertain scientific information and a political responsibility to act to prevent damage to human health and to ecosystems."[26]

The Precautionary Principle, or thinking through the possible consequences of a potential action before committing the act, instructs us to acknowledge mistakes, admit ignorance, doubt the certainty of our knowledge, and act with humility in order to honor our place as an inseparable part of Nature through all generations. Such instruction gives us a way to change our thinking and thus our behavior, personally and collectively, for the greatest good in the present for the present generations and those of the future while simultaneously acting consciously to minimize the potential harm we cause.

The Precautionary Principle's first debut in international law was as Principle 15 of the Rio Declaration of 1992. Principle 15 states: "In order to protect the environment, the precautionary approach shall be widely applied by States according to their capabilities. Where there are threats of serious or irreversible damage, lack of full scientific certainty shall not be used as a reason for postponing cost-effective measures to prevent environmental degradation."[27]

The Rio Declaration does not mandate any specific obligations on states, but the Principle has developed through subsequent treaties, as well as legislation and court decisions in individual nations. As a result, "the precautionary principle has achieved the status of a recognized principle of international environmental law."[28] This recognition is critical because a truly sustainable legal system must be founded on it.

History has proven, however, that legal systems based solely on protecting human values and regulating human desires will always fail the Precautionary Principle when times toughen during economic downturns. Ecosystem integrity is then weighed against human needs and desires, and the balance is inevitably tipped to accommodate human wants.

This problem is painfully obvious in the current regulatory system, which focuses on the health and welfare of the human environment. Without a sustainable, legal precedent wherein the natural environment has a voice of its own, the regulatory system invariably focuses on and chooses the most human-centered, and often the most shortsighted, alternative. When courts create tests of legal balancing to judge between rights or requirements, they are so constructed to benefit the human desires of the moment. Ecosystem integrity may get a nod—or some minimal mitigation—provided it will not harm the perceived march of human "progress."

The only way for the Precautionary Principle to actually curb shortsighted, self-centered human calculus about environmental benefits is for the legal system to be expanded to include the rights of life and health to the Earth, its ecosystems, and every living being. Furthermore, the Rights of Nature language must clearly specify that it is first among all rights, and all balancing tests or reasonableness tests must be crafted under the primary Rights of Nature principles, rather than having Nature's Rights be one set among others all vying for priority.

Revising the legal system's foundation so dramatically is bound to be a long and often contentious process because honoring Nature's Rights cuts directly across the profit-making engine of modern, corporate-controlled society. However, with these provisions in place, it is possible to have the Precautionary Principle work properly. Legal codes and courts of law will then judge the probability of whether a given project does or will harm the right to life of a species or the environment.

The Biophysical Living Trust

Another key concept underlying a Rights of Nature system is the biophysical living trust. Because it is vital to understand this principle, we will digress a little to describe it.[29]

Although most people speak of land "stewardship," the concept of a "living trust" is preferable because "stewardship" does not have a legally recognized "beneficiary"—someone who directly benefits from the proceeds of one's decisions, actions, and the outcomes they produce. Although a "steward," by definition, is someone who "manages" another's property or financial affairs and thereby acts as an agent in the other's stead, there is nothing explicit in the definition about a legal beneficiary. For this reason, "stewardship" is the weaker of the two terms because the fiduciary responsibility of "stewardship" is to the shareholders, whereas the fiduciary responsibility of a "living trust" is to the beneficiaries, none of whom need to be physical shareholders.

A living trust is like a promise, which, in our case, is to keep Nature fully functioning, its ecosystems resilient, biodiverse, and able to flourish. "Promises are scary things," says author Elizabeth Sherrill. "To keep them means relinquishing some of our freedom; to break them means losing some of our integrity. Though we have to make them *today*, promises are all about *tomorrow*—and the only thing we know for sure about tomorrow is that we don't know anything for sure!"[30]

A "living trust," in the legal sense, is a present transfer of property, including legal title, into trust, whether real property (such as forestland or a historical building) or personal property (such as interest in a business). The person who creates the trust—say, the owner of forestland or a historical building—can watch it in operation, determine whether it fully satisfies his or her expectations and, if not, revoke or amend it.

A living trust also allows for the delegation of administering the trust to a professional "trustee," something that is desirable for those who wish to divest themselves of managerial responsibilities. The person or persons who ultimately receive the yield of the trust, for better or worse, are the legal beneficiaries. The viability of the living trust is the legacy passed from one generation to the next.

Though a trustee may receive management expenses from the trust, meaning that a trustee may take what is necessary from the interest, at times even a small stipend, the basic income from the trust, as well as the principle, must be used for the good of the beneficiaries. Yet, natural resources in our capitalist system are *assumed* to be income or revenue, rather than capital.[31] Importantly, a trustee is obligated to seek ways and means to enhance the capital of the trust—not diminish it. Like an apple tree, one can enjoy the fruit, but not destroy the tree. A living trust is about the quality of life offered to the generations of the future; it is *not* about the acquisition of possessions.

"Trusteeship" is a process of building the capacity of people to work collectively in addressing the common interests of all generations in the context of sustainability—biologically, culturally, and economically. A biophysical living trust, in turn, means honoring the productive capacity of an ecosystem within the limitations of Nature's Laws of Reciprocity.[32]

A biophysical living trust is thus predicated on systemic "holism" in which reality consists of an organic and unified whole that is greater than the simple sum of its parts. Consequently, wisdom dictates that we learn to characterize a system by its function, not its parts. The basic initiatives underpinning a living trust—all considered externalities within the current political/economic framework—are as follows:

1. Everything, including humans and nonhumans, is an interactive, interdependent part of a systemic whole.
2. Although parts within a living system differ in structure, their functions within the system are complementary and benefit the system as a whole.
3. The whole is greater than the sum of its parts, because how a system functions is a measure of its biophysical integrity and sustainability in space through time.
4. The biophysical integrity and consequential sustainability of the system are the necessary measures of its economic viability and relative stability.
5. The biophysical integrity of processes has primacy over the economic valuation of components (products).
6. The integrity of the environment and its biophysical processes has primacy over human desires, when such desires would destroy the system's integrity and productivity for future generations.
7. Nature determines the necessary limitations of human endeavors.
8. New concepts need to be tailored specifically to meet current challenges, because old problems cannot be solved in today's world with old thinking.
9. The disenfranchised, as well as future generations, have rights that must be accounted for in present decisions, actions, and potential outcomes.
10. Nonmonetary relationships have value.

In a biophysical living trust, the behavior of a system depends on how individual parts interact as functional components of the whole, not on what each part, perceived in isolation, is doing. The whole, in turn, can be understood only *through* the relationship/interaction of its parts. This understanding gives us a view of systems supporting systems supporting systems, *ad infinitum.*

With the foregoing in mind, the legal system must be set up in ways to prevent immediate human gratification that degrades and despoils the environment for future generations. Future generations are the beneficiaries; thus, current actions must sustain and enhance the biophysical systems on which they will depend. Just like a living trust in law (*inter vivos,* Latin for "between the living") allows a person to put his or her assets in trust for a future beneficiary, a biophysical trust uses the same principle by which living humans respond to Nature as a trust whose beneficiaries are future generations. The difference, of course, is that human beings do not own the "assets" of the biophysical trust, as Nature has its own inalienable rights to survival and integrity. But, as with the legal concept of a trust, the assets for the beneficiaries must be maintained in good order. This is the responsibility of the legal system.

Innovative partnerships are already taking shape at national and international levels, often melding the principles of indigenous peoples with a country's laws or international legal principles. A recent example is the Yasuni Ishpingo Tambococha Tiputini Trust Initiative proposed by Ecuador. This would set up a United Nations administered trust. Ecuador would renounce the rights to oil exploration in a sensitive region and protect the lands intact if industrialized countries contribute at least half the market value of the oil into a UN Development Program–administered trust fund, to be used for renewable energy, reforestation, and social development in Ecuador. This would be a tremendous sacrifice for Ecuador, but the country wants to protect its biodiversity. It also believes that other nations should help share the burden of reducing the atmospheric carbon this sacrifice will create.[33] Creative trust concepts, burden-sharing agreements, governing rules, and funding mechanisms like these hold the future to making the biophysical living trust concept a potent means of establishing and maintaining the Rights of Nature.

The Public Trust Doctrine

We do in fact find this concept deep in the heart of the American legal system: it is called the Public Trust Doctrine. The idea has roots in Western legal thinking as far back as the Roman Empire, and essentially holds that "…a society's governing bodies have an affirmative duty to protect natural resources for the health and well-being of present and future generations. The doctrine has traditionally applied to the rivers, the sea and the coastal shoreline, protecting such activities as navigation, fishing and recreation. The idea is that the unorganized public has sovereign ownership interests, over and above those of the State itself. The State may hold the legal title to the land or water, but the public is the beneficial owner."[34]

American courts have used the Public Trust Doctrine to police governments that try to sell off public assets. For example, the U.S. Supreme Court in *Illinois Central Railroad Co. v. Illinois* (146 U.S. 387 (1892)) barred the Illinois legislature from giving away Lake Michigan shoreline property held in public trust by the State[35] (Figure 7.2). Though not much used since the rise of environmental regulatory systems, the Public Trust Doctrine could easily be expanded to protect a broad array of natural resources, including the atmosphere and the climate. Mary Christina Wood, a professor of Law at the University of Oregon School of Law, and faculty director of its Environmental and Natural Resources Law Center, is a leading proponent of expanding the Public Trust as a remedy against the failures of environmental law: "Failure to recognize these natural resources as sets in the trust simply perpetuates a misguided assumption underlying much of environmental law today—that natural assets are capable of severance and partition."[36]

The Public Trust Doctrine can turn the current environmental regulatory nightmare—which has failed massively to protect the environment as it was designed to—and the "politics of scarcity" into the "politics of abundance," which operate from an ethic of restraint and harmony with natural systems that sustain us.[37] Nevertheless, the Public Trust Doctrine by itself is insufficient to do the job: it is a judicial doctrine designed to safeguard various "commons" resources. It is human-focused, though it hones and sets limits on human desire by protecting resources for the future. But integrated into a comprehensive legal system of Nature's Rights, the Public Trust becomes a powerful means of enforcing Nature's right to flourish free of human interference and for providing the means of the "ethic of restraint" so desperately needed for human society to flourish also. This ethic of restraint must be grounded in Nature's Rights so that the *standard of measurement* for human activity's harm goes beyond human needs and prioritizes Nature's own needs.

FIGURE 7.2
Point Beach is a State Forest in Wisconsin spanning 2,903 acres and includes a park within the borders. It covers 6 miles of Lake Michigan shoreline. Photograph by Yinan Chen (https://commons.wikimedia.org/wiki/File:Gfp-wisconsin-point-beach-state-park-michigan-shoreline.jpg).

What Is Sustainability?

To work properly—even under a Constitution based on the Rights of Nature—the Precautionary Principle and the concept of a biophysical living trust must be clarified in a few fundamental ways, especially to pinpoint what "sustainability" means. This term has been bandied about so frequently that it has effectively lost both its meaning and its moorings, making it nearly impossible to define. Almost any project can be judged sustainable if the criteria for approving it are lax enough. The common term for such sleight of hand is "greenwashing," which can be defined as a company's, a corporation's, a government's, or other group's openly promoting environmental initiatives or images, while actually maneuvering in a way that obscures environmentally damaging activities. Despite such dishonesty, four inviolable concepts underlie true sustainability:

1. True Sustainability Prohibits Human Mitigation—or Substitution—of Any Kind for Political or Monetary Gain

A project that appears sustainable, but is in fact "sustainable" only in the shallowest or most artificial sense, does not meet the standard for a Rights of Nature project, nor can it pass muster under scrutiny by the Precautionary Principle and biophysical living trust concept. Such a project may appear slightly better than the status quo, but in actual fact is worse.

An example would be permitting a new highway to cut through hundreds of acres of forest and wetlands, while seeking to alleviate the destruction by mitigation projects elsewhere that create artificial tree plantations and label them "forests" or create water-filled depressions and deem them "wetlands." Although these projects are legally considered mitigation for the destroyed habitats, they do not balance the destruction or strengthen the ecosystem's integrity. Adding insult to injury, the created, artificial "habitats" are often placed in locations where their natural counterparts would not normally occur, and so they do not flourish. Despite the positive political/monetary gain, the net improvement for ecological integrity (= sustainability) is zero.

2. Sustainable Projects Create Sustainable Levels of Human Use

It is not sustainable to provide the same levels of overconsumption in a way that seems to be more environmentally friendly. Such symptomatic thinking simply appears to create a solution, at least temporarily, that may reduce social guilt about the untenable exploitation without systemically addressing its cause. For example, if a city switches from a coal-fired plant to solar energy without reducing the per-capita level of energy use, there is no true sustainability involved, only a switch from one form of overuse to another.

3. Sustainable Use Shrinks—Not Expands—The Human Footprint on Earth

This is a commonly misunderstood point. Again, energy systems provide an excellent example. In the quest to maintain current levels of energy use in a manner "friendly" to the local environment, vast—and at time distant—areas of the desert, high mountains, and even the sea are being subject to environmentally damaging wind farms,[38] solar farms[39] and (once the technology is perfected) wave-energy farms[40] to provide "clean energy." This is false sustainability, because it greatly enlarges areas of human use, often in regions that had escaped the demands of human necessity thus far.

4. Sustainability Is Flexible but Maintains Nature's Integrity

Sustainability is a flexible and continuous process: what constitutes a sustainable human footprint varies with population, migration, economic necessities/desires, technology changes, and so on. It varies equally as much because of fluctuations in climate, biological cycles of plants and animals, sea-current patterns, and similar factors. Unfortunately "flexibility" is often a code word, in the language of economic expansion, for finding ways to rationalize the desired growth by greenwashing.

To say that sustainability is flexible means that it is a dynamic process. Such flexibility can be accommodated through a Rights of Nature framework in federal and state Constitutions, whose working details are spelled out in subsequent laws. Through such guiding laws, necessary adjustments to maintain sustainability can be made without danger of greenwashing. These modifications are vital to assure that the integrity and productivity of global, ecological systems and their consequent services are recalibrated to guarantee—as much as humanly possible—their ability to maintain their fundamental resilience and biophysical integrity for all generations.

Correcting Course under a Rights of Nature System

Instead of the current system, where human needs and desires are prioritized and Nature's requirements are not even recognized, a Rights of Nature system builds the legal framework first and foremost around the primacy of Nature's biophysical integrity. A legal system that prioritizes Nature's well-being, resilience, and productivity must include the following principles:

1. **A Feedback Mechanism That Corrects Course When Nature, via Changes in the Biophysical System, Makes It Clear Certain Human Actions Are Destructive to the System's Integrity.** This feedback could take several forms, such as a panel of independent scientists—not paid for by any self-interested entities—whose recommendations are then crafted by policymakers into laws curbing conduct that harms Nature. Feedback could also be provided via a "prosecutor for Nature" who hears cases in an independent court system and renders judgments, and whose decisions are forged into laws. To provide checks and balances, both could be used, with the prosecutor's holding being final.
2. **A Legally Enforceable Mechanism for Obtaining Factual Monitored Baseline Data of Sustainable Environmental Function.** In a position so powerful as the one providing factual baseline data on which Rights of Nature laws are based and projects decided, the temptation is very high for political corruption, often through the manipulation of data. The best way to achieve honest, factual ecosystem monitoring may be through a combination of a Nature's advocate or prosecutor with enforcement power, operating in an independent governmental capacity, and citizen enforcement at every level of the planning and decision-making process.

   Leaving this task to centralized bureaucracies has frequently provided biased or compromised data, as well as an unwillingness to use data of integrity to curb harmful practices. The temptations to misuse essential information in a Rights of Nature system, in which harmful economic activity is quickly curbed, would be much greater. One alternative would be so-called "ecosystem trusts" that are accountable to future generations and have property rights, as well as enforceable powers over an ecosystem's ability to flourish.[41] Statutory creativity is desperately needed to find new ways of both providing data and using it to correct course in a Rights of Nature framework of governance.
3. **Direct Citizen Participation in the Legal System to Protect the Rights of Nature Is Essential to Maintaining Legal Integrity.** Scientists are not invariably correct, simply because they do not live on the ground in every locale, are not familiar with every problem, and so cannot be all-knowing. Thus, the legal system must always maintain opportunities for legitimate public participation via meetings, testimony, workshops, rallies, and similar forums. This process must be authentic, however, *not* the "Thank-you-for-sharing" form of participation found so often in the environmental public arena. Too often, an agency considering an action solicits public input, as required by law, tabulates the responses, responds to the comments, and then chooses its predetermined, economically beneficial—but environmentally destructive—alternative.

   Specifically encouraging citizen participation via lawsuits requesting remedies on Nature's behalf, perhaps through a Nature's prosecutor or Nature's tribunal system, would provide an essential watchdog role in the legal Nature's Rights framework to ensure that these primary, all-important rights are enforced through all parts of the economic system.

4. **A Legally Enshrined Set of Environmental Priorities That Will Ensure Nature's Integrity Comes First.** The priorities that shape the new Nature's Rights legal system must include the following, although individuals living in diverse regions may think of others that are essential in their areas:
   A. *Prioritize reintroduction of critical species in areas where they have been extirpated, and cultivation of their habitats.* Example: sea otters are a critical species in the Pacific Northwest nearshore environment because their feeding habits allow kelp forests to flourish. Their populations in California are doing poorly as a result of pollution. In Oregon, sea otters remain extirpated despite a successful reintroduction in Washington. Such critical species are the prime means by which ecosystems can regain much of their biophysical integrity and resilience. Thus, mere reintroduction is insufficient; cultivation or strengthening of their habitats so they can once again flourish and change the environment for the better is also essential. The principles of indigenous cultivation, discussed earlier, need to be modified and adjusted to buttress these important species and the local ecosystem they uphold. This requires moving substantially beyond the current "management" paradigm, in which human desires to extract something or use an area set the rules for ecosystem intervention.
   B. *Prioritize reclamation/restoration of degraded habitats.* This principle would include such things as removing land from agricultural production to recreate wetlands, or ending industrial clear-cutting and tree-farming in order to allow forests, with true age-class diversity, to develop again. It would include curbing land-based pollution and septic leakage into fragile oceans and lakes, or allowing prairies extinguished by livestock overgrazing to revitalize free of range management. For this to happen, the legal system must provide both a mandatory, legal obligation and the means whereby both government and citizens can collaborate in the reclamation of degraded lands.

      One example is removal of harmful roads and damaging rural structures. Unnecessary roads are a great problem throughout the United States, especially in public forests[42] and grasslands of the West. They contribute mightily to erosion and habitat fragmentation.[43] Likewise, many structures, such as culverts, fences, and dams have destroyed thousands of acres of habitat in rivers, streams, rangelands, and forests.[44]
   C. *Prioritize providing large-scale landscapes for ecological resilience.* In wilderness areas and other nonmanaged or minimally managed landscapes, the ecosystem has maximum opportunity to function naturally within in the Laws of Reciprocity. Wilderness areas have often been derided as unnecessary, or as unfair to those who cannot visit them for one reason or another. But the goal in designating such areas is to provide Nature with maximum opportunities for resilience and biophysical integrity. Wilderness protects plants and animals by offering them a refuge, while simultaneously protecting headwater streams and so clean water within the water catchments, as well as contributing to good air quality and a legacy of unspoiled Nature for future generations.[45]

> Providing large-scale opportunities for resilience must become a human priority, with actions taken in such areas designed to be as nonintrusive as possible. The natural areas set aside must be more than the barebones acreage necessary for the ecosystem to function, *and* the intended human uses must be compatible with the areas' biophysical integrity. The value of a wilderness area is greatly devalued if it is surrounded by houses or clearcuts, thereby compromising its sustainable resilience. There must be a gradation of compatible uses adjacent to such sanctuaries of Nature. These areas will, by their very existence, provide priceless and evolving opportunities for natural, environmental, baseline studies about the functional aspects of Nature in various environmental settings.[46]

Ecologically Based Risk Assessment

We introduce thoughts, practices, substances, and technologies into the environment. Before we do so, we are in charge and the proposed action is negotiable. But once introduced, it is effectively out of our control and its effects—good or bad—are forever nonnegotiable.[47]

Western industrialized society seems to find little or no intrinsic value in Nature unless it is demonstrably "good for something" or can be converted into something for which we can find a material value. We therefore use land continually, disallowing it time to itself.

As an example, in the arid western United States, it is time that is being taken away, and, consequently, plants and soil are lost. The "skin" of lichens, fungi, and algae that once formed a protective layer over the surface of the soil of grasslands, deserts, and shrub steppe has been frayed to dust by the hooves of livestock, boots of backpackers, and wheels of dirt bikes and all-terrain vehicles. Today, less than 5% of the skin exists in pristine condition. When undamaged, the skin can cover as much as 80% of the soil and is interspersed here and there only by indigenous grasses and shrubs. This complex crust of "living skin" takes a half-century or more to develop. A fully developed skin, which can be thought of as a tiny ecosystem, may have more than 100 species of minute to small plants and thousands of bacteria and other microorganisms within it.[48]

Our initial introduction into the environment is our pattern of thought, which determines the way we perceive the Earth and the way we act toward it—either as something sacred to be nurtured or only as a commodity to be converted into money. Unfortunately, technology gives us an illusory sense of ever-greater control over our environment and allows us to dupe ourselves into thinking that we are in control of and therefore somehow separated from Nature, rather than being an integral and inseparable part of it.

But if technology can offer a quick fix to our immediate problems, why do we need to assess the imaginary risk in its use? The answer is: As humans develop new technology and draw on the resources of the Earth to do so, they continually generate unprecedented quantities of new, unintended industrial products, such as toxic wastes. Then, because every decision has a risk attached to it, the more we know about the risks of our proposed actions, the better off we will be.

It is therefore imperative that we understand, as best we can, what the effects of our activities will be on the sustainability of our environment as a whole. Over the past few decades, says retired physicist and engineer B. John Garrick, an entire discipline, known as risk assessment or risk analysis, has been formulated around the probabilities, the "what if" of actions. In essence, learning how to *quantify the uncertainties* is a critical part of assessing the risk of introducing something into the environment.[49]

Making people more comfortable with the probabilities, says Garrick, is a matter of changing the terms of the debate. He goes on to say: "During my 40-plus years in the risk [assessment] business, the questions that have come to annoy me most are 'How safe is safe?' or 'How much risk is acceptable?' These are illogical questions. The only answer that makes sense is, 'It depends—on the alternatives available and on the benefits to be gained by making a certain decision.'"[50]

The best possible assessment of potential risk, as Garrick states, requires participation by the public, either directly or through elected representatives. "Governments and the private sector need to develop mechanisms to ensure this input [by the public] without letting the process get bogged down by a few people whose entire agendas may never be expressed and whose actions lead to gross mismanagement of resources. Those who spread false information [on all sides] need to be held accountable—especially since the consequences of their actions can cost billions of dollars,"[51] to say nothing of destroyed ecosystems and the loss of the services they provide, which add to the misery of countless people. However, because government representatives can become beholden to moneyed interests in a democratic system, such as the American one, it is essential that risk assessment be designed as a process that directly involves affected individuals and communities, in addition to elected officials.

"If decisions involving risk are not approached rationally, they will be made on political and emotional bases, which usually is not optimal for society," counsels Garrick.[51] Consequently, assessing risk in a formal manner provides us with a way to better understand the possible effects of the choices we face. Thus, we would be well advised to make the best possible use of what we know about assessing risks, provided we do not become enamored with the outcome. Although it is well beyond the scope of this book to delve more deeply into the complexities of environmental risk assessment, there are excellent, detailed guidelines available.[52]

Many of the issues in environmental risk assessment are familiar to policymakers and conservation leaders grappling with the future effects of proposed projects at a landscape or even regional level. As the size of proposed industrial projects increases, large-scale risk assessment is increasingly urgent. These techniques are also essential in a Rights of Nature system for learning what effects past actions on the environment are likely to have in the future, such as plastics in the ocean or introduction of invasive species to a pristine ecosystem. Risk assessment aids in crafting remedies for ecosystem damage and designing restoration or the cultivation of a critical species on a landscape scale.

Kapustka and Landis, editors of an environmental-risk handbook, explain: "It is possible to do risk assessments at very large scales for a variety of environmental, ecological and human health goals. While there are certainly challenges, there is no conceptual barrier to using risk assessment (or more properly risk analysis) as a tool… over varied time horizons." They do, however, stress the importance of including stakeholders in addition to regulatory authorities.[53]

These risk-analysis techniques can be used for assessment of such issues as measuring the value of wildlands—traditionally a difficult, controversial, and uncertain process. In addition, risk analysis can help elucidate such things as ecological integrity and linked human health, risks of climate change, the effects of invasive species on a landscape scale, escape of pharmaceuticals into the environment, and many other complex issues.

Endnotes

1. Mother Earth Trust. Carbon Credits. http://motherearthtrust.com/home/carbon-credit/?sa=X&ved=0ahUKEwixks3RpcDJAhVN4WMKHZgzAzEQ9QEILTAA (accessed December 3, 2015).
2. United Nations. Framework Convention on Climate Change. http://unfccc.int/land_use_and_climate_change/lulucf/items/3063.php (accessed December 3, 2015).
3. *Sierra Club v. Morton*, 405 U.S. 727 (1972). https://supreme.justia.com/cases/federal/us/405/727/case.html (accessed August 2, 2016).
4. Christopher D. Stone. Should trees have standing?—Toward legal rights for natural objects. *Southern California Law Review*, 45 (1972):450–501.
5. Christopher D. Stone. *Should Trees Have Standing? Law, Morality and the Environment*. Oxford University Press, New York. Third Edition. 2010. p. xiv.
6. Christopher D. Stone. Should trees have standing?—Toward legal rights for natural objects. *op. cit.*
7. *Ibid.* (p. 11).
8. Burns H. Weston and David Bollier. *Green Governance: Ecological Survival, Human Rights and the Law of the Commons*. Cambridge University Press, New York. 2013. p. 69.
9. Christopher D. Stone. Should trees have standing?—Toward legal rights for natural objects. *op. cit.*
10. *Ibid.*
11. *Ibid.*
12. *Ibid.* (p. 63).
13. Christopher D. Stone. Should trees have standing?—Toward legal rights for natural objects. *op. cit.*
14. *Ibid.*
15. Burns H. Weston and David Bollier. *Green Governance: Ecological Survival, Human Rights and the Law of the Commons. op. cit.*
16. Christopher D. Stone. Should trees have standing?—Toward legal rights for natural objects. *op. cit.*
17. *Ibid.* (p. 169).
18. 1000 Friends of Oregon. The Citizen's Guide to Land Use Appeals. https://www.friends.org/sites/...org/.../CitizensGuideToLUBAFinal.pdf (accessed August 3, 2016).
19. Burns H. Weston and David Bollier. *Green Governance: Ecological Survival, Human Rights and the Law of the Commons. op. cit.* (pp. 114–115).
20. Aldo Leopold. https://www.goodreads.com/author/quotes/43828.Aldo_Leopold?page=2 (accessed December 1, 2015).
21. Aldo Leopold. *Round River: From the Journals of Aldo Leopold, 1953*. Oxford University Press, New York. (1972) 286 pp.
22. Tim Newbold, Lawrence N. Hudson, Andrew P. Arnell, and others. Has land use pushed terrestrial biodiversity beyond the planetary boundary? A global assessment. *Science*, 353 (2016):288–291.
23. Robert Thompson. Scientists warn of 'unsafe' decline in biodiversity. http://www.bbc.com/news/science-environment-36805227 (accessed July 15, 2016).
24. *Ibid.*
25. (1) Vorsorgeprinzip. http://www.juraforum.de/lexikon/vorsorgeprinzip (accessed December 1, 2015); (2) Vorsorgeprinzip. http://www.linguee.de/deutsch-englisch/uebersetzung/vorsorgeprinzip.html (accessed December 1, 2015); and (3) Marco Martuzzi and Joel A. Tickner (eds.). The precautionary principle: Protecting public health, the environment and the future of our children. http://www.euro.who.int/__data/assets/pdf_file/0003/91173/E83079.pdf (accessed December 1, 2015).
26. Marco Martuzzi and Joel A. Tickner (eds.). The precautionary principle: Protecting public health, the environment and the future of our children. http://www.euro.who.int/__data/assets/pdf_file/0003/91173/E83079.pdf (accessed December 1, 2015).

27. Edward Guntrip. The common heritage of mankind: An adequate regime for managing the deep seabed? *Melbourne Journal of International Law*, 4 (2003):376–405 (p. 28 of PDF document).
28. *Ibid.*
29. The following discussion of a "biophysical living trust" is based largely on: Chris Maser. *Social-Environmental Planning: The Design Interface between Everyforest and Everycity*. CRC Press, Boca Raton, FL. 2009. 321 pp.
30. Elizabeth Sherrill. The power of a promise. *Guideposts*, August 1998:3.
31. Russ Beaton and Chris Maser. Economics and Ecology: United for a Sustainable World. CRC Press, Boca Raton, FL. 2012. 191 pp.
32. E.W. Sanderson, M. Jaiteh, M.A. Levy, and others. The human footprint and the last of the wild. *BioScience*, 52 (2002):891–904.
33. Burns H. Weston and David Bollier. *Green Governance: Ecological Survival, Human Rights and the Law of the Commons. op. cit.*
34. *Ibid.* (pp. 238–239).
35. United States Supreme Court ILLINOIS CENT. R. CO. v. STATE OF ILLINOIS, (1892) No. 419. http://caselaw.findlaw.com/us-supreme-court/146/387.html (accessed August 3, 2016).
36. Mary Christina Wood. *Nature's Trust: Environmental Law for a New Ecological Age*. Cambridge University Press, New York. 2013. p. 89.
37. *Ibid.*
38. (1) Mark Duchamp. How Much Wildlife Can We Afford to Kill? http://savetheeaglesinternational.org/new/us-windfarms-kill-10-20-times-more-than-previously-thought.html (accessed December 2, 2015); and (2) K. Shawn Smallwood. Comparing bird and bat fatality-rate estimates among North American wind-energy projects. *Wildlife Society*, 37 (2013):19–33.
39. Institute for Energy Research. License to Kill: Wind and Solar Decimate Birds and Bats. http://instituteforenergyresearch.org/analysis/license-to-kill-wind-and-solar-decimate-birds-and-bats/ (accessed December 2, 2015).
40. (1) Northwest National Marine Renewable Energy Center. Effects on the Environment. http://nnmrec.oregonstate.edu/education/effects-environment (accessed December 2, 2015); and (2) Michelle Ma. Concerns emerge about environmental effects of wave-energy technology. *The Seattle Times*. http://www.seattletimes.com/seattle-news/concerns-emerge-about-environmental-effects-of-wave-energy-technology/ (accessed December 2, 2015).
41. Burns H. Weston and David Bollier. *Green Governance: Ecological Survival, Human Rights and the Law of the Commons. op. cit.*
42. Chris Maser. *Our Forest Legacy: Today's Decisions, Tomorrow's Consequences*. Maisonneuve Press, Washington, DC. 2005. 255 pp.
43. (1) Chris Maser. *Earth in Our Care: Ecology, Economy, and Sustainability*. Rutgers University Press, New Brunswick, NJ. (2009) 304 pp.; and (2) Chris Maser. *Social-Environmental Planning: The Design Interface between Everyforest and Everycity*. CRC Press, Boca Raton, FL. (2009) 321 pp.
44. Chris Maser. *Our Forest Legacy: Today's Decisions, Tomorrow's Consequences. op. cit.*
45. (1) Ecological Benefits of Wilderness. http://www.wilderness.net/NWPS/valuesEcological (accessed December 2, 2015); and (2) Pinchot Institute for Conservation. Ensuring the Stewardship of the National Wilderness Preservation System. http://www.wilderness.net/NWPS/documents/brown_report_full.pdf (accessed December 2, 2015).
46. R.A. Mittermeier, C.G. Mittermeier, T.M. Brooks, and others. Wilderness and biodiversity conservation. *Proceedings of the National Academy of Sciences U.S.A.*, 100 (2003):10309–10313.
47. Chris Maser. *Ecological Diversity in Sustainable Development: The Vital and Forgotten Dimension*. Lewis Publishers, Boca Raton, FL. (1999) 402 pp.
48. (1) Jayne Belnap, Roger Rosentreter, Steve Leonard, and others. 2001. Biological Soil Crusts: Ecology and Management. U. S. Department of Interior, Bureau of Land Management and U.S. Geological Survey. *Technical Reference 1730-2*. (2001) 118 pp. http://www.soilcrust.org/crust.pdf (accessed December 17, 2015); (2) Lynell Deines, Roger Rosentreter, David J. Eldridge, and Marcelo D. Serpe. Germination and seedling establishment of two annual grasses on lichen-dominated biological soil crusts. *Plant Soil*, 295 (2007):23–35; (3) Matthew A. Bowker. Biological

soil crust rehabilitation in theory and practice: An underexploited opportunity. *Restoration Ecology*, 15 (2007):13–23; and (4) Wolfgang Elbert, Bettina Weber, Susannah Burrows, and others. 2012. Contribution of cryptogamic covers to the global cycles of carbon and nitrogen. *Nature Geoscience*, 5 (2012):459–462.

49. B. John Garrick: Society must come to terms with risk. *Corvallis Gazette-Times*. Corvallis, OR. November 9, 1997.
50. *Ibid.*
51. *Ibid.*
52. (1) Lawrence A. Kapustka and Wayne G. Landis (eds.). *Environmental Risk Assessment and Management from a Landscape Perspective [USA]*. John Wiley & Sons, Inc., Hoboken, NJ. (2010) 416 pp.; and (2) Áine Gormley, Simon Pollard, and Sophie Rocks. Guidelines for Environmental Risk Assessment and Management [UK]. https://www.gov.uk/government/uploads/system/uploads/attachment_data/file/69450/pb13670-green-leaves-iii-1111071.pdf (accessed January 4, 2016).
53. Lawrence A. Kapustka and Wayne G. Landis (eds.). *Environmental Risk Assessment and Management from a Landscape Perspective [USA]. op. cit.* (p. xii).

Section III

Stumbling Blocks to a Rights of Nature System

[Americans'] one primary and predominant object is to cultivate and settle these prairies, forests, and vast waste lands. The striking and peculiar characteristic of American society is, that it is not so much a democracy as a huge commercial company for the discovery, cultivation, and capitalization of its enormous territory.... The United States is primarily a commercial society ... and only secondarily a nation....

Émile Boutmy (April 13, 1835 to January 25, 1906)[1]
French Political Scientist

Endnote

1. Émile Boutmy. *Études de droit constitutionnel* (*Studies of Constitutional Law*), Macmillan, London. (1891) 183 pp.

8

The Problem of Technology

The role of technology is viewed today as the avenue to all "progress," especially in economic affairs. Since the Industrial Revolution, the combination of human and machine creates the products as output, but without new technology, continual gains in productivity are nullified. Without increases in productivity, an ever-higher standard of living is but a fond imagining.

Technology Is Not the Answer

The combination of technology and free markets has played such a significant role in the economic history of the United States that a distinct ethic, known as "technological optimism," has emerged over the years. In fact, technological optimism is central to the economy versus the environment dilemma. In its simplest form, technological optimism merely contends that technical know-how will bail modern industrial society out of any problem that may come along. In a nutshell, necessity (= desire) is the mother of invention. A simple incentive system emerges when someone, acting in his or her own self-interest, develops a new alternative or finds a new resource and is richly rewarded for his or her ingenuity.

However, technology has seldom been consciously directed toward serving the necessities of human beings for social-environmental sustainability, a thought captured in the reflection of author Havelock Ellis: "The greatest task before civilization at present is to make machines what they ought to be, the slaves instead of the masters of men."[1] Instead, technology has developed willy-nilly inside the unspoken agendas of scientific and engineering. Nevertheless, some technological developments in medicine have admirably served the well-being of people, whereas others have not.

The Industrial Revolution and Luddite Rebellion

In the early stages of the Industrial Revolution in Great Britain, a group of English textile workers and self-employed weavers, known as Luddites, stormed the factories and smashed the machines (between 1811 and 1816), especially in cotton and woolen mills, which they blamed for supplanting their jobs. "Luddism" is often derided as a self-destructive and wrong-headed backlash against the inevitable march of progress. As such, it is dismissed with a shrug and a sneer. But Luddism, and its aftermath, holds the key to understanding the misuse of technology that has landed the peoples of the world in the ecological predicament that looms over us all today.

Luddism is frequently mischaracterized as a protest against technology, but that is inaccurate. "Technology" is defined most simply as "the practical application of knowledge, especially in a particular area"—experimenting with what one knows to solve problems. The use of tools is the broadest example, which has been a part of human life since the dawn of time, and Luddites, like all other humans, used tools. Luddism was a protest against *industrialization* and the severe, social dislocation that resulted, which is vastly different from a protest against technology per se.

Early 19th-century England changed quickly and dramatically from a fundamentally craft and rural laborer economy to a much more heavily populated, urban, factory-based economy. The driver was industrialization: the establishment of needs, or even the perception of needs, which then magically created a market for the goods. The population of England and Wales doubled between 1785 and 1830, from approximately 7.5 million to 16.5 million in 1831. The industrial revolution was poised to create massive amounts of goods for this increased population—plus Britain's overseas colonial empires.

As a result of "enclosures," fencing of the commons for grazing sheep, the entire rural fabric of England and Wales disintegrated. Though rural life was often hard, there were many compensations: knowledge of all one's neighbors; locally made goods sold fairly in local markets; a tradition and history in which all villagers were embedded, often extending back hundreds of years; well-known and honored relationships in dense networks between masters and journeymen, workers and merchants, cottagers and squires.

Enclosures increased the sizes of farms, making them more "efficient," but at the terrible cost of unilaterally eliminating the farms and homesteads of cottagers and tenants, while benefiting a smaller class of rural landowners and gentry. These displaced people, without locally produced income and thus no longer able to provide for their own needs, crowded into the cities and became the cheap workforce needed for the mills that turned out Britain's famed woolen cloth.

Severe, social dislocation accompanied the change from the rural craft and laborer economy to the urban factory economy: steep decline in wages, wretched housing, miserable food, massive populations torn from their social and village roots, families and histories. Approximately 10 percent to 15 percent of Britain's population in the 1813–1818 era and beyond subsisted on some form of poor relief—not counting the millions who had a little work, but not enough to exist outside the most meager margins of sustenance.

Although cotton weavers were the first to be eliminated by machine-based labor in the initial wave of industrialization, other artisans in the weaving trade, such as wool-combers, and traditional agricultural laborers were also eliminated. In addition, the new life in factories was hellish because: (1) a majority of workers were women and children, usually paid less than a third of men's wages; (2) the noise of looms was so deafening that a form of sign language and lip-reading developed for communication; (3) the air was filled with cotton dust that gave workers a lifelong cough; and (4) accidents, even fatal ones, were common because of the unrelenting work among dangerous high-speed machines. Therefore, skilled artisan weavers loathed the factories, and often refused to become wage-laborers except as a last resort before starvation.[2]

How was the British government responding to this tragedy on a massive scale? "[T]he British state worked in myriad ways, by acts of commission and omission, to advance the whole process of industrialization, enrich and protect the manufacturing sector, ensure a compliant labor force, and provide regular and prosperous domestic and foreign markets."[3]

This first wave of industrialization had one more catastrophic effect, one that is highly relevant to our current predicament. Its single-minded focus was to prevail over all limitations of Nature; it "inaugurated an attitude, a deep-seated conviction, that regardless of the inevitable environmental costs, the powers of industrialization could and should be used to control and exploit the forces, species and resources of nature."[4]

Rivers turned black with industrial refuse; wetlands were drained by the thousands; wildlife plummeted nationwide; coal smoke turned the air of towns sooty black; raw sewage ran into waterways; industrial toxic poisons were simply dumped willy-nilly, with no thought about poisoning the land or water; cities were crowded, unhealthy and without the barest amenities. For example, life expectancy in England, as a whole, was approximately 40 years in 1842. But in Manchester and Leeds—two typical industrial towns—life expectancy was 41 years for the middle-class "gentry," but a mere 18 years for the laborer.

The first tidal wave of industrialization appeared to have outdistanced all limits of Nature, community, and tradition—but this was, of course, an illusion. The Luddite rebellions were the first indication that England's laissez-faire economy could go no further. The weavers and knitters of Nottinghamshire were the first to rebel, in 1811, against the insupportable conditions of factory labor that was destroying both livelihood and community[5] (Figure 8.1).

However, there had been isolated incidents of machine-breaking much earlier, such as "a 1675 attack by Spitalfields weavers against a machine they feared could do the work of twenty men—and with the advent of the Industrial Revolution there began a series of attacks aimed at some specific machine or factory that was either putting workers out of their jobs or severely reducing their wages, disputes that had … to do with … the machines themselves, their unwelcome imposition, and their threat to what workers felt was customary, or just, or legitimate."[6]

The first Nottinghamshire raids focused on the destruction of machinery; nighttime attacks; secrecy; member solidarity; and threatening, anonymous letters. As the rebellion

FIGURE 8.1
Luddites, or "frame-breakers," smashing a loom. The British Parliament made destroying machines a criminal act as early as 1721, the punishment for which was exile to a penal colony. However, the continued opposition to mechanization resulted in the Frame-Breaking Act of 1812, and its accompanying death penalty. Artist Unknown (https://commons.wikimedia.org/w/index.php?search=Luddites%2C+or+"frame-breakers%2C"+&title=Special:Search&go=Go&uselang=en&searchToken=4stigddv3z8n4899pudx206vj).

spread, in a disorganized fashion, into Lancashire and Yorkshire, the tactics hardened and began including public demonstrations, arson, burglary, and more frequent raids even in the face of organized troops and police sent to halt the depredations. In Yorkshire, Luddism took its most powerful form, approaching true insurrection, with caches of weapons and outspoken leaders.

Luddism was not an organized phenomenon; it was amorphous, localized, and sporadic, though the values the rebels were seeking to express were always very similar. Historians have often portrayed it as a unified and sophisticated movement, but this is to miss its character: Luddism was an organic response to insupportable local conditions that were being replicated nationwide in rural area after rural area. Despite historians' best efforts, it remains difficult to pinpoint exactly when the Luddite rebellion began, since there were several actions similar in character and close in time in 1811.

Over time, however, the Luddite movement did gain in sophistication, intensify its tactics, and appeared to be more widely organized. Though authorities called out the troops in ever-increasing numbers, they still could not stop the break-ins and destruction of machines; the perpetrators remained hidden. Without the slightest evidence, this sophistication made the authorities certain they were contending with a tightly coordinated and well-commanded organization.

The authorities did eventually catch up with the Luddites. In a particularly notorious example in 1813, the authorities executed in one day 14 Luddites whose average age was around 25. These men were croppers, colliers, woolen spinners, cotton spinners, woolen weavers, and a tailor. At least 15 were killed in action during the government's protracted fight against Luddism, 24 were hanged, 51 sentenced to Australia, and at least two dozen hustled to prison. In the face of what was essentially an occupying army, Luddism faded out, mainly by late 1813.

Nevertheless, it is difficult to pinpoint exactly when Luddism ended: perhaps in 1814, when more than 100 knitting frames were destroyed in Nottinghamshire; or perhaps in the spectacular attack on John Heathcoat's cheap lace factory in Loughborough in June 1816, when more than 50 factory frames were destroyed, ruining the lace production so completely that the factory closed.

Luddism achieved some important goals, despite flaring so briefly, and with the violence that is always a last resort. Wages increased in some areas. There was an end to the use of some of the most despised machines. And, there was a slow, but steady, increase in reformist ideology in the laboring class. Luddism's greatest triumph, however, was that it raised the question that needs to be revisited in every era because technology always has far-reaching consequences. That question is still at the heart of our industrial civilization: What technology should be used, for what ends, at what social-environmental cost, and by whom?[7]

Historians say we are now in the second Industrial Revolution, one dominated by computers: the so-called digital revolution. The effects on human societies and the natural world are incalculable.

> This time around the technology is even more complex and extensive, and its impact even more pervasive and dislocating, touching greater populations with greater speed and at greater scales … automation … an inevitable consequence of computerization … and serves to replace human endeavor in more and more ways in more and more settings … [T]he industrial view of nature … argues, with the full power of industrial science, that most of nature is inert and lifeless…and that other species, without our form of consciousness, are innately inferior. All of those may therefore be considered "resources," for the human species to exploit … and technologies should be designed to make the maximum use of such resources as completely as possible by as many people as possible.[8]

Industrial processes characteristically make unheeding use of natural "resources;" but, as historian Sale points out, "it was not until industrialism grew into its high-tech phase, with the immense power-multiplier of the computer, that this exploitation of resources escalated onto a new plane different not only in *degree*, with exhaustion, extermination, despoliation, and pollution at unprecedented and accelerating rates, but in *kind*, creating that technosphere so immanent in our lives, artificial, powerful and global, and fundamentally at odds with the biosphere."[9]

The Challenge: Technology, Sustainability, and Nature's Rights

This, then, is the task of the generations to come: to revisit the questions first posed by the original Luddite rebellions about the nature and kind of technology, and tailor technology, especially *industrial* technology, to be in the service of the Rights of Nature—not the other way around. Pope Francis in *Laudato Si*, his revolutionary encyclical on ecology, has led the way in pointing to the need for a serious re-evaluation of technology's place in human societies. He pointed out (Chapter 3, Sections I–III) that technology, if developed further in a one-dimensional way, will globalize the technological paradigm and rob societies of their own richness, as well as stunting their ability for communion with Nature. Thwarting this "ecological conversion" will perpetuate the blindness of those who fail to see the magnitude of the problems looming before humanity.[10]

A society based on Nature's Rights must have a technology that is consciously chosen to meld with a sound basis of core human values within the limitations defined by Nature's Laws of Reciprocity. To achieve such an outcome, technological inventions must be conceived, developed, and employed in light of a detached, dispassionate risk assessment of their effects on the available resources and their ecological sustainability. Once again, conscious judgment concerning the basic human values of all generations—present and future—must reign in absolute priority, a sentiment voiced by Aldo Leopold in 1949, "We face the question whether a still higher 'standard of living' is worth its cost in things natural, wild and free. For us of the minority, the opportunity to see geese is more important than television, and the chance to find a pasque flower is a right as inalienable as free speech."[11]

In Nature's scheme of things, the principle of cause and effect is impartial. Technology in and of itself is not the answer to global social-environmental sustainability—no matter how much some people might wish it otherwise.[12]

Technology Is Based on Symptomatic Thinking

For the most part, today's technology is about "more" with respect to resource exploitation for monetary gain and "faster" with respect to computers. There is also the worldwide warfare among ideological factions, all based on a symptomatic fix to something we either want or do not want.

For example, in a small town in the Willamette Valley of western Oregon, the chainsaw can still be used for a quick fix without regard to the consequences a hasty, symptomatic action will cause Nature, or the legacy it will leave the town over the coming years. When it appeared some years ago that drug dealing was taking place in the town's local park, rather than deal with the humans involved (systemic thinking), the town authorities simply clear-cut the wooded area's old trees (symptomatic thinking)—eliminating wildlife habitat, summer shade, recreational opportunities, and the diverse beauty of the park, while allowing the drug dealers to simply move elsewhere.

Engineering to Provide a Quick Fix

We must become students of processes and let go advocacy of positions and battles over narrow points of view, because there are no biological short cuts or technological quick fixes in our current symptomatic thinking that can mend what is broken. *Dramatic, fundamental change in the form of systemic thinking and action is necessary* if we are sincerely committed to attaining environmental sustainability and the social justice it engenders.[13]

In 1998, I (Chris) was honored when Lee Schroeder, then vice president for Finance and Administration at Oregon State University in Corvallis, Oregon, wrote the foreword for my book, *Vision and Leadership in Sustainable Development*. Lee, an engineer, clearly defines the linearity with which most engineers and inventors think—a linearity that defines the technological age in which we find ourselves today:

> I tend to be a linear thinker, as do many engineers. When I have been involved with systems behavior, and the parts of the system interact as they always do, the conclusions reached are often too vague to implement. Since I'm always trying to solve problems and I am driven to get a solution, I tend to revert to the linear approach. Show me a mountain, and I'll move it, but don't make me worry too much about anything more than that.
>
> … We [engineers] focus on the here and now and the immediate future. But we also frequently wonder, in our quiet moments, where all the things we do are leading us and whether we should choose a more sensible course if we are to act truly responsibly. Lily Tomlin once said, 'The trouble with the rat race is that even if you win, you're still a rat.' Her observation gives us cause to think about our actions and the impacts they may have in the future. …
>
> …
>
> As example might be the assumption that placing fish ladders in dams would sustain salmon migrations and hence perpetuate migrant populations. Over time, however, it has been learned that the reservoirs created by the dam (in addition to a host of other human activities) also impact the survival of salmon. Monitoring a single variable, such as counting salmon, has done nothing to clarify the issue or save the fish, which was the objective of the ladders. And a straightforward solution to the problem of saving the salmon is not now available. Things are much more complex than we are usually able to foresee.[14]

The point is that all engineering, from building a dam to fixing a damaged power line after a storm to genetically modifying organisms, is based on finding a way around Nature's processes that pose some outcome we do not want. The linearity of symptomatic thinking, even that which has benevolent intentions, such as electric lights and increased noise of human society, still disregards Nature's Laws of Reciprocity that, over time, can have dramatically undesirable effects, as the following two examples illustrate.

Electric Lights

There was a time in the 1940s and 1950s when the night sky of the Willamette Valley in western Oregon winked with the light of more than a million stars. In those days, the Milky Way was visible from almost everywhere in the south end of the valley, but no more.

Now, only the brightest stars can be seen, even on the darkest of nights, because of light pollution. Thus, we, in the United States, are losing the only portal to the wonder of Nature that is open to virtually everyone.

Glare from lights can simply be uncomfortable or annoying, such as the glare from excessively bright security lights in residential areas. But glare can prevent a motorist from seeing a pedestrian in dark clothing because the driver is blinded by the glare of bright lights from an oncoming vehicle. In addition, the ever-increasing output of lights from commercial and residential settings (such as commercial parking lots and private security systems), as well as from roads, affects our quality of life and our safety.

The most pervasive form of light pollution, however, is "urban sky glow," which is caused by artificial light passing upward, where it reflects off of submicroscopic particles of dust and water in the atmosphere. First noted as a visual problem by astronomers, urban sky glow, which can be seen more than a hundred miles away from large cities, is beginning to seriously destroy our ability to experience the nighttime sky in some of our national parks.[15]

In addition to the diminished wonder and enjoyment engendered by gazing at the stars, light pollution is a rapidly expanding form of human encroachment on other species, particularly in coastal regions, where it alters the behavior of sea nesting turtles.[16] It also affects the foraging behavior of Santa Rosa beach mice, which tend to avoid artificially lit areas.[17] In fact, this artificial light phenomenon appears to be driving some strictly nocturnal species, such as the California glossy snake, toward extinction. According to zoologist Robert Fisher of the U.S. Geological Survey in San Diego, California, "It might be that you can protect the land, but unless you can control the light levels that are invading the land, you're not going to be able to protect some of the species."[18]

The Increase of Noise

Once upon a time, silence could be found throughout much of the world. But as the Industrial Revolution vaulted ahead, silence became a rare and elusive part of Nature. In fact, the world has gotten so noisy, even beneath the ocean, that the ability of many sea creatures to seek food, find mates, protect their young, and escape their predators is severely impaired. The effects of underwater noise can be likened to being trapped in the center of a deafening traffic jam, where the din comes simultaneously from all sides. In deep water, where marine animals rely on their sense of hearing, the surrounding noise is especially harmful.

Noise from supertankers and military sonar equipment, as well as from the explosions of seismic exploration for offshore oil, scrambles the communication signals used by dolphins and whales, which causes them to abandon traditional feeding areas and breeding grounds, change direction during migration, and alter their calls. They also blunder into fishing nets. The global, unintentional, unwanted catch—"bycatch" in today's euphemistic vernacular—of marine mammals is in the hundreds of thousands and likely to have significant demographic effects on many populations. In addition, dolphins and whales can no longer avoid colliding with ships on the open seas, where international shipping produces the most underwater noise pollution, with few regulations to control it.[19]

Both in and beyond the sea, certain levels of noise—unwanted sound—negatively affect our own health, as well as that of our pets, such as zebra finches, which forego fidelity to a mate as sound blares.[20] Some of the other problems associated with noise pollution are loss of hearing; chronic stress; sleep deprivation; high blood pressure; mental distractions, with the resultant loss of enjoyment and productivity—all of which are part of a declining quality of life.

Noise pollution is one of society's growing concerns because it increasingly affects the quality of everyday living—especially if one lives in the flight path of an airport; within a few miles of a railroad crossing; next to an increasingly busy street; near an athletic field, a university fraternity, or ongoing construction. And there seem to be few places one can escape from it.

As urban sprawl claims more and more of a community's landscape, the collective noise of human activities, and their technological devices, increasingly invade the once-quiet sanctuary of private homes—often to the discomfort and frustration of its inhabitants, both human and non-human.

Industrial Profit and Technology Replace Sustainability

One place where it is easy to see how technology has been designed and used to exploit Nature is in the world's oceans, where fishers have long deemed the stocks of fish to be limitless and free for the taking in ever-greater numbers. To this end, various types of fishing gear have been invented specifically to exploit different parts of the ocean in an attempt to take advantage of all possible opportunities for personal gain. As a result, chronic industrialized overfishing has occurred along many continental shelves and coasts, where fishing fleets have been increasingly employed to satisfy the continually growing demand. Today, more than 25 percent of U.S. stocks are overfished; more fish are caught than the ocean can produce, resulting in the collapse of some important fisheries and longstanding fishing communities.

With respect to overfishing the oceans, researchers Daniel Pauly and Dirk Zeller at the University of British Columbia, Canada, contend that official estimates are missing crucial data on small-scale fisheries, illegal fishing, and fish discarded as "bycatch." They argue that the figures submitted to the United Nations Food and Agriculture Organization (which collects global statistics on fishing from countries all over the world) are mainly from large-scale "industrial" fishing and ignore small-scale, commercial fisheries, subsistence fisheries, as well as the discarded bycatch and estimates for illegal fishing. In other words, around 32 million tons of fish go unreported every year, which they bill as more than the entire U.S. population weighs.[20]

According to Daniel Pauly, the senior author, the aggregate of fishing in the ocean was never really sustainable.

> We went through one stock after the other, for example around the British Isles, the stocks in the North Sea were diminished right after the Second World War.
>
> And then British trawlers went to Iceland and did the same thing there, and so on and so did the Germans, the Americans, so did the Soviets.
>
> They had to expand to survive and now the fisheries are in Antarctica.[21]

This intentional exploitation has typically reduced the fish biomass by 80 percent within 15 years on four continental shelves and nine oceanic systems—based on available data from the beginning of exploitation. Moreover, increases in fast-growing species are often reversed within a decade. As a consequence of technology inventions specifically designed for the commercial exploitation of marine fishes, data indicate that the biomass of large predatory fish is currently about 10 percent of pre-industrial levels,[22] thanks in part to two of the most disruptive and widespread, human-induced disturbances to seabed communities worldwide—bottom trawling and bottom dredging, both referred to as "mobile fishing gear."

Another consequence is that Europe's rarest seabird, the Balearic shearwater, faces extinction within 60 years due primarily to being drowned in fishing lines and nets as it dives for its food. Research shows that the global shearwater population is not sustainable in the long term, in part because there are only approximately 3,000 breeding pairs left, each of which lays a single egg per year. Despite the fact that the Balearic shearwater is classified as critically endangered on the International Union for the Conservation of Nature Red List of species, there is a simple solution—setting the demersal longlines at night when the birds are not diving for food.

Demersal longlines (also called bottom-set longlines) are used on the continental shelf and slope to catch a variety of fishes. This line differs from a dropline in that the mainline, with the baited hooks attached, is set along the seabed. One end of the haul-in line has a weight attached to anchor the end of the mainline, and the other has a small, floating buoy with a flag, used to temporarily mark its position at sea. The line is left to fish for up to 6 hours.[23]

Mobile Fishing Gear

Bottom trawling is a method of industrial fishing whereby a large ship, called a "trawler," drags a wide net equipped with rollers, chains, and doors weighing thousands of pounds across the seafloor, scooping up everything in its path. And then there are "rockhopper" nets, which possess large disks that allow the trawling net to jump over and plow through rough terrain without getting snagged, while disturbing and damaging such structures as rock piles and corals (Figure 8.2).

FIGURE 8.2
Celtic Explorer, a bottom trawler, in Galway Bay, Ireland. Photograph by Anilocra at English Wikipedia (https://commons.wikimedia.org/wiki/File:RV_Celtic_Explorer,_Galway_Bay,_Ireland.jpg).

Bottom trawling is both unselective and severely damaging to the seafloor, where the net indiscriminately catches every living thing and object it encounters. In addition, the weight and width of a bottom trawling net can destroy large areas of seafloor habitats that give marine species food and shelter, leaving the marine ecosystem permanently damaged, especially where extensive areas are trawled from 100 percent to 700 percent or more annually.

In addition to trawling, bottom dredges, which are heavy rake-like devices specifically designed to scrape across the seafloor, may be up to 15 feet wide and extensively damage the ocean floor.

In the United States alone, bottom fishing with mobile gear on the Pacific, Atlantic, and Gulf coasts captured more than 800,000,000 pounds of marine life in 2007. Those of no commercial value, euphemistically called a "bycatch," can amount to 90 percent of the total organisms. These unfortunate creatures, such as endangered fishes and deep-sea corals that can live for several hundred years, among many others, are simply thrown overboard dead or dying.[24]

The Environmental Consequences of Bottom Fishing

Bottom-fishing technology is purposefully engineered to destroy the diversity of the seafloor. The combination of large size, heavy weight, and damaging designs makes mobile gear a significant threat to many ocean bottom ecosystems so critical to supporting the abundance and diversity of marine organisms.

Bottom trawls and dredges dramatically alter the ocean floor, thereby causing major changes in biological communities by dispersing boulders, destroying burrows, and removing stationary animals that depend on these structures for breeding, shelter, and feeding. When mobile gear is used over large regions of continental shelves worldwide, it can reduce habitat complexity by smoothing the micro-topography of the bottom (not unlike clear-cutting a forest), removing pebble-cobble substrate with emergent epifauna, and thus eliminating species that produce structures, such as burrows. ("Epifauna" are animals that live on the surface of sediments or soils.)

These considerations have particularly important legal ramifications for any species—particularly long-lived fish, such as cod—that require several types of interconnected habitats to complete their life cycles, some of which are particularly susceptible to damage by trawling and dredging.

Moreover, the effects of mobile fishing gear on biodiversity are most severe in areas least affected by natural disturbance, particularly on the outer continental shelf and slope, where damage from storm waves is negligible and biological processes, including growth, tend to be slow.[25]

A Framework for Ecological Plundering

To understand the present conflict among the status quo of resource overexploitation, Nature's Laws of Reciprocity, and the desperate need for a paradigm switch to a society based on Rights of Nature, we must consider the role of those rationalistic thinkers who first gave frame and body to a reductionist, mechanical worldview over a period of some 300 years.

The Reductionist, Mechanical Worldview

Rationalist thinkers, such as Francis Bacon (1561–1626, English philosopher and essayist),[26] Galileo Galilei (1564–1642, Italian scientist and philosopher),[27] René Descartes (1596–1650, French philosopher and mathematician),[28] John Locke (1632–1704, English philosopher),[29] and Isaac Newton (1642–1727, English mathematician, scientist, and philosopher),[30] legitimized and institutionalized the lust for material wealth over which feudal society had for so long fought. In so doing was born the reductionistic mechanical worldview.[31]

Consider the collective paradigm of these renowned men: "Nature's sole value is in service to the material desires of humanity. But, nature must be tortured before its secrets will be revealed for human use" (Bacon).[32] "Measure what is measurable, and make measurable what is not so" (Galileo).[33] "Because real things are both measurable and quantifiable, they must operate through predictable linear mechanical principles, like an enormous machine" (Newton).[34] "And, like a machine, real things can be understood by disassembling the things themselves into smaller and smaller, more manageable pieces, which can be studied, and then rearranged in an order deemed logical to the human mind" (Descartes).[35]

Reductionistic thinking and acting means taking apart a living system, isolating the components, and rearranging them by keeping only those deemed of value in human terms, reassembling the retained pieces, and expecting the system to function as before. "Mechanical" refers to the common human notion that the world is assembled like a machine, acts like a machine, and thus can be treated like a machine, which has interchangeable parts, like a watch.

Our Western analytical perspective involves a four-part process: (1) dissect the system into its component parts, (2) study each part in isolation, (3) glean a knowledge of the whole by studying its parts, and (4) rearrange the parts in such a way that they satisfy our linear sense of logic.

The implicit assumption is that systems are aggregates of interchangeable parts that function in a linear fashion. Thus, by optimizing each part, we optimize the whole. And so, we continually fragment our problems into smaller, more "manageable" pieces, while our challenges are increasingly systemic. "Linear" in this sense means having only one dimension, that of an ever-extending straight line with no means—or thought of return, such as an economy that is ever expanding.

Both the Enlightenment and the Industrial Revolution were built on the foundation of Bacon's inductive method to control Nature for human gain, and the Cartesian-Newtonian worldview of a great machine subject to methodologies based on reductionism, quantification, and the separation of facts and values.

The Enlightenment, a philosophical movement of the 18th century, was concerned with the critical examination of previously accepted doctrines and institutions from a rationalistic point of view. The Enlightenment's faith was in the wedding of science and technology as a means of ending human scarcity and suffering. Then, in the middle of 18th-century England came the Industrial Revolution with its tremendous social and economic changes brought about by the extensive mechanization of production systems. This resulted in a shift from home manufacturing to large-scale factory production.

The Industrial Revolution's technological advances helped equate "progress" with the satisfaction of material wants, which eventually created a consumer-oriented society.

The Enlightenment and the Industrial Revolution cleared the way for a new paradigm—the expansionist economic worldview. The expansionist economic worldview, parented by capitalism, inaugurated the focus on competing for an ever-expanding hoard of material wealth.[36]

Associated with the European geographical expansion in the Americas, the concept of continuous growth was based on what seemed to be boundless opportunities in a world of limitless natural resources—the personification of the expansionist economic worldview. Once entrenched, this economic paradigm was justified as "social development," which came to be thought of as "social progress."

In this worldview, Nature is merely a vast storehouse of natural resources whose sole value lies in their conversion to commodities for the satisfaction of the ever-increasing material wants of an ever-growing human population. Consequently, material growth is equated with economic growth, which is equated with social development, which is seen as a prerequisite for human happiness and prosperity, the pursuit of which is built into the very fabric of American society and considered inviolate. It is the engine driving the dramatic resource overexploitation so evident now.

Resource Overexploitation and Perceived Loss

Historically, any newly identified resource is inevitably overexploited, often to the point of collapse or extinction. Its overexploitation is based, first, on the perceived rights or entitlement of the exploiter to get his or her share before someone else does and, second, on the right or entitlement to protect his or her economic investment. There is more to it than this, however, because the concept of a healthy capitalistic system is one that is ever growing, ever expanding. But such a system is not biologically sustainable. With natural resources, such non-sustainable exploitation creates a "ratchet effect," where to ratchet means to constantly, albeit unevenly, increase the rate of exploitation of a resource.[37]

The ratchet effect works as follows: During periods of relative economic stability, the rate of harvest of a given renewable resource, say timber or salmon, tends to stabilize at a level that economic theory predicts can be sustained through some scale of time. Such levels, however, are almost always excessive, because economists take existing unknown and unpredictable ecological variables and convert them, in theory at least, into known and predictable economic constants in order to better calculate the expected return on a given investment from a sustained harvest.[38]

Then comes a sequence of good years in the market, or in the availability of the resource, or both, and additional capital investments are encouraged in harvesting and processing because competitive economic growth is the root of capitalism. When conditions return to normal or even below normal, however, the industry, having over-invested, appeals to the government for help because substantial economic capital, and often jobs, are at stake. The government typically responds with direct or indirect subsidies, which only encourage continual over-harvesting.

The ratchet effect is thus caused by unrestrained economic investment to increase short-term yields in good times and strong opposition to losing those yields in bad times. This opposition to losing yields means there is great resistance to using a resource in a biologically sustainable manner because there is no predictability in yields and no guarantee of yield increases in the foreseeable future. In addition, our linear, economic models of ever-increasing yield are built on the assumption that we can, in fact, have an economically sustained yield. This contrived concept fails in the face of the biological limits of a yield's *sustainability*.

Then, because there is no mechanism in our linear, economic models of ever-increasing yield that allows for the uncertainties of ecological cycles and variability or for the inevitable decreases in yield during bad times, the long-term outcome is a heavily subsidized industry. Such an industry continually over-harvests the resource on an artificially created, sustained-yield basis that is not biologically sustainable.[39]

Because the availability of choices dictates the amount of control we feel we have with respect to our sense of security, a potential loss of money is the breeding ground for environmental injustice. This is the kind of environmental injustice in which the present generation steals from all future generations by overexploiting a resource rather than facing the uncertainty of giving up potential income.

The uncertainty of giving up potential income is a mindset that I (Cameron) frequently deal with in working to protect the Oregon coast from ill-advised development. Proposals for golf courses on farmland and coastal shore lands are invariably rationalized as profitable for the developers, landowners, and local community because of construction jobs and subsequent jobs and income from tourism. While these projections can easily be exaggerated and manipulated, the values lost—sustainable farmland and ecologically resilient coastal habitats undisturbed by golf course pollution and increased human traffic—are very difficult to quantify in a public-policy debate. Even though the proposed golf course locations are highly unsuitable and destroy the environment for all generations, the structure of public debate does not allow this overarching problem to become part of the decision-making process, much less center on it.

First, history suggests that a biologically sustainable use of any resource has never been achieved without first overexploiting it, despite historical warnings and contemporary data. If history is correct, resource problems are not environmental problems but rather human ones that we have created many times, in many places, under a wide variety of social, political, and economic systems.

Second, under the prevailing influence of the reductionistic mechanical worldview, it is too easy to dismiss as impractical idealism any attempt to refocus from immediate bread-and-butter issues to long-term processes and futuristic ideas. Further compounding the belief that long-term processes and futuristic ideas are merely impractical idealism is the notion of "conversion potential." For many people, the only value of anything is its "conversion potential" from a resource into a commercial product. This is the concept that Nature, having no intrinsic value, must be converted into money before any value can be assigned to it. All Nature is thus seen only in terms of its conversion potential.

Third, the fundamental issues involving resources, the environment, and people are complex and process-driven. The integrated knowledge of multiple disciplines is required to understand them. These underlying complexities of the physical and biological systems preclude a simplistic approach of management through technology. In addition, the wide, natural variability and the compounding, cumulative influence of continual human activity mask the results of overexploitation until they are severe—the consequences of which all generations must ultimately face.

Genetically Modified Organisms: The Corporate Bid to Control the Building Blocks of Life

The corporate bid to control the building blocks of life began with "biopiracy," which is the act of corporations pursuing the ancient wisdom of indigenous peoples to locate and understand the use of their traditional medicinal plants, and then exploit them commercially—unbeknownst to and at the expense of the indigenous people and the nation in which they live.[40]

Biopiracy

Hoping to find cures worth billions of dollars, scientists from the United States and Europe have even taken blood, hair, and saliva from indigenous peoples, which means biopiracy of the human body, for which there is neither moral justification nor recompense.

Jeremy Rifkin, in his 1999 article entitled "God in a Labcoat," points out that many of today's best-known molecular biologists are imprisoned in the hubris of Baconian tradition.[41] Francis Bacon, who lived in the 17th century, is considered to be the father of modern science through the scientific method of inquiry. He is often said to have viewed Nature as a "common harlot" and thus urged future generations to "tame," "squeeze," "mold," and "shape" Nature in order to be in control of the physical world—a claim questioned by Nieves H. De Madariaga Masthew.[42]

Scientists who embrace this traditional Baconian worldview through the reductionist lens, according to Rifkin, deem themselves, "grand engineers [who are] continually editing and recombining the genetic components of life into compliant organisms designed for human service."[43] This worldview could also easily describe corporations that compete for greater profit margins through increasingly centralized control of global resource and markets. Their competition has even led to a "gene rush" by so-called biological pirates.

Legalizing Biopiracy

Biological prospecting or "bioprospecting," as Kimbrell calls it, has the potential to be a veritable gold mine for industry and science because the untapped genetic materials found in the non-industrialized countries of the world may well yield cures for some of society's ills, as well as a bounty of cash. The prospect of potential monetary wealth from Nature's genetic materials has already sent "biopirates" into countries such as India to steal living things and potential drugs under the guise of improving "unoccupied" knowledge. Biopiracy is not solely the product of scientists wed more thoroughly than in the past to corporate greed for their research dollars; it is also the result of a new law based on a 1980s U.S. Supreme Court decision, *Diamond v. Chakrabarty*.[44] By granting legal permission for bioprospecting with the intent to commit biopiracy, the effect of this little-known court decision is, according to Kimbrell, "one of the most important judicial decisions of the twentieth century."[45]

The case began in 1971, when microbiologist Ananda Mohan Chakrabarty, an employee of General Electric, genetically engineered a bacterium that could digest oil. That same year, General Electric applied to the U.S. Patent and Trademark Office for a patent on Chakrabarty's oil-eating bacterium. The Patent and Trademark Office deliberated for several years and finally rejected the application under the traditional legal doctrine that life-forms, which are products of Nature, are not legally patentable.

However, the case was eventually referred to the U.S. Supreme Court, which handed down a surprise five-to-four decision in June 1980, ruling that the patent was to be granted because the "relevant distinction is not between living and inanimate things but whether living products could be seen as 'human-made inventions.'"[46] Allowing life to be seen as "human-made inventions" and thus patentable is a notion on ethically thin ice. Nevertheless, the Patent and Trademark Office ruled in 1985 that genetically engineered plants are patentable and in 1987 extended patenting to all genetically altered or engineered animals.[47] Within a few years, microbes, plants, animals, and human cells, cell lines, and genes were being patented.

By a margin of one vote, the U.S. Supreme Court unilaterally handed over to private ownership the genetic building blocks of life on planet Earth. So began the race to create "genetically modified organisms" by such corporate giants as Monsanto.

Genetic Engineering and Its Hidden Risks

Most supporters of "genetic engineering" downplay the difference between genetically engineered organisms and the techniques of time-honored selective breeding. They claim, for example, that the only difference between genetically engineered plants and traditional crops is that genetic manipulation is not only more precise but also faster and cheaper. Although this claim would seem to be good news, it is ecologically misleading, genetically irresponsible, and totally contrary to any Rights of Nature philosophy.

"Experiments have shown," writes Ricarda Steinbrecher, a genetic scientist and member of the British Society for Allergy, Environmental, and Nutritional Medicine, "that a gene is not an independent entity as was originally thought." Genetic engineers increasingly want to "transform plants and animals into designed commodities," while ignoring the many unknown hazards.[48]

Steinbrecher cites the example of a 1990 experiment in Germany, where the gene for the color red was taken from corn and transferred, together with a gene for antibiotic resistance, into the flowers of white petunias. The only expectation was a field of 20,000 red-flowering petunias. However, the genetically engineered petunias not only turned unexpectedly red but also had more leaves and shoots, a higher resistance to fungi, and lower fertility. These unforeseen results were completely unrelated to the genes for color and antibiotic resistance. Such unanticipated results have been termed "pleiotropic effects," which by their very nature are totally unpredictable. ("Pleiotropy," from the Greek *pleion*, meaning "more," and *tropos*, meaning "way," occurs when a single gene influences two or more seemingly unrelated genetic traits. Consequently, a mutation in a pleiotropic gene may have an effect on some—or all—biophysical traits simultaneously.)

In this case, says Steinbrecher, the pleiotropic effects were both clearly visible and easily identified without molecular analysis. But what happens, asks Steinbrecher, if pleiotropic effects are not so obvious, if they in secret affect the composition of proteins, the expression of hormones, or the concentration of nutrients, toxins, or allergens? Who is going to monitor all the possible pleiotropic effects before a genetically engineered plant is introduced into the environment or placed on our dinner plates? And there are neither regulations nor voluntary guidelines and practices with which to check for pleiotropic effects.[49]

So what, one might say, the possibility of genetic engineering remains. We will simply design organisms to meet our needs despite the supposed ecological limiting factors. Yes, we can work on this, but it is a gigantic gamble with the entire biophysical future of the Earth.

To examine this gamble, let us begin with a new buzzword, "transgenic technology," which is taking one part of an organism's genes and placing it into the genes of another organism. Transgenic technology is, after all, far from a precise practice because genes are not machines that can simply be snipped from their host and placed somewhere else without opening the possibility of totally unpredictable, unwanted, and potentially uncontrollable results. Consider that a trait, which is not evident in one organism and is said to be suppressed, may come out of hiding, as it were, and become visibly expressed when working in concert with the full set of chromosomes present in a normal reproductive germ cell of another organism.

Roughly two-thirds of the transgenic research in agricultural crops is aimed at creating plants that can resist stronger than normal dosages of pesticides and herbicides. The reason is that most crops developed through selective breeding can compete only weakly with weeds. Corporate symptomatic reasoning might unfold something like this: If the crop plants are easily out-competed by weeds and our chemical herbicides are detrimental to both the weeds and crop plants, then we must create genetically engineered crop plants that can withstand more chemical herbicides. In this way, the crops will be protected, we sell more herbicides, kill the weeds, and the world is better fed.

In 1997, Monsanto made its first commercial planting of "Roundup Ready" cotton, which was created to withstand high dosages of the powerful herbicide Roundup. By October of that year, more than 30,000 acres planted to Roundup Ready cotton were in serious trouble because the cotton bolls were falling off the plant throughout the South. In fact, the cotton bolls were deformed in such a characteristic way that the farmers named the syndrome "parrot beak."

Although the farmers lost millions of dollars, Roundup Ready cotton had been deregulated, which meant that the U.S. Department of Agriculture did not require Monsanto to report the incident. According to Kelly Wiseman, co-op general manager of the Community Food Co-op of Bozeman, Montana, the company tried to claim that unfavorable weather conditions were at fault, despite the fact that only Roundup Ready cotton exhibited parrot beak. Wiseman went on to say that Monsanto was investigating what happened, meaning the results would most likely remain a closely guarded secret under U.S. patent laws.[50]

What if the genetic engineering of the world's crop plants were to lead to "superweeds," as there already are "superinsects" and "superbacteria"—all genetically resistant to the chemicals that are supposed to kill them? After all, many so-called weeds are close relatives of crop plants, especially in the mustard family. And there is good evidence that cross-species mutations can occur within a few generations, which can lead to herbicide-resistant weeds.[51]

Consider, as previously stated, that a trait not evident in one organism because it is genetically suppressed may come out of hiding, as it were, and become visibly expressed, or dominant, when working in concert with the full set of chromosomes present in a normal reproductive or germ cell of another organism. For example, DNA from genetically engineered corn has shown up in samples of indigenous corn in four fields in the Sierra Norte de Oaxaca in southern Mexico. This finding is "particularly striking," said University of California, Berkeley, researchers Ignacio Chapela and David Quist in 2001, because Mexico has had a moratorium on genetically engineered corn since 1998.

Corn genetically engineered to resist herbicides or to produce its own insecticides threatens to reduce the variety of plants in that region of Mexico because it may be able to out-compete the indigenous species. According to Quist and Chapela, the probability is high that diversity is going to be crowded out by these genetic bullies. This type of unwanted genetic transference is termed "genetic pollution." In addition, the herbicide resistance could jump into weedy relatives and create super weeds that are beyond control. Furthermore, plants that have been genetically engineered to produce their own insecticide can have serious, deleterious effects on indigenous insects and microbes in the soil and thus have negative effects on indigenous plants.[52]

Cloning in Forestry

Artificial cloning is nothing new. It has been a common practice in forestry for some time, but it has grave ecological problems waiting in the wings. To see these problems, however, one must get past the short-term, economic, linear thinking that makes the notion of cloning attractive in the first place.

Genetic manipulation of forest trees was born in the concept of short-term economic expediency. By necessity, this process ignores long-term ecological ramifications to and within the forest as a whole, because tenable management practices for short-term profits must be based on *predictable, uniform* results. "Predictable" and "uniform" are the operative words in cloning. The term "clone," invented by J.B.S. Haldane from the Ancient Greek word *klōn* ("twig"), refers to the process whereby a new plant can be created from a twig.

Although grafting was first used in tree breeding in 1937 in Europe by Syrach Larsen, the art of grafting was developed by the Chinese thousands of years ago. With respect to forestry, grafting was originally done solely in greenhouses, but experiments with grafting in the open areas gave better results. In forest-tree propagation, grafting and cuttings are of fundamental importance because, through these operations, forest companies are able to keep exactly the same visible features found on a selected tree. In other words, they establish a "clone."

A young, selected tree—a clone—can be moved to a place where it can be worked with easily, as well as brought together with other clones. In addition, it be can saved for the future in graft depots, where it can get picked up the day it is to be planted. And lastly, it is easier to test a selected tree, both as a clone and as a progeny, through free and controlled pollination. Moreover, grafting makes it possible to save time because the twig is able to retain its biophysical characteristics when grafted into a young, rooted stem, so grafts from seed-bearing trees can be expected to start flowering two to five years after the grafting.[53]

Some people consider "genetic improvements," usually based on cloning in some facet of the operation, to be the panacea of forestry. In the short term, this is seen as maximizing return on investments while creating a predictable and uniform base of raw materials for the benefit of the timber industry. Beyond the short-term economic benefit to timber companies, however, it is imperative to address the immense importance of long-term biotic, genetic, and functional diversity in relation to the adaptability of ecosystems at the scale of the landscape. Current criteria for the economic evaluation of tree farms (even-aged plantations of trees as a crop) do not adequately account for these and other crucial issues that accrue from the manipulation of whole forests for short-term profit.

Genetic manipulation, mainly cloning—industrial panacea or not—contains hidden costs. Three hidden, clearly evident costs can be shown for tree farms consisting of cloned trees, particularly tree farms of commercial scale, also thought of as "industrial forests."

The first hidden cost is lack of predictability; we cannot predict any results, because no one has yet grown a "genetically improved" tree farm for even one full rotation of a century of more. We are therefore playing genetic roulette with future forests.

The second hidden cost is that by manipulating the genetics of the trees, the ecological processes in the entire tree farm are being altered by changing how the individual trees function. For example, if they grow faster, they will have larger cells, more sapwood and less heartwood, which changes the way they recycle in the soil. In turn, all other connected biological functions are altered. And, we do not even know what these functions are, let alone what difference they will make in the long-term health of the forest ecosystem as a whole through time. We can, however, make some educated guesses.

A central thrust of modern agricultural technology (including forestry under this umbrella) has been to: (1) isolate such individual organisms, as a particular species of tree, which possess desirable, economic characteristics; (2) enhance their desired characteristics through breeding—increasingly genetic engineering; and (3) replicate them on a massive scale. This type of plant is termed an "ideotype," which literally means a form denoting the idea for the *ideal model* of the desired, economic plant.

Any management, but particularly that which focuses on individual ideotypes, unavoidably changes other properties of the ecosystem in addition to productivity, which is the target of the change. Genetic diversity and its fundamental, structural relationships, the architectural aspects of indigenous plant communities, are altered as well. Although numerous ecologists and foresters have raised concerns about this for some time, these concerns are too often viewed as grounded in unproven criticism, which has little or nothing to do with perceived economic need.

Quite to the contrary, system-level properties are likely to play a seminal role in the biophysical integrity of the ecosystem, and productive, sustainable forests provide values that are no less real because they cannot be traded in the marketplace. Society will continue to demand wood, and forests will provide it. But forests not only produce wood; they also play a central role in the dynamics of global climate, in addition to which they harbor immense biotic diversity that in turn maintains soil fertility.

The third hidden cost is like a numbered Swiss bank account, which has a complete denomination of its own, particular currency. The currency in the forest (called "stored genetic variability") is unseen, as is the currency in the bank account. One does not have to see the currency in order to get the correct change, however. If there is an unlimited amount of money in the bank account, as there is genetic variability in a natural, unmanipulated forest, then the exact change can be received for any chosen denomination, even from an automatic, mechanical teller.

This means a forest can, within limits, adapt genetically to changes in climate from natural climatic changes, a human-caused greenhouse effect, increasing air pollution, and so on. When genetic variability (a particular denomination of Nature's currency) is withdrawn from the forest's genetic account, the ability of the forest to adapt to changing conditions—something it must be able to do in order to survive—becomes artificially limited. Without the one-cent coin, for example, exact change for some transactions can no longer be received from the bank account; it becomes limited and loses flexibility in terms of possible transactions.

Let us take this one step further. A large forest is cut down across the landscape from northern Washington to southern Oregon and from the Pacific coast to the dry eastern side of the Cascade Mountain crest. Genetically engineered Douglas fir seedlings, based on cloning in some part of the engineering process, are then planted to grow quickly. In addition, seedlings planted in northern Washington are selected to withstand cold, those in the south to withstand heat, those in the west to withstand wet weather, and those in the east to withstand dry weather and a short growing season. In order to gain genetic selectivity, foresters have artificially adapted the trees to their commercial set of values. In so doing, they have short-changed the trees' flexibility or "genetic plasticity"—the trees' inherent ability to adapt to changing conditions.

Through such genetic engineering as cloning, genetic diversity—and therefore adaptability—is increasingly limited over time, as genetically handicapped, identical trees are planted in a progressively changing environment. The more the environment changes, the more the cloned trees, which have been robbed of some portion of their adaptability, appear to remain the same, like a historic relict out of the past. Such relicts—geologically termed "living fossils"—are holdovers that cannot adapt fast enough to keep pace with the changing conditions of their environment. Extinction, therefore, is always just around the corner.

Animal Cloning

Cloning in forestry, as an increasingly standardized method of manipulating crop trees, was followed by Dolly the sheep. Dolly was the first mammal to be successfully cloned from an adult cell, which took place at the University of Edinburgh's Roslin Institute in Midlothian, Scotland. Dolly started her life, as with all other cloned animals, in a test tube. Once normal development was confirmed at six days, the embryo that was eventually to become Dolly was transferred into a surrogate mother whose pregnancy was confirmed by an ultrasound scan at about 45 days of gestation. Thereafter, the flawless pregnancy was closely monitored for the remaining 100 days, and Dolly was born on July 5, 1996. Her birth was not announced, however, until February 22, 1997. She lived at the Roslin Institute until her death on February 14, 2003, at six years of age[54] (Figure 8.3).

According to Ian Wilmut, one of the scientists who cloned Dolly, the purpose of cloning is, in the end, economic. Someday, said Wilmut, a dairy farmer might clone a few cows that are especially good at producing milk, resisting disease, and having calves, but the farmer would not want an entire herd of cloned cows because populations require genetic diversity to prevent a lethal disease from eliminating the whole herd. He went on to say that what scientists are learning about cloning will allow a much more efficient way to insert genes into livestock, which in turn can be used to make animals secrete such things as valuable drugs in their milk.[55] Moreover, as of 2010, the U.S. Federal Drug Administration

FIGURE 8.3
Dolly was a cloned sheep—the first mammal successfully cloned from an adult cell. Dolly in the National Museum of Scotland, Edinburgh. Photograph by Tim Vickers (https://commons.wikimedia.org/wiki/File:Dollyscotland_(crop).jpg).

has approved the sale of food products from cloned animals and their offspring, saying that such food items will not require labels.[56]

Human Cloning

Where in all this can one find any respect for the life of the animals that are being violated through such biopiracy? And what about humans? Are we next? The answer is yes, according to Chicago scientist Richard Seed, who, in 1998, said he would clone a human despite the overwhelming ethical considerations. Seed vowed to clone himself, with his wife carrying the embryo that is to be created by combining the nucleus of one of his own cells with a donor egg.[57]

With respect to humans, the aim of cloning is to create a genetically identical copy of a particular human. The term, as it is generally used, refers to artificial human cloning, which is the artificial reproduction of human cells and tissues. It does not refer to the natural conception and delivery of identical twins, triplets, and so on.

Although Seed's declaration prompted President Clinton to call for a federal ban on the cloning of human beings, drug researchers pointed out that the Food and Drug Administration already required anyone who wanted to perform research that uses the technique of cloning to file for permission. Members of the drug industry said they would prefer a federal ban that only *forbids* the cloning of a whole human being—if people felt there must be a law based on social morality.

Despite the debate, scientists already used cloning as a technique to test how identical cells react to different substances. Cloning had already led to drugs used for cystic fibrosis and strokes. For this reason, two trade groups launched a state-by-state campaign to fight 50 anti-cloning bills that were being considered in legislatures from California to Connecticut. But they were too late to block the first state law against cloning. Pete Wilson, governor of California, signed a bill on October 4, 1997, that made it a crime to clone a human or to purchase fetal cells to do so. Fines for violating this law can be as much as $1 million.[58]

Today, however, the battle to allow the creation of "designer human babies" rages on. According to a group of scientists, ethicists, and policy experts from The Hinxton Group, an International Consortium on Stem Cells, Ethics & Law Headquartered in Manchester, UK, it is "essential" that the genetic modification of human embryos be allowed. A Hinxton Group report says that editing the genetic code of early-stage embryos is of "tremendous value" to research. Moreover, a range of novel techniques allows a scientist to travel to a precise location in human DNA with a pair of "molecular scissors" guided by a molecular *sat-nav* (abbreviation for "satellite navigation") that tells the scissors where to cut what the scientist wants. In essence, the techniques have not only transformed research in a wide range of fields, but also means genetically modified human babies are ceasing to be a future prospect and fast becoming a possibility.

The report adds that while genetically modified human babies should not be allowed to be born at the moment, it may become "morally acceptable" under some circumstances in the future. Whereas disease-free children or "designer babies" may be on the horizon, the more immediate uses are far less controversial and could restore the reputation of gene therapy in adults and children.[59]

According to cloning pioneer Dr. Tony Perry, who was part of the team to clone the first mice and pigs, the prospect was still fiction, but science was rapidly catching up to make elements of it possible. He also indicated that designer babies—genetically modified for beauty, intelligence, or to be free of disease—were no longer in H.G. Wells territory.[60] Real-world human genetic engineering could no longer be considered mere science fiction.

Perry said in an interview with the British Broadcasting Company, "My view is this is such a wonderful opportunity to remove horrible diseases that it would be unethical not to explore it. I think it is a sin of omission, if you have a method where you can prevent someone suffering and you don't take that opportunity then it is wrong, it is unethical." But, he added, "that [it] needs to be in context of a full debate."[61]

The Director of the U.S. National Institutes of Health, Dr. Francis Collins, has made it clear his organization will not fund such research. "The concept of altering the human germline in embryos for clinical purposes has been debated over many years from many different perspectives, and has been viewed almost universally as a line that should not be crossed. [A "germline" is the sex cells (eggs and sperm) that are used by sexually reproducing organisms to pass along their genes from generation to generation.] Advances in technology have given us an elegant new way of carrying out genome editing, but the strong arguments against engaging in this activity remain.[62]

Nevertheless, scientists in the United Kingdom have been given permission—by the fertility regulator—to genetically modify human embryos. It is the first time a country has approved the DNA-altering technique in human embryos. The research, to take place at the Francis Crick Institute in London, aims to provide a deeper understanding of the earliest moments of human life, but thus far it will be illegal to implant the modified embryos into a woman.

DNA is the blueprint of life—the "instruction memo" for building the human body. That being the case, genetic editing allows the precise manipulation of human DNA. Needless to say, this field of investigation and its motives are attracting controversy over concerns it is opening the door to genetically modified designer babies.[63]

Although the use of cloning as a scientific technique is clearly an ethical issue for society to resolve, there is also a practical aspect to the debate, about which little is heard. The practicality is the fact that genetically engineered parts, individuals, and species are, in the commercial sense—and even more so in the ecological sense—untested products, which are likely to have unwanted, hidden consequences because the people who create them are working, at best, with a great deal of uncertainty. As Edward Tenner, a visiting professor at Princeton University, says, "It is part of the nature of radically new ideas that they are not the kinds of ideas we thought they would be."[64] In essence, whatever the outcome of such symptomatic thinking, the technology of genetic engineering is diametrically opposed to the Rights of Nature, which protects Earth's right to flourish outside of human interference.

Biomimicry and Technology—Are They Compatible with Nature's Rights?

Biomimicry is a technological approach to innovation that seeks sustainable solutions to human challenges by emulating Nature's evolved patterns and strategies, in essence a new approach to living. The technological goal is to create products and processes that are well adapted to life on Earth over the long term.

The question, however, is whether biomimicry and technology are really compatible with Nature's laws of reciprocity, and Nature's Rights. The answer is "yes," but only within strict limitations. One example is solar cells to convert the sun's radiation to electricity. Another is the technological duplication of functional body parts.

Nature's Chloroplasts and Technology's Solar Cells

It has long been understood that green plants use chlorophyll molecules that are enclosed in chloroplasts (specialized subunits), where the chlorophyll absorbs sunlight and produces energy in a process known as photosynthesis.

People tinkered with the idea of a solar cell and solar panel to artificially mimic the photosynthetic process. As far back as 1839, it was discovered that some treated substances had the ability to generate electricity when light fell on them—four decades prior to the invention of the first workable electric light bulb.

But it was not until the late 1800s that an invention, using selenium, could be described as a solar cell. Then, in the 1950s, it was further discovered that silicon performs much better, and the way was paved to create a truly usable solar cell or photovoltaic cell (an artificial cell that is able to generate a current or voltage when exposed to visible light).

These solar-voltaic cells are a technological device that mimics a chloroplast and its chlorophyll by collecting sunlight during the daylight hours and converting it into electricity within the cell for human use. In turn, these cells are arranged in a grid-like pattern on the surface of the solar panel, an artificial rendition of a plant's green leaf, the function of which is to mimic the photosynthetic process.[65]

While solar panels are passive in their collection of sunlight, they cause damage, frequently large-scale damage, by disrupting the ecological rhythms of the land on which they are installed. The current industrial-scale model of alternative energy is itself

FIGURE 8.4
Information about the solar panels on Oregon State University's property. Photograph by Chris Maser.

FIGURE 8.5
One of several arrays of solar panels on Oregon State University's property. Photograph by Chris Maser.

misusing the positive potential of biomimicry. Placing solar panels across thousands of acres of fragile desert habitat hitherto unexploited by humans disrupts ecological processes in the name of sustainable energy. It violates a key principle discussed earlier in this book: true sustainability does not increase the human footprint. Solar panels erected on a small scale (Figures 8.4 and 8.5) or installed on individual buildings (Figure 8.6) fit the framework of Nature's Rights much more closely. But such is not the case with wind power.

Turbines and Wind Farms

Although wind power is considered to be one of the most environmentally friendly energy sources, it has been shown to cause enormous bird and bat fatalities. In some cases, wind turbines may pose a threat to local avian populations and to certain special-status species that have been given legal protection. A recent study reported that 1,000 or more bird fatalities may occur at Altamont Pass, California every year, with approximately 50 percent of these being raptors, and nearly all protected by the Migratory Bird Treaty Act, the Bald Eagle and Golden Eagle Protection Act, and/or the Endangered Species Act. Raptors are also protected under the California Fish and Game Code, which makes it illegal to "take," possess, or destroy any raptor. The U.S. Fish and Wildlife Service considers any injury or mortality of any raptor from a collision with a wind turbine, or ancillary facilities to be a

FIGURE 8.6
Solar panels on the roof of a business. Photograph by Chris Maser.

"take" and, therefore, a violation of the law. Violations can result in fines from $100,000 to $500,000.[66]

Moreover, in 2012, breaking the European omerta on wind-farm mortality, the Spanish Ornithological Society not only reviewed actual carcass counts from 136 monitoring studies but also concluded that Spain's 18,000 wind turbines are killing 6–18 million birds and bats annually.[67] ("Omerta" is the code allegedly applied to members of the Mafia, requiring them to remain silent concerning any crimes of which they have knowledge.)

In the United States, according to K. Shawn Smallwood, 573,000 birds and 888,000 bats are killed by wind turbines in the United States annually, but this is, in all probability, grossly understated because the areas searched under wind turbines are limited to 200-foot radiuses, even though the huge, modern turbines catapult 90 percent of the bird and bat carcasses much further. In addition, wind-farm owners, operating under voluntary guidelines of the U.S. Fish and Wildlife Service, commission studies to search areas that not only are much too small but also conduct the searches only once every 30–90 days, ensuring that scavengers remove most carcasses, while ignoring wounded birds that happen to be found within search perimeters.[68]

> [Moreover] since the early 1980s, the industry has known there is no way its propeller-style turbines could ever be safe for raptors. With exposed blade tips spinning in open space at speeds up to 200 mph, it was impossible. Wind developers also knew they would have a public relations nightmare if people ever learned how many eagles are actually being cut in half—or left with a smashed wing to stumble around for days before dying.

FIGURE 8.7
Palm Springs Windmill Farms, CA. Photograph by and courtesy of Irma Britton.

> To hide this awful truth, strict wind farm operating guidelines were established—including high security, gag orders in leases and other agreements, and the prevention of accurate, meaningful mortality studies.
>
> For the industry, this business plan has succeeded quite well in keeping a lid on the mortality problem. While the public has some understanding that birds are killed by wind turbines, it doesn't have a clue about the real mortality numbers. And the industry gets rewarded with subsidies and immunity from endangered species and other wildlife laws[69] (Figure 8.7).

Clearly, wind farms—as opposed to sustainably constructed and properly placed solar panels—are a symptomatic, technological fix for more electricity that both disregards both Nature's Laws of Reciprocity and ultimately the social-environmental impact of losing the ecological services of so many birds and bats. Beyond that, the loss of their manifold ecological services will compound not only with increasing loss of each individual killed but also with the loss of each potential generation that cannot be produced now. This brings us to the creation of technological body parts.

The Technological Creation of Body Parts

The technological creation of body parts (such as artificial knees, hips, shoulders, and so on) requires careful, systemic simulation of how they function within a human body without causing damage to the body's functional dynamics. In other words, these kinds of carefully designed and constructed body parts fit within Nature's Laws of Reciprocity as they affect the health and vitality of the patient.

Then there is the creation and implantation of such technological constructs as medical stents, which do not simulate a body part but are nonetheless essential for body function in the case of partial breakdown. A stent is a thin, flexible tube (catheter), usually of metal mesh, that is used to treat narrow or weak arteries, which are blood vessels that carry blood away from the heart to other parts of the body. The stent is inserted through the skin in the upper thigh or arm into the artery to improve the flow of blood and help prevent the artery from bursting.

Although stents are usually metal, some are made of a specialized fabric, termed "stent grafts," for use in larger arteries. In addition, some stents are coated with medicine that is slowly and continuously released into the artery, where it helps prevent the artery from becoming blocked again. These are "drug-eluting stents."

While no artificial, medical device is free of potential problems for the patient, each is designed, as carefully as humanly possible, to fit systemically within a patient's body and restore it to a healthy function. Whereas the creation of artificial knees, hips, shoulders, and so on, carefully and faithfully follows Nature's design, a stent does not, and so the risk of error is greater. In such cases of medical technology, Nature's Laws of Reciprocity are studied, honored, and followed as closely as possible, because each human body is a microcosm of Nature. And, because such an unnatural device as a stent is being introduced into a living system, where human health and vitality take precedence over technology, each such fabrication is not only tested over and over but also subjected to extensive risk analysis and continual monitoring. Then there are the initial, human test cases, where the risk is high. Such tests can only be conducted with the agreement of the patient, who seldom has a feasible alternative but to risk the near certainty of his or her death without the new invention.[70]

Conclusion

Things can go wrong under the best of circumstances with such things as artificial body parts or stents. All technological inventions carry with them an unknown and unknowable risk when introduced into any part of Nature's realm. Such risk is astronomically increased, however, when politics and greed hide the unwanted outcomes of theoretically sustainable or "alternative" technology for personal reasons, as evidenced by wind farms and the continual slaughter of birds and bats caused by the huge turbines.

And wind farms are not alone. In many other common instances, technology is designed specifically to overexploit ecosystems regardless of consequences, such as massive equipment for tree-felling or hydraulic fracturing. Little attempt is made to reduce the negative consequences of such technology, save for publicity statements aimed at hiding the devastating effects by refocusing public attention on the economic benefits of allowing the ecological damage to continue.

As always, and unfortunately, the self-centered behavior of the minority, both corporate and individual, affects the well-being of all generations and thus forces particular attention to the wisdom with which laws are written and the diligence with which they are implemented and upheld. Writing statutes to redesign technology to a human-centered scale, focused on restraining human needs rather than large-scale profit, maximum consumerism, and linear efficiency would help tremendously to reduce the dangers technology poses to ecosystem survival. This is the task of a Rights of Nature legal and social system. At the core, the need for return to technology's human scale was the message of the Luddite Rebellion.

Endnotes

1. Havelock Ellis. http://www.brainyquote.com/quotes/quotes/h/havelockel137638.html (accessed December 21, 2015).
2. The preceding seven paragraphs are based on: Kirkpatrick Sale. *Rebels against the Future: The Luddites and Their War on the Industrial Revolution*. Addison-Wesley Publishing Co., Boston, MA. (1996) 320 pp.
3. *Ibid.* (p. 49).
4. *Ibid.* (pp. 53–54).
5. The preceding two paragraphs are based on: *Ibid.*
6. *Ibid.* (pp. 60–70).
7. The preceding six paragraphs are based on: *Ibid.*
8. *Ibid.* (pp. 209–212).
9. *Ibid.* (p. 229).
10. Pope Francis and his *Laudatro Si* is based on: Encyclical Letter Laudato Si of The Holy Father Francis on Care for Our Common Home [in English]. http://w2.vatican.va/content/francesco/en/encyclicals/documents/papa-francesco_20150524_enciclica-laudato-si.html (accessed August 19, 2015).
11. Aldo Leopold. *A Sand County Almanac and Sketches Here and There*. Oxford University Press, Inc., New York, NY. (1949) 228 pp. (Foreword).
12. The foregoing discussion is based on: Chris Maser, Russ Beaton, and Kevin Smith. *Setting the Stage for Sustainability: A Citizen's Handbook*. Lewis Publishers, Boca Raton, FL. (1998) 275 pp.
13. Chris Maser. Do we owe anything to the future? 1992. pp. 195–213. *In*: *Multiple Use and Sustained Yield: Changing Philosophies for Federal Land Management?* Proceedings and summary of a workshop convened on March 5 and 6, 1992, Washington, DC. Congressional Research Service, Library of Congress. Committee Print No. 11. US Government Printing Office, Washington, DC.
14. Lee Schroeder. Foreword, p. xiii. *In*: Chris Maser. *Vision and Leadership in Sustainable Development*. Lewis Publishers, Boca Raton, FL. (1998) 235 pp.
15. (1) Robert F. Baldwin and Stephen C. Trombulak. Losing the dark: A case for a national policy on land conservation. *Conservation Biology*, 21 (2007):1133–1134; and (2) Ben Harder. Light all night. *Science News*, 169 (2006):170–172.
16. Elizabeth Howell. Light Pollution Deters Nesting Sea Turtles. http://www.livescience.com/37278-light-pollution-sea-turtle-nesting.html (accessed December 21, 2015).
17. Brittany L. Bird, Lyn C. Branch, and Deborah L. Miller. Effects of coastal lighting on foraging behavior of beach mice. *Conservation Biology*, 18 (2004):1435–1439.
18. Ben Harder. Light all night. *op. cit.*
19. This paragraph is based on: (1) P.J. Bryant, C.M. Lafferty, and S.K. Lafferty. Reoccupation of Laguna Guerrero Negro Baja California, Mexico, by Gray Whales. pp. 375–386. *In*: *The Gray Whale Eschrictius robustus*. (M.L. Jones, S.L. Swartz, and S. Leatherwood, eds.) Academic Press, Orlando, FL. (1984); (2) M. Andre, C. Kamminga, and D. Ketten. Are Low-Frequency Sounds a Marine Hazard: A Case Study in the Canary Islands. Underwater Bio-sonar and Bioacoustics Symposium, Loughborough University, UK, December 16–17, 1997; (3) A.B. Morton and H.K. Symonds. Displacement of *Orcinus orca* (L.) by high amplitude sound in British Columbia. *Journal of Marine Science*, 59 (2002):71–80; (4) P.J.O. Miller, N. Biasson, A. Samuels, and P.L. Tyack. Whale songs lengthen in response to sonar. *Nature*, 405 (2000):903; (5) K.C. Balcomb and D.E. Claridge. A mass stranding of cetaceans caused by naval sonar in the Bahamas. *Bahamas Journal of Science*, 8 (2001):1–12; (6) R.D. McCauley, J. Fewtrell, and A.N. Popper. High intensity anthropogenic sound damages fish ears. *Journal of the Acoustical Society of America*, 113 (2003):638–642; (7) Andrew J. Read, Phebe Drinker, and Simon Northridge. Bycatch of Marine

Mammals in U.S. and Global Fisheries. *Conservation Biology*, 20 (2006):163–169; and (8) Lars Bejder, Amy Samuels, Hal Whitehead, and others. Decline in relative abundance of bottlenose dolphins exposed to long-term disturbance. *Conservation Biology*, 20 (2006):1791–1798.

20. (1) Daniel Pauly and Dirk Zeller. Catch reconstructions reveal that global marine fisheries catches are higher than reported and declining. *Nature Communications*, 7 (Article number: 10244, published January 19, 2016) doi:10.1038/ncomms10244 (accessed January 19, 2016); (2) D. Zeller, S. Harper, K. Zylich, and D. Pauly. Synthesis of underreported small-scale fisheries catch in Pacific island waters. *Coal Reefs*, 34 (2015):25–39; and (3) D. Zeller, P. Rossing, S. Harper, and others. The Baltic Sea: Estimates of total fisheries removals 1950–2007. *Fisheries Research*, 108 (2011):356–363.
21. Matt McGrath. Global fishing catch significantly under-reported, says study. http://www.bbc.com/news/science-environment-35347446 (accessed January 19, 2016).
22. (1) Ransom A. Myers and Boris Worm. Rapid worldwide depletion of predatory fish communities. *Nature*, 423 (2003):280–283; and (2) Destructive Fishing. http://www.marine-conservation.org/what-we-do/program-areas/how-we-fish/destructive-fishing/ (accessed December 26, 2015).
23. The preceding two paragraphs are based on: (1) Meritxell Genovart, José Manuel Arcos, David Álvarez, and others. Demography of the critically endangered Balearic shearwater: The impact of fisheries and time to extinction. *Journal of Applied Ecology*, March 8, 2016 DOI: 10.1111/1365-2664.12622; (2) Helen Briggs. Europe's rarest seabird 'faces extinction.' http://www.bbc.com/news/science-environment-35778655 (accessed March 12, 2016); and (3) http://www.fish.gov.au/fishing_methods/Pages/hook_and_line.aspx (accessed March 12, 2016).
24. The foregoing discussion of mobile fishing gear is based on: (1) Destructive Fishing. http://www.marine-conservation.org/what-we-do/program-areas/how-we-fish/destructive-fishing/ (accessed December 26, 2015); (2) Bottom trawling impacts on ocean, clearly visible from space. *ScienceDaily*, http://www.sciencedaily.com/releases/2008/02/080215121207.htm (accessed December 26, 2015); and (3) Bottom Trawling and Dredging. http://act.oceanconservancy.org/site/DocServer/fsTrawling.pdf?docID=213 (accessed December 26, 2015).
25. The preceding three paragraphs are based on: (1) Jonna Engel and Rikk Kvitek. Effects of otter trawling on a benthic community in Monterey Bay National Marine Sanctuary. *Conservation Biology*, 12 (1998):1204–1214; (2) Peter J. Auster. A conceptual model of the impacts of fishing gear on the integrity of fish habitats. *Conservation Biology*, 12 (1998):1198–1203; (3) Cynthia H. Pilskaln, James H. Churchill, and Lawrence M. Mayer. Resuspension of sediment by bottom trawling in the Gulf of Maine and potential geochemical consequences. *Conservation Biology*, 12 (1998):1223–1229; and (4) Les Watling and Elliott A. Norse. Disturbance of the seabed by mobile fishing gear: A Comparison to forest clearcutting. *Conservation Biology*, 12 (1998):1180–1197.
26. (1) Anthony M. Quinton and Baron Quinton. Francis Bacon, Viscount Saint Alban. http://www.britannica.com/biography/Francis-Bacon-Viscount-Saint-Alban-Baron-Verulam (accessed December 22, 2015); and (2) Carolyn Merchant. "The Violence of Impediments," Francis Bacon and the Origins of Experimentation. (accessed December 22, 2015).
27. J.J. O'Connor and E.F. Robertson. Galileo Galilei. http://www-groups.dcs.st-and.ac.uk/~history/Biographies/Galileo.htmls (accessed December 22, 2015).
28. René Descartes. http://www-history.mcs.st-and.ac.uk/Biographies/Descartes.html (accessed December 22, 2015)
29. William Uzgalis. John Locke. http://plato.stanford.edu/entries/locke/ (accessed December 22, 2015).
30. Eric W. Weisstein. Newton, Isaac (1642–1727). http://scienceworld.wolfram.com/biography/Newton.html (accessed December 22, 2015).
31. Chris Maser. Cultural values versus science. *Trumpeter*, 12 (1995):116–118.
32. Jeremy Rifkin. *Biosphere Politics: A Cultural Odyssey from the Middle Ages to the New Age.* HarperSanFrancisco, San Francisco, CA. (1991) 388 pp.
33. Quotations by Galileo Galilei. http://www-groups.dcs.st-and.ac.uk/~history/Quotations/Galileo.html (accessed December 22, 2015).

34. Chris Maser. Part one: Global challenges: Cultural values versus science. *Journal of Sustainable Forestry*, 4 (1997):3–4.
35. Luke Mastin. The Basics of Philosophy—Reductionism. http://www.philosophybasics.com/branch_reductionism.html (accessed December 22, 2015).
36. The preceding two paragraphs are based on: Duncan M. Taylor. Disagreeing on the basics. *Alternatives*, 18 (1992):26–33.
37. Donald Ludwig, Ray Hilborn, and Carl Walters. Uncertainty, resource exploitation, and conservation: Lessons from history. *Science*, 260 (1993):17, 36.
38. Russ Beaton and Chris Maser. *Economics and Ecology: United for a Sustainable World*. CRC Press, Boca Raton, FL. (2012) 191 pp.
39. *Ibid.*
40. Andrew Kimbrell. Breaking the law of life. *Resurgence*, 182 (1997):10–11.
41. Jeremy Rifkin. God in a labcoat. *Utne Reader*, May–June (1999):66–71.
42. Nieves H. De Madariaga Masthew. Francis Bacon, Slave-Driver of Servant of Nature? http://www.sirbacon.org/mathewsessay.htm (accessed February 9, 2016).
43. Jeremy Rifkin. God in a labcoat. *op. cit.*
44. Diamond *v.* Chakrabarty, (1980) No. 79-136. (accessed February 9, 2016).
45. Andrew Kimbrell. Breaking the law of life. *op. cit.*
46. Diamond *v.* Chakrabarty, (1980) No. 79-136. *op. cit.*
47. Douglas Robinson and Nina Medlock. Diamond v. Chakrabarty: A retrospective on 25 years of biotech patents. *Intellectual Property & Technology Law Journal*, 17 (2005):12–15.
48. Ricarda Steinbrecher. What is wrong with nature? *Resurgence*, 188 (1998):16–19.
49. The preceding two paragraphs are based on: *Ibid.*
50. The preceding two paragraphs are based on: Chris Maser. *Ecological Diversity in Sustainable Development: The Vital and Forgotten Dimension*. Lewis Publishers, Boca Raton, FL. (1999) 402 pp.
51. Jeff Barnard. Genetic engineering may spawn 'super weed.' *Corvallis Gazette-Times*, Corvallis, OR. September 3, 1998.
52. The preceding two paragraphs are based on: (1) David Quist and Ignacio H. Chapela. Transgenic DNA introgressed into traditional maize landraces in Oaxaca, Mexico. *Nature*, 414 (2001):541–543; (2) Anita Manning. Gene-altered DNA may be 'pollutin' corn. *USA Today*, November 29, 2001; and (3) and Mark Stevenson. Mexicans angered by genetically modified corn. Corvallis (OR) *Gazette-Times*, December 30, 2001.
53. The preceding two paragraphs are based on: N.G. Jacobson. Grafting of Douglas Fir and Establishment of Seed Orchards. npn.rngr.net/.../grafting-of-douglas-fir-and-establishment-of-seed-orchards/.../file (accessed February 9, 2016).
54. Dolly the Sheep. http://www.roslin.ed.ac.uk/public-interest/dolly-the-sheep/a-life-of-dolly/ (accessed February 10, 2016).
55. Malcolm Ritter. Scientists clone adult mammal for first time. *Corvallis Gazette-Times*, Corvallis, OR. February 24, 1997.
56. Fast Facts about Animal Cloning. http://endanimalcloning.org/factsaboutanimalcloning.shtml (accessed February 10, 2016)
57. (1) James Webb. Scientist says he'll clone a human. *Corvallis Gazette-Times*, Corvallis, OR. January 8, 1998; and (2) *Boston Globe*. Wife to carry clone. *Corvallis Gazette-Times*, Corvallis, OR. September 7, 1998.
58. The preceding two paragraphs are based on: John Hendren. Cloning debate moves to states. *Corvallis Gazette-Times*, Corvallis, OR. March 18, 1998.
59. The preceding two paragraphs are based on: (1) James Gallagher. GM embryos 'essential', says report. http://www.bbc.com/news/health-34200029 (accessed February 7, 2016); and (2) What Is Cloning? http://learn.genetics.utah.edu/content/cloning/whatiscloning/ (accessed February 7, 2016).
60. Toru Suzuki, Maki Asami, and Anthony C.F. Perry. Asymmetric parental genome engineering by Cas9 during mouse meiotic exit. *Scientific Reports*, **4**, Article number: 7621 (2014) doi:10.1038/srep07621.

61. James Gallagher. Embryo engineering a moral duty, says top scientist. http://www.bbc.com/news/uk-politics-32633510 (accessed February 7, 2016).
62. *Ibid.*
63. The preceding two paragraphs are based on: (1) 23andMe's 'build-a-baby' patent criticized. http://www.bbc.com/news/technology-24381149 (accessed February 7, 2016); (2) James Gallagher. 'Designer babies' debate should start, scientists say. http://www.bbc.com/news/health-30742774 (accessed February 7, 2016); and (3) James Gallagher. Scientists get 'gene editing' go-ahead. http://www.bbc.com/news/health-35459054 (accessed February 7, 2016).
64. (1) Mark Harris. To be or not to be? *Vegetarian Times*, June (1998):64–70; and (2) David Susuki and P. Knudtson. *Genethics, the Ethics of Engineering Life*. Stoddart Publishing, Toronto, Canada. (1988) 384 pp.
65. (1) Jeremy Battista. Chloroplast: Definition, Structure, Function & Examples. http://study.com/academy/lesson/chloroplast-definition-structure-function-examples.html (accessed December 23, 2015); (2) What Are Solar Panels? http://www.solarpanelinfo.com/solar-panels/what-are-solar-panels.php (accessed December 23, 2015); (3) Photovoltaic Panel Construction. http://www.solar-facts.com/panels/panel-construction.php (accessed December 23, 2015); (4) How Do Photovoltaic Panels Generate Electricity? http://www.solar-facts.com/panels/how-panels-work.php (accessed December 23, 2015); and (5) A Guide to Solar Energy. http://www.viridiansolar.co.uk/Solar_Energy_Guide_5_3.htm (accessed December 23, 2015).
66. David Sterner, Linda Spiegel, Kelly Birkinshaw, and others. A roadmap for PIER research on avian collisions with wind turbines in California. *Public Interest Energy Research*, P500-02-070F. (2002) 54 pp.
67. Spanish Ornithological Society: Spain's wind turbines kill 6 to 18 million birds and bats yearly. http://savetheeaglesinternational.org/releases/spanish-wind-farms-kill-6-to-18-million-birds-bats-a-year.html (accessed December 24, 2015).
68. (1) K. Shawn Smallwood. Comparing bird and bat fatality-rate estimates among North American wind-energy projects. *Wildlife Society Bulletin*, 37 (2013):19–33; and (2) Jim Wiegand. Big Wind & Avian Mortality (Part II: Hiding the Problem) https://www.masterresource.org/cuisinarts-of-the-air/wind-avian-mortality-ii/ (accessed December 24, 2015).
69. Jim Wiegand. Hiding the Slaughter Committed by Big Wind. http://www.epaabuse.com/11676/editorials/hiding-the-slaughter-committed-by-big-wind/ (accessed December 24, 2015).
70. The preceding three paragraphs are based on: (1) National Heart, Lung, and Blood Institute. What Is a Stent? http://www.nhlbi.nih.gov/health/health-topics/topics/stents (accessed December 24, 2015); and (2) How Are Stents Used? http://www.nhlbi.nih.gov/health/health-topics/topics/stents/used (accessed December 24, 2015).

9

Corporations and the Rights of Nature

The corporation is an invention of the British Crown through the creation of the East India Company by Queen Elizabeth I in 1600, which, being the original transnational corporation, set today's precedence for big businesses. The East India Company, "found India rich and left it poor," says author Nick Robin. The corporate structure of the East India Company was deemed necessary to allow the British to exploit their colonies in such a way that the owner of the enterprise was, for the first time, separated from responsibility for how the enterprise behaved. The corporation is thus a "legal fiction" that lets the investors who own the business avoid personal responsibility whenever the business dealings are unethical or even blatantly illegal, despite the fact that such unscrupulous behavior profits them enormously.[1]

For the first 100 years of American history after the Revolution, there were few corporations because early American governments limited them strictly to business roles, fearing their ability to dominate trade and wield great, but anonymous, powers, as voiced by Thomas Jefferson, who lived just long enough to witness the early development of the corporate system in the United States. Seeing the handwriting on the wall, he said, "Banking establishments [his term for corporations] are more dangerous than standing armies." Jefferson warned in his last years that money and banking establishments (corporations) would destroy liberty and restore absolutism, effectively nullifying the ideals for which the American Revolution was fought.[2]

The privilege of incorporation in early America was usually granted only to further activities beneficial to the public, such as the construction of a canal or the building of a road. Charters were granted for a limited time; corporations could only engage in activities related to their corporate purpose; could not own stock; could be, and often were, terminated if they caused public harm. Owners and managers were held responsible for the corporation's criminal acts, and a corporation could not make any political or charitable contributions in its own name.

Under the common European model of the time, the corporate charter created a liability shield against debt and harm. But American legislators distrusted the corporate ability to aggregate power and resources, and designed corporations for easy dissolution if their privileges were abused. Corporations, however, sought more control, and in 1819, the U.S. Supreme Court tried to give it to them by its ruling in *Trustees of Dartmouth College v. Woodward* (4 Wheat. 518 [1819]). The Court held that New Hampshire could not revoke a charter to Dartmouth College by King George III, because it had no revocation clause. This ruling outraged citizens and lawmakers nationwide; over several decades, beginning in 1844, 19 states amended their constitutions to give their legislatures power to alter or revoke corporate charters.[3]

But the stakes were high. To control the powers of the corporate charter was also to control resources, labor, wages, and even political sovereignty. Corporations abused and stretched their powers more and more frequently to become conglomerates and trusts.

The American Civil War era was a turning point in the rise of corporate power, partly attributed to government spending in the war effort. Legislators routinely granted corporations limited liability, decreasing citizen authority over them, and extended (rather than limited) charters. So it was that numerous corporations were chartered (= created) to supply the Union Army, much to the foreboding of President Abraham Lincoln, who wrote in a letter to Colonel William F. Elkins, dated November 21, 1864:

> We may congratulate ourselves that this cruel war is nearing its end. It has cost a vast amount of treasure and blood.... It has indeed been a trying hour for the Republic; but I see in the near future a crisis approaching that unnerves me and causes me to tremble for the safety of my country. As a result of the war, corporations have been enthroned and an era of corruption in high places will follow, and the money power of the country will endeavor to prolong its reign by working upon the prejudices of the people until all wealth is aggregated in a few hands and the Republic is destroyed. I feel at this moment more anxiety for the safety of my country than ever before, even in the midst of war. God grant that my suspicions may prove groundless.[4]

The Beginning of Corporate Control

In the next three decades, the era of the robber barons took hold, along with corruption and vast fortunes amassed by a few, well-connected people. But perhaps the greatest milestone in increasing corporate power was the 1886 U.S. Supreme Court case of *Santa Clara County v. Southern Pacific Railroad*, in which the Court is reputed to have held, for the first time, that corporations are "persons" that enjoy protections of the 14th Amendment, which protects personal rights. The question of corporate personhood was raised in the arguments of the case, but the Justices did not declare in any written Opinion that corporations are persons. However, the Court reporter, J.C. Bancroft Davis, summarizing the case in what are now called "headnotes," quoted the Chief Justice as saying that the Court did so hold:

> The defendant Corporations are persons within the intent of the clause in section 1 of the Fourteenth Amendment to the Constitution of the United States, which forbids a State to deny to any person within its jurisdiction the equal protection of the laws.... Before argument MR. CHIEF JUSTICE WAITE said: "The court does not wish to hear argument on the question whether the provision in the Fourteenth Amendment to the Constitution, which forbids a State to deny to any person within its jurisdiction the equal protection of the laws, applies to these corporations. We are all of opinion that it does."[5]

Prior to this case, the railroad corporations had lost every time they sought to have the corporation protected as a person. But though this case was about the ways Santa Clara County was taxing the railroad (which it was suing for back taxes), the corporate attorneys spent much time arguing about the need for personhood—though Congress had not passed a law permitting this.[6] However, though the Justices never issued an Opinion on corporate personhood and did not rule on federal constitutional issues at all, the headnotes to the case were cited as precedent for this dangerous idea, and is thus the fountainhead of the "corporate personhood" doctrine.

It is especially noteworthy that this case, as summarized, gave corporations personhood under the Fourteenth Amendment, which meant that states could no longer deny corporations due process or equal protection of the laws. It was an easy step from there to the judicial grant of Fifth Amendment rights of due process to corporations in 1893, which provided them the means to challenge federal, as well as state, laws that limited their corporate goals.

To counter the rise of the robber barons, Congress, in the late 19th century, passed a suite of progressive legislation aimed at regulating the socially deleterious effects of corporations. The first was the Interstate Commerce Act of 1887: "Approved on February 4, 1887, the Interstate Commerce Act created an Interstate Commerce Commission to oversee the conduct of the railroad industry. With this act, the railroads became the first industry subject to Federal regulation."[7]

This was followed by the Sherman Antitrust Act of 1890, which was formulated to prevent large firms from controlling a single industry. Section 1 states:

> Every contract, combination in the form of trust or otherwise, or conspiracy, in restraint of trade or commerce among the several States, or with foreign nations, is declared to be illegal. Every person who shall make any contract or engage in any combination or conspiracy hereby declared to be illegal shall be deemed guilty of a felony, and, on conviction thereof, shall be punished by fine not exceeding $10,000,000 if a corporation, or, if any other person, $350,000, or by imprisonment not exceeding three years, or by both said punishments, in the discretion of the court.[8]

The main objective of the Progressives was reining in government corruption. Some of the largest corporations and trusts ever seen were created in this era, including Standard Oil; J.P. Morgan's U.S. Steel (formed 1901); Northern Securities Company, a railroad trust also formed in 1901 to control the Northern Pacific Railway, Great Northern Railway, Chicago, Burlington and Quincy Railroad and associated lines; and many others. These gigantic corporations created industrialized factory systems, company towns, and wholly owned subsidiaries controlling vast swaths of resources.

Then, in 1912, Theodore Roosevelt, a devout conservationist, led the progressive elements out of the Republican Party because President William H. Taft was removing Gifford Pinchot as chief of the U.S. Forest Service—and the Progressive Party was born. It disappeared again in 1916 when Roosevelt returned to the Republican camp.[9] The Progressive Party's platform of November 5, 1912 included:

> **A Covenant with the People**
>
> This declaration is our covenant with the people, and we hereby bind the party and its candidates in State and Nation to the pledges made herein.
>
> **The Rule of the People**
>
> The National Progressive party, committed to the principles of government by a self-controlled democracy expressing its will through representatives of the people, pledges itself to secure such alterations in the fundamental law of the several States and of the United States as shall insure the representative character of the government.
>
> In particular, the party declares for direct primaries for the nomination of State and National officers, for nation-wide preferential primaries for candidates for the presidency; for the direct election of United States Senators by the people; and we urge on the States the policy of the short ballot, with responsibility to the people secured by the initiative, referendum and recall.

Amendment of Constitution

The Progressive party, believing that a free people should have the power from time to time to amend their fundamental law so as to adapt it progressively to the changing needs of the people, pledges itself to provide a more easy and expeditious method of amending the Federal Constitution.[10]

Nevertheless, it appeared to a congressional committee in 1913 that certain corporations had continued to grow bigger, despite the trust-busting activities of the administrations of Presidents Theodore Roosevelt and William Howard Taft under the Sherman Act, the vague language of which had provided numerous loopholes to large corporations. Consequently, the control of money and credit was in the hands of a few men, giving them the power to plunge the nation into a financial panic. Thus, when President Woodrow Wilson asked for a drastic revision of existing antitrust legislation, Congress responded by passing the Clayton Antitrust Act, in 1914, to clarify and strengthen the Sherman Antitrust Act of 1890, closing the loopholes that had heretofore enabled corporations to create restrictive business arrangements that, while not actually illegal, thwarted open competition.[11] Section 2 of the Clayton Act was then amended in 1936 by the Robinson–Patman Act,[12] and Section 7 was amended by the Celler–Kefauver Act in 1950[13]—both to reinforce their respective provisions.

The Progressive Party was reinstated in 1924 because the liberals in both major political parties were frustrated with conservative control of the day. They formed the League of Progressive Political Action (the Progressive Party), at which time Senator Robert Marion La Follette, Sr., a Republican from Wisconsin, accepted the party's presidential nomination. Senator La Follette, who emerged as one of the leaders of the Progressive Movement, wrote in his autobiography:

> Within the last dozen years trusts and combinations have been in nearly every branch of industry. Competition has been ruthlessly crushed, extortionate prices have been exacted from consumers, independent business development has been arrested, invention stifled, and the door of opportunity has been closed, except to large aggregations of capital ... it is everywhere recognized that trusts and combinations are to-day the gravest danger menacing our free institutions[14] (Figure 9.1).

The Growth of Corporate Personhood

Personhood has greatly eased the path of corporate ability to maximize profits free of ecological, social, or moral restraints. Congress did not create corporate personhood; no law mandates it or sets limits to it, nor has it been subject to political debate among lawmakers and citizenry. It is a judicially created doctrine.

Among the milestones creating corporate personhood are the following cases: In 1893, a corporation first claimed 5th Amendment protection (*Noble v. Union River Logging Railroad Company*[15]). In 1933, the courts prohibited chain store taxes, because they violated the rights of corporate "due process" (*Louis K. Liggett Co. v. Lee*[16]). In 1970, a jury trial—guaranteed to individuals by the 7th Amendment—was granted to corporations (*Ross et al., Trustees v. Bernhard et al.*[17]). In 1976, corporations were granted the 5th Amendment right against double jeopardy: a corporation argued successfully that it was double jeopardy to require a retrial in an antitrust case (*United States v. Martin Linen Supply Co.*[18]). In 1978, the Court extended 4th Amendment protection for corporations, requiring federal inspectors to obtain a search warrant for safety inspections on corporate property (*Marshall v. Barlow*[19]).

FIGURE 9.1
Senator Robert M. La Follette, Sr. Photograph from the Library of Congress (https://commons.wikimedia.org/wiki/File:Robert_M_La_Follette,_Sr.jpg).

The United States Supreme Court has continued the initial trend of granting complete equality between corporations and people, as well as equating money with speech. In the 1976 case *Buckley v. Valeo*, the Court ruled that political contributions are equivalent to speech.[20] In other words, First Amendment protections extend to financial contributions made to candidates or political parties. In *Randall v. Sorrell*, the Court in 2006 struck down a Vermont law limiting how much money a single person could donate to candidates for state-elected offices.[21]

In some cases not dealing directly with corporations, the Court again greatly expanded the opportunity for corporate influence. For example, a political action committee called Citizens United wanted to broadcast TV ads criticizing presidential candidate Hilary Clinton, but could not because of a 2002 law (the Bipartisan Campaign Reform Act[22]), which barred corporations and unions from paying for media that mentioned candidates in an election period. Citizens United sued the Federal Election Commission, which enforced campaign finance laws and sets election rules.

In 2010, the key U.S. Supreme Court case *Citizens United v. Federal Election Commission* built on more than a century of shaky, corporate-personhood decisions. Though this case did not itself grant corporations the rights of people, it greatly extended the territory in which such arguments could be made.[23]

In a close 5–4 decision, the Supreme Court ruled in *Citizens United v. Federal Election Commission* that the First Amendment protects not only a person's right to speak but also *the act of speech itself*, regardless of the speaker. Therefore, the First Amendment protects the speech of corporations and unions, regardless of whether they are considered "people." The Court also held that barring independent political spending is a form of prohibiting free speech protected by the First Amendment.

Limited Liability, Subsidiaries, and Transnationals

Side by side with increasing corporate personhood, and related to it, corporate limited liability has expanded. "Limited liability" simply means that a company's shareholders are not personally liable for the company's acts or debts; only the company itself is. Limited liability can be expanded to company directors and managers as well, to shield them individually from the company's misdeeds—as opposed to a general partnership, wherein the partners are liable for all the debts of the company. Limited liability, common in the United States by the mid-19th century, greatly encouraged the move to large-scale industry, since an individual's wealth was not liable if the company itself were to fail. As a result, companies began to accumulate large pools of financial capital for the first time.[24]

"Piercing the corporate veil," as it is called, is a heavily litigated area of United States law. However, there is no clear and concrete standard dictating when the corporate veil may be pierced to hold shareholders, directors, or managers liable. Furthermore, every U.S. corporation is chartered in a "home state," and states vary in the extent to which limited liability can be removed in cases of corporate wrongdoing. Courts end up juggling various tests relating to wrongful conduct, proximate cause, and ownership. Since the law is unsettled, courts also have to decide—on a case-by-case basis—whether and when there is adequate proof to warrant piercing the corporate veil.[25]

An especially pernicious problem linked to limited liability is the foreign subsidiary. The multinational corporation's ability to form foreign subsidiaries has dramatically increased the corporate reach. Why does a corporation, chartered (let us say) in a state of the United States, form a subsidiary?

> A multinational can form a subsidiary by negotiating with a foreign government to open an office or production facility in the country. In return, the multinational may receive financial incentives to locate in a certain area. The main reason for subsidiaries is economics... The country may offer the business a lower rate or a number of years without national taxes to aid in establishing the subsidiary. Corporations also create subsidiaries for the specific purpose of limiting their liability in connection with a new product or just for operating in a new country. The parent and subsidiary have separate legal identities, which means one is not liable for the actions of the other.[26]

In other words, the subsidiary, combined with limited liability and corporate personhood, creates a vast, faceless entity able to undertake enormous projects abroad that often have staggering environmental consequences. Yet the corporation, or its subsidiary, faces few consequences, and the corporate directors or managers who made the decisions suffer no liability at all. The subsidiary, as a separate legal entity, is not usually liable under the environmental laws of its home country and/or state, and often operates in countries with weaker environmental laws.

Many multinational subsidiaries are difficult to trace to a home country and make the company accountable for its actions, despite a corporation's originating in, or being partly controlled by interests in, a particular country. Multinationals do not necessarily exhibit loyalty to or adherence to the laws of the original home country, with many or even most operations far-flung via independent subsidiaries.

Even more disturbing, transnational corporations do not identify at all with a national home, wherever they may have originated. Whereas multinational corporations are chartered in a country and have foreign subsidiaries, *transnational* corporations—though originally chartered in a particular country—spread their operations across many countries to make decisions from as global a perspective as possible. They do not wish to identify with any one national home. Nestlé, a well-known corporation, is a transnational:

> When you buy the firm's products, do you have any idea that it is a registered Swiss company? Executives at Nestlé ... view the entire world as their domain for acquiring resources, locating production facilities, marketing goods and services, and establishing brand image. They seek total integration of global operations, try to make major decisions from a global perspective, distribute work among worldwide points of excellence, and employ senior executives from many different countries.[27]

Multinationals, subsidiaries, and especially transnational corporations can become virtually ungovernable via any democratic process:

> Companies are accountable to the general public mainly through the governments of the countries where the companies conduct their activities. This is at least the 'standard' situation ... However, it can become quite problematic in the case of TNCs [transnational corporations]. The congruence between the two sides of the accountability relationship can be put into question by the mismatch between the growing integration of world markets and the fragmented character of world politics.... Broadly speaking, there are four sources of accountability gaps in the relationship between TNCs and citizenries: the collusion between government officials and the directors of TNCs; the consequence of regulatory competition; the problem of weak and collapsed states; and subversive activities by TNCs.[28]

The more corporations can claim the rights of people, the more they vigorously defend those "rights," whereby they abuse both people and resources for the sake of profit, as well as spending massive amounts of money to keep regulatory laws at a minimum. How does all this relate to environmental sustainability? Protection of individual rights to life, liberty, and property by due process of law and equal protection of the laws—the language of the 5th Amendment curbing abuse of federal power, and the 14th Amendment curbing abuse of state power—once granted to corporations, shields these fictional entities from governmental power to enforcement to restrain abuse of the environment. In addition, the corporate liability shield, especially combined with formation of multinational subsidiaries and the globalization that encourages transnational corporations, is an exceedingly powerful tool that allows corporations to cause massive human, cultural, and environmental damage—but suffer no consequences. Corporate limited liability is nearly universal, and neither the judicial nor legislative branches have come to grips with its abuses by corporations shielding themselves from the consequences of their environmental and human abuse, whether in their home state, other states, or elsewhere in the world via forming foreign subsidiaries or taking a transnational form.

Local, state, or national environmental laws, as well as health and safety laws protecting individuals, are jeopardized by the limited liability shield, corporate globalization, and increasing corporate personhood, because corporations argue that restrictions on such things as open pit mining, requirements for post-mining restoration, limitations on clear-cutting forests, requirements to monitor water quality near large-scale farming operations, protecting worker safety from environmental harm, lax industrial standards that reduce profits, and similar measures will affect their "rights" as corporate persons to use ecological resources and people in ways socially responsible individuals would not. Moreover, corporate "rights" has become a cornerstone not only of U.S. law but also of international law in the last quarter century.

Corporate Disregard for the Environment—and People

Corporations—with few exceptions—have used and continue to use an inventive array of legal theories to get supportive court rulings, as illustrated by a few historic and recent examples of their abuse of both natural resources and people in the United States and abroad.

Ludlow Mining Massacre

On April 20, 1914, the Colorado Fuel and Iron Company and the Colorado National Guard attacked a tent colony of 1,200 striking coal miners and their families, leading to at least 24 deaths. The Ludlow Massacre was the finale of a series of widespread strikes against Colorado coalmine owners protesting inhumane working conditions, organized by the United Mine Workers of America. The three largest companies involved were the John D. Rockefeller-owned Colorado Fuel & Iron Company, the Rocky Mountain Fuel Company, and the Victor-American Fuel Company.[29]

Bhopal Disaster

In December 1984, a Union Carbide India Limited pesticide plant in Bhopal, India, released at least 30 tons of a highly toxic gas called methyl isocyanate, as well as other poisonous gases. The reason for the leak has not been conclusively determined, but the plant had a history of leaks and negligent maintenance, dating back to the mid-1970s. The gases stayed low to the ground, leading to exposure by upward of 600,000 people in the shantytowns surrounding the pesticide plant. It is estimated that 15,000 people were killed over the years from exposure to the toxic gases. The site has never been adequately cleaned up; Dow Chemical has owned Union Carbide, which had a majority stake in the plant, since 2001, and has further slowed cleanup efforts. Tons of contaminated wastes remain buried underground, and drinking water is contaminated, as several investigations have shown.[30]

Climate Change Science

The Center for International Environmental Law, in concert with many governmental and nonprofit researchers, has been able to document that the American Petroleum Institute and the fossil fuel industry—collectively one of the largest segments of American corporate activity—were already debating climate science, and the possible effects of burning fossil fuels, in the 1950s. Leading fossil fuel companies (including Union Oil, Standard Oil of California, Esso, and Shell) created the Smoke and Fumes Committee in 1946 to fund scientific research into air pollution, and use the scientific findings to block or weaken environmental regulations.[31]

By 1958 at the latest, the industry was actively bankrolling research into the relationship between fossil fuels and rising atmospheric carbon dioxide. They understood the risks by the 1960s. But beginning in the 1970s and 1980s, they used scientific research, public opinion manipulation, and active lobbying at state and national levels to create a body of misinformation and create uncertainty about whether any real emergency existed from burning fossil fuels, both for air pollution and climate change.[32]

In 1957, two scientists at Scripps Institute, Roger Revelle and Hans Suess, found that CO_2 resulting from burning fossil fuels would remain in the atmosphere. The Stanford Research Institute, in a 1968 report to the American Petroleum Institute, specifically cited their research: "...although there are other possible sources for the additional CO_2 now being observed in the atmosphere, none seems to fit the presently observed situation as well as the fossil fuel emanation theory. In summary, Revelle makes the point that man is now engaged in a vast geophysical experiment with his environment, the earth. Significant temperature changes are almost certain to occur by the year 2000 and these could bring about climatic changes."[33]

The industry continued to fund research into climate science, including the origins of hurricanes, and investigated other theories to put forward for rising temperatures, as well as researching carbon sinks that could contain carbon levels without the need to reduce emissions.[34] These included forests, soils, and marine macrophytes (from the Greek *makros* ["long"] and *phyton* ["plant"]—in this case plants large enough to see with the naked eye living in aquatic or semiaquatic environments).[35]

They also studied weather-modification techniques that would use fossil fuels products, such as using asphalt to spur rainfall or spreading oil on the ocean to weaken hurricanes.[36]

Perhaps most startling is that the industry, as far back as 1958, investigated the use of "ice clouds" created by petroleum combustion, or other methods, to melt Arctic ice. The petroleum industry was aware of the vast oil reserves under the Arctic ice pack by the early 1940s, and by the 1970s began designing special platforms for Arctic drilling, and ships and tankers to withstand icy waters and Arctic conditions. Chevron, for example, patented a design for Arctic drilling platforms in 1974.[37] Esso (now ExxonMobil) patented an oil tanker design for use in Arctic waters in 1974.[38] What is more, the oil industry was refuting the science showing the dangers of Arctic oil spills in that same period, arguing that to cause Arctic sea ice melt over a large area, an oil spill would have to be on a massive, and therefore unlikely, scale.[39]

Building on this and other research, the oil companies, by the 1980s at the latest, had a sophisticated and wide-ranging understanding of climate science. They used the science not only to shape their own decisions but also to promote climate skepticism in the public and among lawmakers. Attorneys General in three states thus far—New York, Massachusetts, and California—and the Virgin Islands have begun investigations into what the fossil-fuel industry, especially Exxon Mobil, knew about climate change, what facts or public policy debates they manipulated, what risks they failed to take into account or about which they failed to warn consumers (Figure 9.2). Attorneys General from 17 states have formed a coalition to collaborate on legal efforts to research and investigate whether fossil-fuel companies misled investors and the public on the impacts of climate change.[40] In retaliation, Exxon Mobil filed a lawsuit against the Attorney General of Massachusetts for that state's climate science investigation and Exxon's role in possible misrepresentation and cover-up. The company subsequently expanded its lawsuit to bring suit against the Attorney General of New York State as well. New York has focused its investigation on Exxon's potential

FIGURE 9.2
Commercial oil tanker *AbQaiq* readies itself to receive oil at Mīnā' al-Bakr, an offshore Iraqi oil installation, where it will take on an estimated 2 million barrels of crude oil. Photograph by Photographer's Mate 2nd Class Andrew M. Meyers, U.S. Navy (https://commons.wikimedia.org/wiki/File:Tanker_offshore_terminal.jpg).

lies to the public and investors about the risks of climate change. Exxon argued the two states' investigations were politically motivated, conducted only "in a coordinated effort to silence and intimidate one side of the public policy debate on how to address climate change." Investigations such as this, spearheaded by Attorneys General, have the potential to change both the nature of the debate over climate change, and the case law on damages allowed for corporate climate change-inducing activities.[41]

Regulation of Genetically Modified Organisms (GMOs) in Hawaii

Hawaii and Kauai Counties in Hawaii have both passed ordinances to curb the planting of genetically modified crops, concerned about the massive pesticide use on genetically modified crops and the so-called "GMO drift," in which genetically modified pollen contaminates nearby nonmodified crops and indigenous plants. Both ordinances have been faced with corporate legal challenges from such organizations as the Biotechnology Industry Association, the Nursery Association, and the Cattlemen's Council, as well as such biotech companies as DuPont's Pioneer Hi-Bred International. The corporations argued that such laws violated their corporate, constitutional "rights" to engage in commerce and grow genetically modified crops, meaning that municipalities have no right to pass restrictive ordinances.[42] Not surprisingly, a federal judge subsequently ruled that the Kauai County ordinance—requiring large-scale agricultural operations to disclose the use of pesticides and genetically modified crops—was preempted by state law and therefore unenforceable.[43]

Growth Hormones in the Dairy Industry

Monsanto Company has tried for more than 10 years to censor informed decision-making on the use of synthetic growth hormones in cattle to increase milk production, because it manufactures the only such hormone available in the United States. It successfully sued the state of Vermont to get a food labeling law struck down (*International Dairy Foods Association v. Amestoy*). The court held that manufacturers could not be compelled to disclose such information, "absent…some indication that this information bears on a reasonable concern for human health or safety…"[44]

Monsanto has also pressured other dairies to drop labels stating that milk from their cows is free of artificial growth hormones on grounds that such labels imply milk from non-treated cows is superior. For example, Oakhurst Dairy in Maine labels its milk: "Our farmers' pledge: no artificial hormones." Monsanto's lawsuit contends the label implies that milk from Oakhurst's cows is somehow better than that from cows treated with rBST (= bovine somatotrophin), thereby unfairly harming Monsanto's business.[45]

Mountaintop-Removal Mining

The most environmentally destructive form of mining ever developed is mountaintop-removal mining to extract coal. Mountaintop mining is essentially strip-mining the top of a mountain, where the top 200 to 600 feet of a 2,000 to 4,000-foot mountain is blown up. The mountains are never restored to their original heights, and the hundreds of millions of tons of rubble are dumped into the mountain's canyons and streams, as well as the valleys that separate the steep mountains. The mining company then extracts a 24-inch to 50-inch seam of coal.

Ted Hapney (of the United Mine Workers of America) contended, "mountaintop mining can be done in an environmentally sensitive way." However, U.S. District Judge Charles Haden ruled in 2000 that the federal Clean Water Act prohibits regulators from approving new mines whose waste would damage streams that run 6 months or more of the year. According to Bill Raney, president of the West Virginia Coal Association, "we know mountaintop removal ain't pretty. But the land recovers, and it's usually put to better economic use" (Figures 9.3 and 9.4).

"The whole country should be concerned about this [Judge Haden's ruling]," chimes in Andy [no other name given] of the West Virginia Department of Environmental Protection, which grants mining permits. "You've got a handful of environmentalists stopping a crucial energy source. The nation made a decision in the 1960s and '70s to move away from nuclear power and use coal. We created a monster, and now we want to crucify it."[46]

The George W. Bush administration was in favor of changing Clean Water Act rules so this type of mining could continue filling stream channels with debris; nevertheless, Judge Haden blocked such activity for a second time. Judge Haden wrote that the U.S. Army Corps of Engineers issued permits in 2000 that allowed coal companies to fill 87 miles of streambeds with waste from mining. "Past … permit approvals were issued in express disregard of the Corps' own regulations and the [Clean Water Act]" wrote Haden. "As such, they were illegal." Haden also ruled that a new regulation issued by the Corps was merely "an attempt to legalize their long-standing, illegal regulatory practice …. The rule change was designed simply for the benefit of the mining industry and its employees."[47]

During the two decades 1989–2009, mountaintop-removal mining in Appalachia has destroyed or severely damaged more than a million acres of forest and buried nearly

FIGURE 9.3
Mountaintop-removal mine in Pike County, Kentucky, just off U.S. 23. Photograph by Matt Wasson, Appalachian Voices (https://commons.wikimedia.org/wiki/File:Mountaintop_removal_mine_in_Pike_County,_Kentucky.jpg).

2000 miles of streams.[48] By the end of 2012, approximately 2,200 square miles of the Appalachian Mountains had been permanently ruined by mountaintop-removal mining.[49]

Why is mountaintop-removal mining so popular with coal companies? It allows them to extract more coal with fewer worker hours, thus increasing profits. Principal companies involved include Massey Coal Company (acquired by Alpha Natural Resources), Arch Coal, CONSOL Energy, International Coal Group, James River Coal Company, and Patriot Coal.[50]

Clear-Cut Logging

Timber companies frequently consider clear-cutting (a form of removing trees for timber) to be the most economical. It consists of cutting all the trees in a given area, sometimes a few acres, but often in large swaths. The effects on soil stability, water quality, and ecological integrity have been devastating in many parts of the world.

In western Oregon, a heavily forested region, only about 10 percent of the original forest remains. In the Amazon basin, where perhaps half of the original forest is still intact, it is very much in danger of continued depredation. Moreover, replanting clear-cut areas almost always means planting "tree farms" of similar age and species, so as to create a uniform crop of trees, rather than the diversity age classes and species characteristic of a true forest. This also has serious implications for genetic diversity, as well as that of other plants and all the forest animals. In some areas, such as South America, clear-cuts are turned into farmland, whereas in Malaysia, they usually end up in palm-oil or rubber plantations.[51]

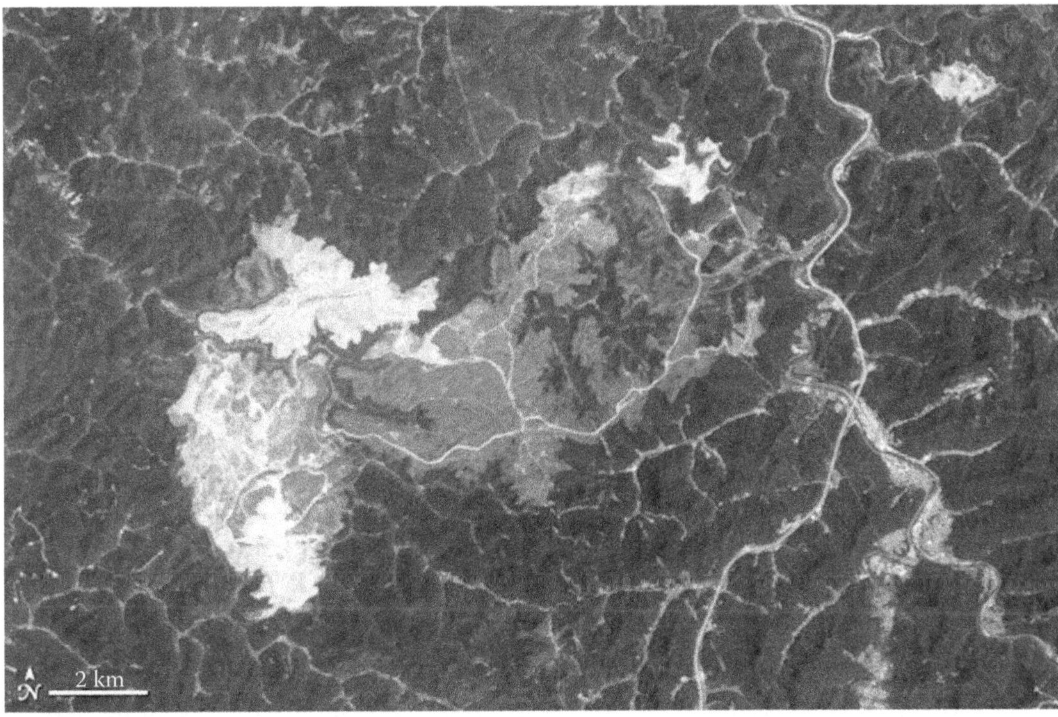

FIGURE 9.4
Hobet mountaintop-removal mine in Boone County, West Virginia. Photograph by U.S. National Aeronautics and Space Administration, 2009 (https://commons.wikimedia.org/wiki/File:Hobet_Mountaintop_mine_West_Virginia_2009-06-02.jpg).

Sierra Pacific Industries, for example, is the largest landowner in California, a privately owned company owning more than 1.5 million acres of land within the state; it also bids on nearly 40 percent of the timber logged from the National Forests in California. Sierra Pacific Industries practices clear-cutting, having clear-cut hundreds of thousands of acres in the Sierra Nevada, and then replanted the land to homogenous, pesticide-laden tree farms. Between 1992 and 1999, Sierra Pacific Industries further increased its clear-cutting rate, by an astounding 2,426 percent, devastating wildlife, water catchments, and environmental integrity in the Sierra Nevada region. California's forest practices statutes and timber company-friendly regulators have failed to rein in the destructive activity by Sierra Pacific Industries, as well as other timber companies.[52]

Where Do We Go from Here?

There continue to be nationwide efforts to rein in corporate environmental abuses at many levels of government. While there is disagreement as to whether federal, state, or local efforts are the most effective to regulate corporations, the truth is that none will accomplish this goal by themselves because it is in the corporate interest to corrupt any regulatory mechanism.

A Rights of Nature legal system must go hand in hand with ending corporate personhood, liability shields, and corporate capacity to create international subsidiaries or form transnational entities. The history of corporate depredations on ecosystem resilience and Nature's ability to function makes it clear these are essential steps. This will require a multipronged approach in the United States and abroad because transnational corporations, multinational corporations, and foreign subsidiaries have created environmental havoc in many countries, where environmental laws are even weaker than those in the United States. To reach this goal, concerned citizens will need to do the following:

1. Place Rights of Nature as the primary rights holder in the United States Constitution and all state Constitutions, so that protecting Nature and its right to flourish free of degradation, destruction, and pollution comes first in legal disputes.
2. End corporate liability shields via federal legislation, so that corporate shareholders, directors, and managers are personally liable for misdeeds of the company in any state or nation in which it operates. This will greatly restrain corporate abuses.
3. Strengthen state powers, by federal legislation if necessary, to revoke charters of their home state corporations upon evidence—using the Precautionary Principle as the standard—of causing harm to the environment or people, either in the home state, other states, or other nations.
4. Eliminate the power of American corporations to form foreign subsidiaries. Subsidiaries are a prime cause of environmental degradation caused by corporate projects in foreign countries. In the United States, this would likely require both federal legislation and changes to state corporate-charter rules, as subsidiaries take many technical forms.
5. Eliminate the ability of American corporations to take transnational forms, by restricting the corporation's operations to its home state or particular product line in the state or regional area.
6. Research whether or not American corporations should be allowed to form subsidiaries (or otherwise do business) under narrowly limited circumstances in American states other than the home state. If so, it should only be for very circumscribed purposes. Corporate activities in other states—whether directly or via subsidiary—often cause tremendous environmental harm, because large corporations have vast resources at their command. Multi-state operations are effective tools for increasing corporate powers and personhood outside the reach of individual state laws. Multi-state subsidiaries and operations also give corporations the opportunity to compete for tax subsidies and other benefits state by state, and favor location in the state offering the greatest benefits. This stunts local economies, community structure, and the local tax base.
7. Research the best legal means to create local associations necessary to accomplish specific purposes an individual cannot carry out—such as an infrastructure project—and set up the legal possibility to create local or regional associations chartered for specific purposes, but created for limited periods and for limited projects only. Transition the current legal structure allowing national, multinational, and transnational corporations to this more limited structure.
8. Expand the constitutional right of citizens to sue for social-environmental violations under the Rights of Nature provisions, with legal protections to prohibit

corporate backlash or countersuit. This will require creating Nature Tribunals with exclusive jurisdiction over violations of Nature's Rights.

9. Build and strengthen the rights of residents of affected or targeted areas to make their voices heard against corporate projects through democratic forums, legal cases, and non-violent gatherings to let leaders know that they, the people, care about the environment on which they depend as an inseparable part of Nature. Such involvement recently led the gigantic Southern Copper Company to cancel the proposed $1.4 billion Tia Maria open pit copper mine in southern Peru's Arequipa region.[53]
10. Amend the U.S. Constitution to prohibit corporate personhood. Such an amendment must be proposed either by a two-thirds vote of Congress or else by a Constitutional Convention convened when the legislatures of two-thirds of the states request one be convened. Ratification of a proposed amendment then requires three-quarters of the states approve it.

"Move to Amend" is an organization building a coalition with the intent of amending the U.S. Constitution to abolish all corporate personhood.[54] This will have to be a multi-year, grassroots campaign. A similar campaign to abolish corporate personhood—or any facsimile serving the same function—may be needed in other countries or international forums, such as the European Union. Such amendments will be necessary to prevent corporations from fleeing to foreign jurisdictions, where they could undertake irresponsible activities to the everlasting detriment of worldwide social-environmental integrity.

It must be noted that there is nothing wrong in and of itself in forming associations of various kinds to undertake necessary projects or design and manufacture products that individuals alone do not have resources for. The question lies in how to allow—and, if necessary, control—business associations so they do not branch into faceless, powerful corporations. We have sought, in this chapter, to begin the discussion of how to undertake the desperately needed, large-scale changes (both nationally and internationally) to bring worldwide corporate abuse under control.

That corporate abuse occurs is also of concern to some in the corporate community, and there is a movement to reform and create socially and environmentally responsible corporations. These are called the B Corporations: for-profit companies certified by the nonprofit B Lab that they meet strict standards of social and environmental performance, accountability, and transparency. There are currently more than 1,600 B Corporations worldwide. They seek to use business as a force for good in their communities, for their workers, and for the environment.[55]

Combining the successes of the B-Corporation movement with needed legal reforms can end corporate abuse of the environment and communities. It will take concentrated effort to return associations to the human-centered, local entities they once were—in a new form bettered by the experience of corporate abuse the United States and the world have undergone in the last century and a half. It can be done, and for the sake of the world's future, it must be done.

Endnotes

1. The foregoing paragraph is based on: (1) Jim Hightower. Chomp! *Utne Reader,* March–April (1998):57–61, 104; (2) Nick Robins. Loot. *Resurgence,* 210 (2001):12–16; and (3) David C. Korten. 2001. What to do when corporations rule the world. *Yes! A Journal of Positive Futures,* Summer (2001):48–51.
2. (1) Noam Chomsky. How free is the free market? *Resurgence,* 173 (1995):6–9; and (2) Thomas Jefferson. (This quotation is often cited as being in an 1802 letter to Secretary of the Treasury Albert Gallatin, and/or "later published in The Debate Over the Recharter of the Bank Bill [1809].") http://www.monticello.org/site/jefferson/private-banks-quotation (accessed July 7, 2015).
3. (1) Trustees of Dartmouth College *v.* [William H.] Woodward. *Encyclopaedia Britannica.* (Last updated January 15, 2015) http://www.britannica.com/event/Dartmouth-College-case (accessed July 6, 2015); and (2) Dartmouth College *v.* Woodward. The Oyez Project at IIT Chicago-Kent College of Law. 07 July 2015. http://www.oyez.org/cases/1792-1850/1818/1818_0 (accessed July 6, 2015).
4. Abraham Lincoln, Archer Hayes Shaw, and David Chambers. *The Lincoln Encyclopedia: The Spoken and Written Words of A. Lincoln Arranged for Ready Reference.* Macmillan, New York. (1950) p. 40.
5. Santa Clara County *v.* Southern Pac. R. Co. (1886). http://caselaw.findlaw.com/us-supreme-court/118/394.html (accessed July 6, 2015).
6. Thom Hartmann. *Unequal Protection: The Rise of Corporate Dominance and Theft of Human Rights (Second Edition).* Berrett-Koehler Publishers, Inc., San Francisco, CA. (2010) 373 pp.
7. Interstate Commerce Act (1887). http://www.ourdocuments.gov/doc.php?flash=true&doc=49 (accessed July 6, 2015).
8. Sherman Antitrust Act (1890). http://www.stern.nyu.edu/networks/ShermanClaytonFTC_Acts.pdf (accessed July 6, 2015).
9. Progressive Party. http://www.u-s-history.com/pages/h1755.html (accessed July 6, 2015).
10. Gerhard Peters and John T. Woolley. Progressive Party Platform of [November 5,] 1912. http://www.presidency.ucsb.edu/ws/?pid=29617 (accessed July 7, 2015).
11. Clayton Antitrust Act. http://www.britannica.com/event/Clayton-Antitrust-Act (accessed July 8, 2015).
12. Ross E. Elfand. The Robinson–Patman Act. American Bar Association. http://www.americanbar.org/groups/young_lawyers/publications/the_101_201_practice_series/robinson_patman_act.html (accessed July 8, 2015).
13. American Antitrust Institute Public Interest Advocacy Workshop on Mergers. National Press Club, Washington, DC (October 11, 2013) 2 pp. http://www.antitrustinstitute.org/sites/default/files/Section%207.pdf (accessed July 8, 2015).
14. Robert Marion La Follette. *La Follette's Autobiography: A Personal Narrative of Political Experiences.* The Robert M. La Follette Co., Madison, Wisconsin. (1911, 1913) p. 684.
15. Noble *v.* Union River Logging R Co, (1893). http://caselaw.findlaw.com/us-supreme-court/147/165.html (accessed July 8, 2015).
16. Louis K. Liggett Co. *v.* Lee, 288 US 517—Supreme Court 1933. http://scholar.google.com/scholar_case?case=7798071385596153798&q=Liggett+v.+Lee&hl=en&as_sdt=6,38&as_vis=1 (accessed July 8, 2015).
17. Ross *v.* Bernhard, 396 US 531—Supreme Court 1970. http://scholar.google.com/scholar_case?case=3226359321665935534&hl=en&as_sdt=6&as_vis=1&oi=scholarr (accessed July 8, 2015).
18. United States *v.* Martin Linen Supply Co., 430 US 564—Supreme Court 1977. http://scholar.google.com/scholar_case?case=15633467112423421514&hl=en&as_sdt=6&as_vis=1&oi=scholarr (accessed July 8, 2015).
19. Ross *v.* Bernhard: The uncertain future of the Seventh Amendment. *The Yale Law Journal,* 81 (1971):112–133.

20. Buckley *v.* Valeo. https://www.law.cornell.edu/supremecourt/text/424/1 (accessed July 8, 2015).
21. Randall et al. *v.* Sorrell et al. https://www.law.cornell.edu/supct/html/04-1528.ZS.html (accessed July 8, 2015).
22. H.R.2356—Bipartisan Campaign Reform Act of 2002. https://www.congress.gov/bill/107th-congress/house-bill/2356 (accessed July 8, 2015).
23. Citizens United *v.* Federal Election Commission (No. 08–205: argued March 24, 2009; reargued September 9, 2009; decided January 21, 2010) https://www.law.cornell.edu/supct/html/08-205.ZS.html (accessed July 8, 2015).
24. Limited liability. https://en.wikipedia.org/wiki/Limited_liability (accessed August 1, 2016).
25. Piercing the corporate veil. https://en.wikipedia.org/wiki/Piercing_the_corporate_veil (accessed August 1, 2016).
26. Monica Sanders. Relations between International Companies and Their Subsidiaries http://smallbusiness.chron.com/relations-between-international-companies-subsidiaries-24591.html (accessed August 1, 2016).
27. John R. Schermerhorn, Jr. *Exploring Management*. John Wiley & Sons, New York. (2010) p. 387.
28. Mathias Koenig-Archibugi. Transnational Corporations and Public Accountability. http://dspace.africaportal.org/jspui/bitstream/123456789/8719/1/Transnational%20Corporations%20and%20Public%20Accountability.pdf?1 (pp. 5–6) (accessed August 1, 2016).
29. Militia slaughters strikers at Ludlow, Colorado. http://www.history.com/this-day-in-history/militia-slaughters-strikers-at-ludlow-colorado (accessed July 8, 2015).
30. (1) Edward Broughton. The Bhopal disaster and its aftermath: A review. *Environmental Health*, 4 (2005):6 pp. http://www.ehjournal.net/content/pdf/1476-069X-4-6.pdf (accessed July 14, 2015); (2) Ingrid Eckerman. Bhopal gas catastrophe 1984: Causes and consequences. *Encyclopedia of Environmental Health* (2013):302–316; (3) Ingrid Eckerman. *Asian-Pacific Newsletter on Occupational Health and Safety*, 13 (2006):48–49; and (4) Hannah Osborne. Bhopal disaster 30th anniversary: Facts about the world's worst industrial tragedy. *International Business Times*, December 1, 2014. http://www.ibtimes.co.uk/bhopal-disaster-30th-anniversary-facts-about-worlds-worst-industrial-tragedy-1477489 (accessed July 8, 2015).
31. (1) Smoke and Fumes Committee. https://www.smokeandfumes.org/fumes/moments/2 (accessed August 1, 2016); (2) Chris D'Angelo. Big Oil Could Have Cut CO_2 Emissions In 1970s—But Did Nothing. http://www.huffingtonpost.com/entry/big-oil-emissions_us_573c9d81e4b0aee7b8e8a046 (accessed August 1, 2016); and (3) Matt Smith. The Oil Industry Was Warned about Climate Change in 1968. https://news.vice.com/article/the-oil-industry-was-warned-about-climate-change-in-1968 (accessed August 1, 2016).
32. Smoke and Fumes Committee, *op. cit.*
33. E. Robinson, R.C. Robbins. Sources, Abundance, and Fate of Gaseous Atmospheric Pollutants. Final report and supplement. United States: Stanford Research Institute, Menlo Park, CA. (1968) 125 pp.
34. Roberet E. Inman, Royal B. Ingersoll, and Elaine A. Levy. Soil: A natural sink for carbon monoxide. *Science*, 172 (1971):1229–1231.
35. S.V. Smith. Marine macrophytes as a global carbon sink. *Science*, 211 (1981):838–840.
36. (1) James F. Black and Barry L. Tarmy. The use of asphalt coatings to increase rainfall. *Journal of Applied Meteorology*, 2 (1963):557–564; and (2) H. Wexler. Modifying weather on a large scale: Current proposals are either impractical or likely to produce cutes that are worse than the ailment. *Science*, 128 (1958):1059–1063.
37. T. Hudson and G. Strickland. (1974). U.S. Patent No. 3,831,385. Washington, DC: U.S. Patent and Trademark Office.
38. W. Devine. (1973). U.S. Patent No. 3,745,960. Washington, DC: U.S. Patent and Trademark Office.
39. (1) R.C. Ayers, J.L. Glaeser, S. Martin, and W.J. Campbell. Oil spills in the Arctic Ocean: Extent of spreading and possibility of large-scale thermal effects. *Science*, 186(1974):843–846; and (2) Smoke and Fumes. https://www.smokeandfumes.org/fumes/moments/10 (accessed August 1, 2016).

40. (1) Ben Jervey. State Investigations into What Exxon Knew Double, and Exxon Gets Defensive. http://www.desmogblog.com/2016/04/01/more-state-attorneys-general-investigate-exxon-exxon-gets-defensive (accessed August 1, 2016); and (2) Smoke and Fumes. https://www.smokeandfumes.org/fumes/moments/13 (accessed August 1, 2016).
41. (1) Brendan DeMelle. Exxon Sues MA Attorney General in Retaliatory Attempt to Intimidate 'Exxon Knew' Climate Accountability Movement. http://www.desmogblog.com/2016/06/16/exxon-sues-ma-attorney-general-retaliatory-attempt-intimidate-exxon-knew-climate-accountability-movement (accessed August 1, 2016); (2) Tom Korosec. Exxon Adds New York to Suit in Bid to Stop Climate Change Probe. https://www.bloomberg.com/news/articles/2016-11-12/exxon-adds-new-york-to-suit-in-bid-to-stop-climate-change-probe (accessed November 28, 2016); and (3) Justin Gillis and Clifford Krauss. Exxon Mobil Investigated for Possible Climate Change Lies by New York Attorney General. http://www.nytimes.com/2015/11/06/science/exxon-mobil-under-investigation-in-new-york-over-climate-statements.html (accessed November 28, 2016).
42. Corporate Rule Examples. https://movetoamend.org/corporate-rule-examples (accessed July 9, 2015).
43. Nestor Garcia. Federal judge declares new Kauai GMO, pesticide law invalid. http://khon2.com/2014/08/25/federal-judge-declares-new-kauai-gmo-pesticide-law-invalid/ (accessed July 9, 2015).
44. International Dairy Foods Association et al. *v.* Jeffrey L. Amestoy et al. http://www.leagle.com/decision/19951144898FSupp246_11108.xml/INTERNATIONAL%20DAIRY%20FOODS%20ASS'N%20v.%20AMESTOY (accessed July 9, 2015).
45. (1) Kristen Philipkoski. Monsanto *v.* Oakhurst Dairy: Does Monsanto Corporation Have the Right to Keep You from Knowing the Contents of Your Food? http:/reclaimdemocracy.org/monsanto-v-oakhurst-dairy/ (accessed July 9, 2015); and (2) Oakhurst Dairy. http://www.oakhurstdairy.com/ (accessed July 9, 2015).
46. Dennis Cauchon. Once again, a fierce fight on Blair Mountain. *USA Today*. March 29, 2000.
47. The discussion of Judge Haden's ruling on mountaintop-removal mining is based on: (1) Brian Farkas. Judge blocks Appalachian coal mining permits. *The Oregonian*, Portland, OR. May 9, 2000; and (2) Traci Watson. Mountaintop mining halted. *USA Today*. May 9, 2000.
48. Leveling Appalachia: The Legacy of Mountaintop Removal Mining. http://e360.yale.edu/feature/leveling_appalachia_the_legacy_of_mountaintop_removal_mining/2198/ (accessed July 9, 2015).
49. (1) Top four U.S. coal companies supplied more than half of U.S. coal production in 2011. (October 2, 2013) http://www.eia.gov/todayinenergy/detail.cfm?id=13211 (accessed July 9, 2015); (2) Mountaintop Removal Mining Companies Banks Should Avoid. http://d3n8a8pro7vhmx.cloudfront.net/rainforestactionnetwork/legacy_url/1299/ran_mtrcompaniestoavoid_2012.pdf?1402698590 (accessed July 9, 2015); and (3) Rob Perks. Appalachian heartbreak: Time to end mountaintop removal coal mining. *Natural Resources Defense Council*. 12 pp. https://www.nrdc.org/land/appalachian/files/appalachian.pdf (accessed July 9, 2015).
50. Rob Perks. Appalachian heartbreak: Time to end mountaintop removal coal mining. *op. cit.*
51. (1) Chris Maser. *The Redesigned Forest*. R&E. Miles Publishers, San Pedro, CA. (1988) 234 pp.; (2) Chris Maser. *Sustainable Forestry: Philosophy, Science, and Economics*. St. Lucie Press, Delray Beach, FL. (1994) 373 pp.; (3) Chris Maser. *Our Forest Legacy: Today's Decisions, Tomorrow's Consequences*. Maisonneuve Press, Washington, DC (2005) 255 pp.; and (4) Chris Maser. *Earth in Our Care: Ecology, Economy, and Sustainability*. Rutgers University Press, New Brunswick, NJ. (2009) 304 pp.
52. Industrial Logging in the Sierra Nevada. *Sierra Forest Legacy*, 2012. http://www.sierraforestlegacy.org/FC_FireForestEcology/FFE_IndustrialForestlands.php (accessed July 9, 2015).
53. (1) Robert Kozak. Southern copper scraps Tia Maria copper project in Peru. *The Wall Street Journal*, March 27, 2015. http://www.wsj.com/articles/southern-copper-scraps-tia-maria-copper-project-in-peru-1427471246 (accessed July 10, 2015); and (2) Mitra Taj. Peruvian foes

of Tia Maria copper mine expand month-long protest. *Reuters*, April 22, 2015. http://www.reuters.com/article/2015/04/23/peru-mining-protests-idUSL1N0XJ2PB20150423 (accessed July 10, 2015).
54. End Corporate Rule, Legalized Democracy. http://movetoamend.org/ (accessed July 10, 2015).
55. B Corp Community. https://www.bcorporation.net/b-corp-community (accessed August 1, 2016).

Section IV

Rights of Nature for Land, Water, and Air

Fundamentally, the Rights of Nature must allow the land, water, and air of the planet to flourish, maintain their processes, and regenerate according to their own principles, the Laws of Reciprocity detailed earlier in this book. This section of the book assumes that Rights of Nature has already been placed in federal and state Constitutions, so that we can lay out a blueprint that extends beyond the near horizon. Obviously, changing the governing system to make Rights of Nature primary must come first.

We begin this section with a discussion of land—that part of Nature most subject to the fiction of ownership. We focus first on the needs of urbanized lands. We then turn our attention to the problems and Rights of Nature in the three major, terrestrial ecosystems in our national landscape: forests, grasslands, and wetlands. Water is not exactly owned in the United States, usually being considered as belonging to the people or the state, but there is a complex system of water rights governing its use. Air, the least subject to questions of ownership, is most obviously a global commons—despite declared airspaces for reasons of national security. But, all three, in their different ways, are vulnerable to abuse and misuse, as part of the commons on which all life depends.

Each chapter begins with a discussion of the principles of the Rights of Nature, as taken from the Ecuadorian Constitution, the Bolivian Law of Pachamama, and the Universal Declaration of the Rights of Mother Earth. In each instance, we focus on those parts of the law most relevant to land, water, or air, as the case may be.

We then give an overview of the social governance of these three elemental aspects of life, look at some aspects of their biology, and give a summary of the legal system governing their use in the United States. The biological overviews are confined to the U.S., but even so can only be general in nature, since ecosystems vary tremendously across landscapes and have multiple biophysical differences by region.

The legal overviews concentrate primarily on federal laws, because in the hierarchical American legal system there are usually state and local laws as well. To describe even some of these is beyond the scope of this book. The one exception is the discussion of private urban and rural land, because use of private land in the United States is governed principally via local zoning codes.

Each chapter then has a list of management considerations, which are principles that need to be set into place to create and safeguard a Rights of Nature legal system that governs use. The details are necessarily vague, but the principle is urgent: only by retooling our use of the land, water, and air can we create a deeply and continuously sustainable society in which Nature can flourish and thus provide all that is necessary for human societies to do likewise.

10

Land

Land Is Not a Commodity

Land, which must be used by all generations of people worldwide, as well as all the life-forms and processes of Nature through all of time, is not a commodity that can be owned in fee simple, albeit one can today purchase a parcel of land and obtain a title and deed to justify in one's mind that such ownership exists. But fundamentally, land is not a commodity, and by many cultures is considered a "commons" for use by all under specific communal rules.

This more communal view of land use originated with nomadic or semi-nomadic peoples. Hunter-gatherers were, by nature and necessity, nomadic—a traditional form of wandering in which people moved their encampment several times a year as they either searched for food or followed the known seasonal order of their food supply.

Although hunter-gatherers had the right of personal ownership, it applied only to mobile property, that which they could carry with them, such as their hunting knives or gathering baskets. On the other hand, things they could not carry with them, such as land, was to be shared equally through rights of use, but could not be personally controlled to the exclusion of others or abused to the detriment of future generations.[1]

Thus, the "commons usufruct law" arose as a natural outgrowth of various hunter-gatherer cultures because almost all hunter-gatherers, including nomadic herders and many village-based societies as well, shared a land-tenure system based on the rights of common usage that, until recently, were far more common than regimes based on the rights of private property. In traditional systems of common property, the land is held in a kinship-based collective, while individuals owned movable property. Rules of reciprocal access made it possible for an individual to satisfy life's necessities by drawing on the resources of several territories. Today, the legal definition of usufruct in the United States is:

> Usufruct is a right in a property owned by another, normally for a limited time or until death. It is the right to use the property, to enjoy the fruits and income of the property, to rent the property out and to collect the rents, all to the exclusion of the underlying owner. The usufructuary has the full right to use the property but cannot dispose of the property nor can it be destroyed.
>
> The extent of usufruct is defined by agreement and may be for a stated term, covering only certain stated properties, it could be set to terminate if certain conditions are met, such as marriage of a child or remarriage of a spouse, it can be granted to several people to share jointly, and it can be given to one person for a period of time and to another after some stated event occurs.[2]

Land: The Rights of Nature

A legal system that enshrines the Rights of Nature in its founding documents must focus consciously and carefully on human land use. Without land, people could not survive, and all terrestrial ecosystems would become extinct.

The new Ecuadorian Constitution says that "Nature, or Pacha Mama, where life is reproduced and occurs, has the right to integral respect for its existence and for the maintenance and regeneration of its life cycles, structure, functions and evolutionary processes."[3] Further, in recognition of the fact that human "ownership" of land often combines with the legal authority to abuse ecosystem sustainability, the Constitution states, "Nature also has the right to be restored, and its integral restoration is held to be independent of natural persons or legal entities."[4]

Turning now to the provisions of Bolivia's new law, it states that Mother Earth is guaranteed the right of regeneration, "…that the diverse living systems of Mother Earth may absorb damage, adapt to shocks, and regenerate without significantly altering their structural and functional characteristics…."[5] Most importantly, the Bolivian law directly confronts the question of "property rights" versus Mother Earth's requirements, stating in Section 5, "Neither living systems nor processes that sustain them may be commercialized, nor serve anyone's private property."[6]

The Universal Declaration of the Rights of Mother Earth, which sprang directly from the philosophies embodied in the Ecuadorian and Bolivian provisions, provides the rights of Mother Earth, and all beings of which she is composed, in Article 2. These include the right to life and to exist, the right "to continue their vital cycles and processes free from human disruptions;" and the right "to maintain its identity and integrity as a distinct, self-regulating and interrelated being."[7] Clearly these rights could not be sustained without a land base in which ecosystem processes could flourish uninterrupted.

Placing broadly similar provisions in the United States Constitution and the state Constitutions would allow states and local governments to experiment with how best to address the requirement that land and ecosystems have the inalienable right to exist, flourish, and regenerate outside of human interference. Adopting Rights of Nature ordinances at the local level has occurred in the United States in a growing number of instances, but it is not enough. The American legal system is hierarchical, and local ordinances that conflict with state or federal laws will be struck down by the courts. The only way for Nature's Rights to become the fundamental basis of *local* land-use planning is for Rights of Nature to be placed in federal and state Constitutions, so Nature is the first rights-holder in the legal framework from the national level to the most local.

The Urban Landscape: Regulating Humans in Community

Land use is the body of law that regulates how an owner may use his or her land. The most commonplace tool in land use is a legal requirement for some kind of community planning, combined with mandatory zoning, which restricts uses allowed on various parcels. Zoning without a hand-in-hand planning system leads to narrow, exclusive zones, and lacks an overall vision for the community. Zoning by itself solves only half of a community's effort to chart its own future. But, once a planning system is in place, a parcel zoned for residential use cannot, for example, be used for an industrial purpose without a public process. In the public forum, the applicant, surrounding owners, and other interested parties discussed the change and local decision makers render a verdict. Nationwide, variations of this scenario take place daily. Fittingly, most land-use law is found either at the state or, more frequently, at the local level since use (and misuse) of land occurs locally, place by place.

In order to investigate how to devise land-use regulations that uphold the Rights of Nature and thus Nature's inalienable right to exist intact and unspoiled, we must first investigate a bit of historical land-use regulation in the Western paradigm, as it manifests in the United States. Note at the outset that all forms of land-use planning and restrictive zoning are, by definition, communal because they broaden the sphere of influence the landowner must accept, or face punishment from the legal authority of the community. Thus, planning and zoning, when taken together, can easily fit into a framework of Nature's Rights, where they broaden and deepen the depth of communal responsibility to include the ecosystem, but do not change it in kind.

The Origins of Land Use Regulation in England

At the time the American republic was formed, England had no national, legal framework regulating land use. Nevertheless, England did have local controls narrowing a landowner's right to unfettered use of land. Most of these were slow-evolving, common-law doctrines initially designed and then enforced by the courts.

After the Norman Conquest in 1066, William the Conqueror imposed a system of feudal tenure on England, in which his kin, allies, advisors, or other parties were granted rights of the land. But they held the land right of the King, not as an individual, private ownership. The grantee owned services to the grantor, beginning with the King, as explained in the earlier chapter on the feudal system. The feudal system was already well developed on the European continent at this time, before it was imported to England. Individual ownership in England grew out of the system imposed by the Norman conquerors, initially through the Statute of Quia Emptores of 1290. This law forbade subinfeudation, a practice that allowed one tenant to grant land to another, with the grantor then becoming the grantee's lord. The statute was critical in stopping the expansion of the feudal pyramid, which created endless chains of obligation and unclear rights of ownership. The 1290 law gave owners the right to sell land to other private parties, although with the state retaining some rights.[8]

Once this critical threshold was passed, English common law developed to protect the rights of landowners to a very generous degree. William Blackstone, an early and perhaps the best known of English common law commentators, described property rights in 1782, as "that sole and despotic dominion which one man claims over the external things of the world, in total exclusion of the right of any other individual in the universe."[9]

Initially, there were few restrictions to unfettered use. The landowner could control trespass via action in common-law courts; intentional trespass was always actionable even if no actual damage occurred. The only balancing factor was the developing law of nuisance, which forbade owners from using their property in a way that injured the land of others. The doctrine was limited, however, to barring land uses that could *actually* be proven to have harmed adjacent lands or its occupants. The doctrine grew slowly and haphazardly, depending solely on private court cases of one party against another. The kinds of nuisances complained of were such things as smoke, dust, or noise that in some way restricted the normal uses of the affected properties.

The land use conflicts under a doctrine of unfettered, private land use in rural England were inconsequential compared to those that developed in newly flourishing urban areas: overcrowding, polluted water sources, traffic congestion, and highly unsanitary conditions.[10] London was the primary catalyst for change. The earliest building constraints probably date about 1189, regulated things like siting privies, problems with gutter runoff, and house encroachment onto the street. Thatched roofs were banned in London after a major fire in 1212 in which about a thousand people died. London's Mayor and aldermen had an Assize of Nuisance to settle nuisance cases between neighbors, and such experts as masons and carpenters advised the court. (An *Assize of Nuisance* is a common-law writ for nuisance with the twofold purpose of abating the activity and of compensating for any harm done by the activity.) Other cities followed London's lead; Bristol, for example, had building inspectors as early as 1391.

The real catalyst for change was the tragic Great Fire of London in 1666, which destroyed 80 percent of the city. King Charles II published a proclamation subsequent to an investigation after the Great Fire. The Act for the Rebuilding of London of 1667 put new land use restrictions into place, requiring brick exteriors, wider streets, and access to Thames water for firefighting. Noxious activities, such as breweries and tanneries, also came to be banned in central London. The municipality itself could regulate building size, height, materials, and placement.[11]

A Brief History of Land Use Regulation in the United States

Land use regulation in England at the time of the U.S. republic was strictly municipal, and principally limited to health, safety, and nuisance abatement. The notion of strong, individual property rights was well established in the U.S. colonial period and thereafter, but colonial authorities, via land charters, imposed basic restrictions on such things as building size or location. As early settlements evolved into cities and towns, the powers of land use regulation devolved to them, but only as creatures of the state. For example, the 1866 Metropolitan Health Law of New York provided the first legal grounds for comprehensive control by municipalities of sanitary conditions in cities of New York State. As a result, New York City formed a Metropolitan Board of Health. It was the first modern municipal health authority in the United States, and had power to enforce sanitary laws and inspect properties, among other duties.[12]

The first case that upheld zoning authority against Constitutional challenge was *Village of Euclid, Ohio v. Ambler Realty Co.*, 272, U.S. 365 (1926). Euclid was a suburb of Cleveland, and Ambler owned nearly seventy acres in the village. Concerned about being swallowed by the expansion of industrial Cleveland, Euclid passed a zoning ordinance that restricted uses in order to prevent development from overtaking the village. Ambler Realty sued on grounds that the zoning ordinance substantially reduced their property value. The United States Supreme Court upheld the village's zoning ordinance, on grounds that the city's police power allowed ordinances zoning lands for different uses, as they have a substantial relation to public health, safety, and the general welfare.[13] Zoning was rather innovative at this time, and this decision encouraged the passage of zoning ordinances nationwide.

In the case of *Berman v. Parker*, 348 U.S. 26 (1954), the US Supreme Court, interpreting the U.S. Constitution's Fifth Amendment Takings Clause, unanimously held that private property could be taken via eminent domain for a public purpose with just compensation. Subsequent cases have often revolved around the question of what a "public purpose" is, how broadly it should be defined, and when eminent domain may be used against an owner for a larger purpose, such as a redevelopment plan.[14]

Berman v. Parker was scrutinized again in the notorious 2005 case of *Kelo v. City of New London* 545 U.S. 469 (2005), in which the Supreme Court clarified its earlier ruling and allowed eminent domain in a situation where the city's goal was economic development rather than removing blight. A closely divided court reasoned that community benefits from economic growth made private redevelopment plans permissible as a "public use" under the Takings Clause. The Court essentially held that the City of New London, Connecticut, could "take" private property—homes, in this instance—and sell it to a private developer, presumably under a redevelopment plan that would produce greater economic benefits.[15]

Fear of possible corporate abuse from *Kelo*'s expansive interpretation was based on the facts of the case. The City of New London condemned private property so it could be used as part of a comprehensive redevelopment plan. The city's goals were to increase tax revenues and to replace lower-middle-class homeowners with higher-income residents. But despite the favorable decision, the redeveloper could not obtain financing, so the redevelopment project collapsed, and the entire property was still vacant as of 2014. Pfizer, the large corporation that was to anchor the "urban village" redevelopment, merged with another company, combined its facilities, and closed its New London research campus, resulting in the loss of over a thousand jobs.[16]

Public reaction to the *Kelo* case was highly unfavorable, coming from both politically liberal and conservative sides of the spectrum. There was widespread fear that a continuation of such legal thinking would benefit large corporations at the expense of small homeowners or communities weak in political power, such as the elderly and minorities. Subsequent to the decision, 44 states amended their laws to prohibit use of eminent domain for economic development, except to prohibit blight.

Zoning and Regulation: Diverse Approaches toward a Goal

There are several kinds of zoning in the United States. The most prevalent is probably the so-called Euclidean or "building block" zoning, which segregates land uses into districts or regions, and restricts the development allowed on lots in a given district. For example, a district might be zoned for "single family residential," where building is restricted to detached houses on lots of one acre or less, while high-rise developments or clustered housing is not allowed. This was the initial kind of zoning that took hold in the United States, beginning with the country's first comprehensive zoning ordinance in New York City in 1916. It divided the city into multiple land-use districts, with specific uses allowed in each district.[17] The law originated in 1915 when the Equitable Building was built, rising 42 stories above the street and casting a seven-acre shadow over the neighborhood. Clearly, effective height and setback controls were needed. The original law has been drastically amended many times since to keep pace with a quickly changing city.

"Performance" zoning uses goal-oriented criteria to establish parameters for proposed projects. This allows the municipality and the landowners to hone a proposed development toward meeting established zoning goals. It can be flexible and accountable, which explains its popularity. Unlike traditional zoning, performance zoning is seldom arbitrary. On the other hand, its flexibility can make it tough to implement, or it can simply dissolve into meeting the perceived needs of the developer.

"Incentive" zoning, on the other hand, was developed to provide a rewards-based system to encourage land uses that could meet established urban development goals, such as affordable housing. It, too, provides much flexibility, and so can be very complex to administer, because the incentives to meet the goals often need to be regularly revisited

FIGURE 10.1
Philadelphia skyline. Photograph by ErgoSum88 (https://commons.wikimedia.org/wiki/File:Philadelphia_skyline.JPG).

and frequently revised to provide appropriately tailored incentives to developers. This kind of zoning was designed for developers in such large cities as Chicago and New York.

Most recent is "smart growth" zoning, which aims to overcome problems of urban sprawl, deterioration of inner cities, and the fragmentation of local habitats and ecosystems. Master planning of an area is a frequent tool of smart growth zoning and is used to develop a community consensus. Mixed use, rather than the old segregation of uses, is also increasingly common, aiming to revitalize commercial and neighborhood areas. Smart growth operates on principles of affordability, clustered development, and protection of critical environmental areas by focusing on community livability. This, in turn, requires identifying environmentally sensitive areas *before* development begins, frequently as part of the initial planning process. Such identification can be done via environmental impact assessments in states having a State Environmental Policy Act,[18] or through the requirements of wetland delineation, riparian zone mapping, and similar tools.

Communities can then protect sensitive areas via regulation, acquisition, conservation easements, transfer of development rights, and other tools. Properly implemented, this kind of sustainable zoning protects natural resources and provides more livable built environments. There are more and more local environmental laws, and requirements for environmental protection in comprehensive plans, watershed plans, and subdivision plats, as well as the creation of special "conservation districts," where one or another type of environmental protection takes precedence over other land uses. Open space and park requirements are now frequently found in local and state land use requirements nationwide. The interaction between traditional zoning and these environmental land use laws is a growing and important area of land use regulation[19] (Figure 10.1).

The Problem of Environmental Pollution and Local Regulation

A new kind of problem arose, however, when large-scale environmental regulation became necessary, because environmental problems do not stop at the borders of towns, counties, or states. Before this, local governments and states with oversight power—which Oregon, for example, created with its comprehensive land use planning system of 1973—regulated all zoning. But in order to control the pollution of air and water, hazardous wastes, and other trans-boundary environmental problems, a longer governmental reach was necessary. Provisions of the U.S. Constitution provided the means.

The federal government has powers that touch on the question of private ownership and the communal good. Article I, Section 8 of the U.S. Constitution gives the federal government the power to regulate interstate commerce.[20] As defined by decades of court cases, this power allows the federal government to regulate issues between two or more states concerning trade and navigation—though always with regard for the welfare of the public. A large number of bedrock national environmental laws are based on the power of the Commerce Clause,[21] beginning with the National Environmental Policy Act of 1969.[22] Subsequently, Congress passed the Clean Water Act of 1972,[23] the Coastal Zone Management Act of 1972,[24] the Endangered Species Act of 1973,[25] and the Comprehensive Environmental Response, Compensation and Liability Act of 1980.[26] This last statute was passed to erect a structure for cleaning up hazardous waste sites and thereby prevent future contamination, not only by assigning liability to relevant parties but also by requiring them to pay for the cleanup.

As a result of the new layer of federal environmental regulation, states, counties, and municipalities developed a detailed, complex network of federal guidance, federal grants, local control with federal oversight and governmental assistance contingent on the requirements of local zoning being put into place. For example, the National Flood Insurance Program of 1968[27] provides federal insurance for property owners damaged by flooding—but only if the supervising local government adopts and enforces building regulations in federally delineated floodplains.[28] Even though often criticized for being too lax, National Flood Insurance Program regulations have helped limit unregulated building in dangerous flood-prone areas.

The Coastal Zone Management Act of 1972 provides grants to states for planning, thereby encouraging local governments to adopt regulations and plans that comply with federal principles of resource protection. The Oregon Coastal Management Program,[29] for example, is funded primarily through federal monies supplied by the Coastal Zone Management Act. Again, though the federal standards are imperfect, they have had some effect in curbing inappropriate uses in fragile, vulnerable coastal zones. In all these statutes, there is some federal preemption of state and local land-use controls, just as states have partially preempted local ordinances with statewide comprehensive land-use laws, such as Oregon's.

I (Cameron) work in protecting the Oregon coast. Requirements of the Coastal Zone Management Act are met by Oregon's land use system of state laws implemented by the coastal counties and cities, through local ordinances and comprehensive planning. However, the federal standards for development projects are not very robust, which hampers efforts to halt highly damaging projects, such as the liquefied natural gas export facility that was proposed for the southern Oregon coast nearly a decade ago, originally as an import facility.

The Jordan Cove Energy Project in Coos County would be built on a sand spit, which is in both a tsunami zone and the earthquake-prone Cascadia subduction zone. The sand spit protects the towns of Coos Bay and North Bend—more than 40,000 people—from the open ocean.

No worse place could be imagined for a liquefied natural gas facility. The project also includes a proposed pipeline of more than 230 miles to pass under Coos estuary, over the Coast Mountain Range and the Cascade Mountains, to end in the far-distant town of Malin (near Klamath Falls in south central Oregon), where it would link to another pipeline.

Yet, the Oregon Department of Land Conservation and Development (the state land use oversight agency) considers compliance with the Coastal Zone Management Act as nothing other than making sure the company has all its many permits. Basically, the Act has degenerated into a giant checklist of requirements. This view is enormously frustrating to activists and residents who are looking for a comprehensive review of the potential massive effects to the coastline and coastal communities from the project.

Rights of Nature and Zoning: The New Paradigm We Must Build

Pope Francis, in his encyclical *Laudato Si*, clearly outlined some of the main tenets of a land use system that would be humane, and protect the Earth, our common home. He wrote:

> The extreme poverty experienced in areas lacking harmony, open spaces or potential for integration, can lead to incidents of brutality and to exploitation by criminal organizations. In the unstable neighbourhoods of mega-cities, the daily experience of overcrowding and social anonymity can create a sense of uprootedness. ...There is also a need to protect those common areas, visual landmarks and urban landscapes which increase out sense of belonging ... within a city which includes us and brings us together. ...For this same reason, in both urban and rural settings, it is helpful to set aside some places which can be preserved and protected from constant changes brought by human intervention ... An integral ecology is inseparable from the notion of the common good.[30]

The encyclical goes on to explore the notion of the common good in detail, including respect for the human person, the family, and social peace and stability. Furthermore, "The notion of the common good also extends to future generations. ...We can no longer speak of sustainable development apart from intergenerational solidarity."[31]

Caring for the necessities of future generations is central to both *Laudato Si* and Nature's Rights. Indeed, the Rights of Nature constitute the bedrock from which to achieve the vision laid out in the *Laudato Si*, because it is the sole framework where humans and all of Nature are respected and placed in the only social system that can create the ultimate, common good whereby humans are allowed to flourish.

Major Issues in Zoning, Planning, and the Rights of Nature

Let's look at a few major issues in zoning and planning, in both rural and urban areas, that can be redesigned in light of an overarching framework of Nature's Rights. With Nature's right to flourish as the fundamental goal, many zoning problems become much easier to solve.

Building in Areas with Urban Infrastructure

In past eras, a community's history was delivered from one generation to the next so the community would know itself through time. But, with the stage set by the postwar housing industry, things began to change noticeably, as corporate depersonalization commenced its insidious cancer-like growth into the heart of community. Roads became bigger, straighter, faster, and increasingly went through prime agricultural land to connect shopping malls. Then came larger and larger subdivisions with cheaper and cheaper ticky-tacky tract housing, some of which was constructed in floodplains or on unstable soils.

I (Chris) remember, for example, looking at a house in Las Vegas, Nevada, in 1990, that mirrored what took place in my hometown after the war. As I looked around the kitchen of a newly constructed house, I noticed that the counters were literally pulling away from the wall. I asked the builder if the situation was going to be corrected. "Nah," he said, "somebody will buy it."

Centralization had arrived on the landscape as it had earlier in corporations. Driving on superhighways became a necessity, and with it came pollution of air and water, which increased with every extra mile that had to be driven and every additional automobile on the road. And the gentle motion and the relaxed pace of the traditional street gave way to ever-increasing speed. As author Jean Chesneaux observed, "The street as an art of life is disappearing in favour of traffic arteries. People drive through them on the way to somewhere else."[32] There is no word in English with a positive connotation for going slowly or lingering on streets as a way to participate in a sense of community.

One landmark of the past resides in historic buildings. Moreover, an area can be designated a historic district when enough old buildings remain. The buildings record a snapshot of a particular era in the community's history. Buildings erected in those earlier days were meant to endure and paid homage to history through their design, including elegant solutions to the age-old problems posed by the cycles of weather and light. They paid respect to the future because they were consciously built to endure beyond the lifetimes of the people who designed and constructed them,[33] in what James Howard Kunstler calls "chronological connectivity."[34]

We have rejected both the past and the sustainable future since 1945, says Kunstler, a repudiation that is plainly manifest in our graceless buildings, each constructed to disintegrate within a few decades. This consciously built-in decline—euphemistically termed "design life"—may last 50 years. Since today's buildings are expected to serve only our era, we seem unwilling to expend money or effort for their beauty, maintenance, or their service as storytellers to the generations of the future.[35]

Beyond that, few so-called "developers" care about those elegant solutions to the problems created by weather and light. This process of disconnecting from the time continuum of the past, through the present, into the future and from weather and light diminishes us spiritually, impoverishes us socially, and destroys the time-honored cultural patterns we call community.

What we neglect we lose, be it a house, a street, a downtown, or community itself, along with the sustainability of its surrounding landscape. Communities are neither meant to be disposable nor designed in terms of planned obsolescence, which translates into increasingly shorter life span for a "throw-away" society. Only to the extent true community is built and cherished does social-environmental sustainability become possible.

Limiting Urban Sprawl: A Key Aspect of Nature's Rights

Zoning law and the policing powers represented by zoning ordinances are a means for letting people know where they can build, what they can build, and what sorts of activities are allowed. Zoning to limit development from encroaching into open space is based on the general understanding that the more contiguous and well-designed an open space system is, the better it can be protected in a relatively natural state as part of the landscape matrix of the region, such as riparian areas, floodplains, wetlands, grasslands, and forestlands.[36]

The need for such protection is exemplified by the effect indigenous species experience between the degree of habitat connectivity and the degree of habitat fragmentation in a given landscape. Fragmentation of ecosystems is perhaps the most common, worldwide alteration of landscapes, one that generally results in a mosaic of habitat "islands" of different sizes surrounded by and within a matrix of agricultural and urban development.

The ongoing incursion of this cultural matrix into undeveloped habitats causes fluctuations in the landscapes' receptivity to solar radiation, as well as the movement of wind, water, and nutrients. These changes have variable influences on the biota within and among remnant areas and between the edge of a habitat island and the surrounding matrix. Among birds, for example, the rate of nest predation is nearly twice as high along abrupt, permanent edges of all types as it is along more gradual edges, where plant succession is allowed to occur.[37]

The retention of open space can also be justified by understanding the human need for contrast in one's environment, such as quiet spaces to serve as a welcome relief from the built environment. Additionally, a provision of open spaces within and around an urban setting typically enhances property values.[38]

Birds in urban landscapes generally occupy parks, some of which may be analogous to forest fragments, whereas tree-lined streets form linear corridors that connect the fragments within the urban matrix. To understand the species-habitat dynamics of an urban setting, a study conducted in Madrid, Spain, examined the effects of street location within the urbanscape, the vegetative structure along the streets, and the human disturbance (pedestrian and automotive) on bird-species richness within the street corridors. In addition, the birds' temporal persistence, density of feeding and nesting guilds, and the probability of a street's being occupied by a single species were also accounted for.

The number of species increased from the least suitable habitats (streets without vegetation) to the most suitable habitats (urban parks) with tree-lined streets being an intermediate landscape element. Tree-lined streets that connected urban parks positively influenced the number of species within the streets' vegetation, as well as species persistence, population density, and the probability that the individual species would continue to occupy the streets. Nevertheless, human disturbance did exert a negative influence on the same variables.

Wooded streets could potentially function as corridors that would allow certain species to live well by supporting alternative habitats for feeding and nesting, particularly those birds that feed on the ground and nest in trees or tree cavities. Local improvements in quality and complexity of the vegetation associated with certain streets, as well as a reduction in the disturbance caused by people, could exert a positive influence on the regional connectivity of streets, as a system of urban corridors for birds. Because of the differential use of corridors by species with various habitat requirements, streets as habitat corridors could be further improved by taking the requirements of different species into account.[39]

First, there is the resident community or that group of birds inhabiting the area to which they have a strong sense of year-round fidelity. In order to stay throughout the year, year after year, they must be able to meet all of their ongoing requirements for food, water, shelter, open space, and privacy. Then there are the summer visitors that overwinter in the southern latitudes and fly north to reproduce. They arrive in time to build their nests, rear their young, and in so doing must fit in with the yearlong residents without competing severely for food, water, shelter, space, or privacy—especially during nesting.

There are also winter visitors that spend the summer in northern latitudes, where they rear their young, and fly south in the autumn to overwinter in the same area as the yearlong residents, but after the summer visitors have left. They must also fit in with the yearlong residents without severely competing for food, water, shelter, space, and privacy during times of harsh weather and periodic scarcities of food.

FIGURE 10.2
Tree-lined street in Savannah, Georgia. Photograph by Censusdata at English Wikipedia (https://commons.wikimedia.org/wiki/File:Treelined_street_in_Savannah,_Georgia.jpg).

On top of all this are the migrants that come through in spring and autumn on their way to and from their summer nesting grounds and winter-feeding grounds. They pause just long enough to rest and replenish their dwindling reserves of body fat by using local food, water, shelter, and space, to which they have only the passing fidelity necessary to sustain them on their long journey.

The crux of the issue is the carrying capacity of the habitat for the yearlong resident community. If food, water, shelter, space, and privacy are sufficient to accommodate the yearlong resident community, as well as the seasonal visitors, winter visitors, and migrants, then all is well. If not, then, in effect, each bird in addition to the yearlong residents causes the area of land and its resources to shrink per resident bird (Figure 10.2).

New Roads Inside and Outside City Limits

Writers James O. Wilson and James Howard Kunstler argued, in the online magazine *Slate*, that, "we have transformed the human ecology of America, from sea to shining sea, into a national automobile slum."[40] Bill Bishop, editorial page columnist for the *Herald-Leader* in Lexington, Kentucky, meanwhile, wonders if we just "can't remember any other way to live?"[41]

If a city chooses to design its transportation system around ecological constraints, it will place the components of the system where they will best honor the integrity and connectivity of the available habitats, including the city's interface with its surrounding landscape. When a transportation system is planned around ecological *effectiveness*, the probability of being able to have a relatively good system of open spaces is greatly increased.

On the other hand, if a city chooses to design its transportation system around economic *efficiency*, available habitats will suffer far greater fragmentation than if an open space system itself had driven the city's planning and implementation. Under the efficiency mode, open space, as a sustainable system, will be foregone because fragmentation of habitat is inevitably maximized, as are noise and light pollution.

Nevertheless, the choice between ecological constraints (with its emphasis on quality of life) or economic constraints (with its emphasis on profit margins) is seldom posed. Peter Headicar, Transport Planning at Oxford Brookes University in England, states this perpetual lack of choice eloquently. He says the basic questions about planning a transportation system within an urban setting are not often asked because "they are both politically uncomfortable and tractable only over the longer term—hence conveniently forever deferrable in the present."[42]

Protecting Sensitive Lands

Protecting sensitive lands in urban and rural settings requires an inventory of known sensitive lands, as well as an understanding of how they contribute to the landscape's biophysical functioning in that region. With that in mind, let us explore some land use principles in greater detail, and then focus on a few of the better-known sensitive areas. Last, it is necessary to describe some of the tools available to protect these lands.

Constraints on Building in Sensitive Areas

The land surrounding a community's urban area gives the community its ambience. The wise acquisition of open spaces in the various components of the surrounding landscape, whether Nature's or the culture's ecosystem, protects, to some extent at least, the uniqueness of the community's setting and hence the uniqueness of the community itself.

Moreover, a well-implemented system of open spaces helps ameliorate the cumulative effects of a concentrated human population on its immediate surroundings. Open space also connects people with the land and its variety of habitats and life-forms. Most importantly, open space, as the nonnegotiable constraint around which a community plans and carries out its development, allows both roads and people to be placed in the best locations from a sustainable point of view, both environmental and socially.

Riparian Zones

Riparian areas can be identified by the presence of vegetation that requires flowing or standing water and conditions more moist than normal. These areas may vary considerably in size and the complexity of their vegetative cover.

Riparian areas have the following things in common: (1) they create well-defined habitats within much drier surrounding areas, (2) they make up a minor portion of the overall area, (3) they are generally more productive than the remainder of the area in terms of the biomass of plants and animals, (4) wildlife use riparian areas disproportionately more than any other type of habitat, and (5) they are a critical source of diversity within an ecosystem.

In addition, riparian areas supply organic material in the form of leaves and twigs that become an important component of the aquatic food chain. They also supply large woody debris in the form of fallen trees that, in turn, form a critical part of the land/water interface, the stability of banks along streams and rivers, and instream habitat for a complex of aquatic plants, as well as aquatic invertebrate and vertebrate organisms.[43] Moreover, riparian areas close to coastal shores supply large driftwood critical to the oceanic ecosystem.[44]

Riparian areas are also an important source of large woody debris (= driftwood) for the stream or river whose banks they protect from erosion.[45] Further, these areas are periodically flooded in winter, which, along with floodplains, is how a stream or river dissipates part of its energy. Such dissipation is an important function. Without it, floodwaters would cause significantly more damage than they already do in settled areas.

Floodplains

A floodplain is a low, flat area of land that borders a stream or river and is subject to flooding. Unfortunately, a river's floodplain is sometimes viewed as completely separate from a river's active channel.

Like riparian areas, floodplains are critical to maintain as open areas because, as the name implies, they frequently flood. These are areas where storm-swollen streams and rivers spread out, decentralizing the velocity of their flow by encountering friction caused by the increased surface area of their temporary bottoms, both of which dissipate much of the floodwater's energy and prevent erosion along the river's banks.

In addition, when inundated, floodplains act as natural filters that remove excess sediment and nutrients that would otherwise degrade the water quality and increase treatment costs for municipalities that get their potable water from the river. As well, they allow water to percolate into the ground, where it can replenish the underground water (or aquifers), which serve not only as a primary source of water for many communities but also are critical for the irrigation that grows much of the world's food crops[46] (Figure 10.3).

Wetlands

Wetlands are those areas where water is at or near the surface of the soil for varying periods during the year, including the growing season. They vary widely because of local and regional differences in soils, topography, climate, hydrology, water chemistry, vegetation, and other factors, including human disturbance.

It is the soil's saturation with water ("hydric" soils, from the Greek *hydr* ["wet"]) that largely determines the types of plant and animal communities that can live in and on it and how the resulting wetland develops. In addition, the prolonged presence of water both promotes the development of a characteristic wetland and creates the conditions for specially adapted plants (hydrophytes, from the Greek *hydr* ["water"] and *phyt* ["plant"]) that grow in waterlogged soil or partly or totally submerged in the water.

Wetlands are important features in the urban landscape wherever they occur. Some of the functions they provide include protecting and improving water quality, providing fish and wildlife habitats, storing floodwaters, and maintaining surface-water flow during dry periods. These valuable functions are the result of the unique, natural characteristics of wetlands, and are thus vital to the landscape matrix.[47]

FIGURE 10.3
Aerial photograph of Red River's meandering course and floodplains in Lafayette and Miller counties, Arkansas. Photograph by USDA Farm Service Agency (https://commons.wikimedia.org/wiki/Category:Floodplain #/media/File:RedRiverMeandersArkansas1.jpg).

Geohazard Areas

"Geohazard" is the term for a geological phenomenon that has the potential to create widespread damage. Typical geohazards include earthquakes, volcanoes, landslides, mudflows, floods, snow avalanches, sinkholes, and tsunamis.

For example, a catastrophic landslide, such as the one near Oso, Washington state, in March 2014, which killed 43 people living in a rural residential community on the North

Fork Stillaguamish River, is a particularly acute reminder that an unstable geological landscape combined with triggering human activities can have deadly effects. In this case, an entire hillside sloughed off and deposited tons of debris 30 to 70 feet deep over a square mile in the river valley (Figure 10.4). This unstable area is subject to massive landslides around every 140 years, and the combination of the geological instability, heavy rainfall, and clear-cut logging on the hillside produced the deadly slide.[48]

Every geohazard differs, based on such things as the geological composition of the land in terms of rock types or tectonic activity, as well as the socioeconomic impact an activated geological event would have. Some geohazards can have unexpected consequences from distant locations, such as a wide-reaching pyroclastic flow (from the Greek *pyr* ["fire"] and *klastóç* ["broken"]) from an erupting volcano, which can cause far more widespread damage than many people would expect.

In populated areas of known geohazards, the counsel of a geotechnical engineer is essential for the benefit of all the townspeople, to assess the potential risks and protect a community from known geohazards such as periodic flooding.[49] Some regions of the country, such as several coastal counties in Oregon subject to strong coastal erosion and landslides, require geohazard reports of developers before deciding whether to allow a development in a geohazard area.

FIGURE 10.4
A view of the Oso area mudslide of Snohomish County, Washington, that killed 43 people on March 22, 2014. The slide covered a 1-square-mile area in a rural community about 55 miles northeast of Seattle. Photograph by the U.S. Navy (https://commons.wikimedia.org/wiki/File:Oso_Mudslide_22_March_2014_Aerial_view.jpg).

Tools for Protecting Sensitive Areas

There is an array of tools available to protect land, ranging from nonregulatory options to laws that provide incentives or outright funding.[50] Three examples are: (1) outright purchase of land, (2) conservation easements, and (3) purchase of development rights.

1. *Purchase of Land:* The outright purchase of land from a willing landowner by a government agency, such as a wildlife refuge or a national park, is generally for the purpose of protecting natural resources, and may include providing public access. Land may also be purchased by a not-for-profit conservation organization, such as The Nature Conservancy, whose mission is "to conserve the lands and waters on which all life depends. Our vision is a world where the diversity of life thrives, and people act to conserve nature for its own sake and its ability to fulfill our needs and enrich our lives."[51] On the other hand, a landowner may choose to donate some amount of land to a government agency or a not-for-profit conservation organization and reap a tax advantage in return.
2. *Conservation Easements:* Conservation easements protect land for future generations while allowing owners to retain many uses of their private property as long as they live on it. Landowners have found that conservation easements offer great flexibility, while providing a permanent assurance that the land they donate will be protected from development. For example, an easement on property containing rare wildlife habitat might prohibit any development, whereas one on a farm might allow continued farming and the building of additional agricultural structures. Moreover, an easement may apply to only a portion of person's property without requiring public access.

   A landowner may sell or donate a conservation easement. If the easement is donated and the donation benefits the public by permanently protecting important resources. If it meets other requirements of the federal tax code, it can qualify as a tax-deductible, charitable donation. Perhaps most importantly, a conservation easement is legally binding, whether the property is sold or passed to heirs. By removing the potential for commercial development, the easement lowers the market value of the land, which in turn lowers the estate tax. Whether the easement is donated during a person's life or through a will, it can make a critical difference in the heirs' ability to keep the land intact.[52]
3. *Purchase or Transfer of Development Rights:* Purchase of Development Rights is an incentive-based, voluntary program with the intent of permanently protecting productive, sensitive, or aesthetic landscapes, while retaining private ownership and caretaking. In this program, a landowner sells the development rights of a parcel of land to a public agency, land trust, or unit of government. A conservation easement is recorded on the title of the property that permanently limits commercial development. While the right to develop or subdivide the land is permanently restricted, the landowner retains all other rights and responsibilities associated with the land. In some programs, the development rights, instead of being purchased, can be transferred to another more suitable parcel where development can be intensified. Programs can be designed to provide both options.

   Purchase of development rights and conservation easements do not necessarily require public access, though it may be granted as part of the agreement, or as a requirement of the funding program. Possible uses of these initiatives include the permanent protection of: (1) farmland, (2) forestland, (3) open space, (4) rural character, (5) critical habitat, (6) riparian corridors, (7) wetlands, (8) areas for groundwater recharge, and (9) storage of stormwater.[53]

Management Considerations for a Rights of Nature System

It is clear that adding the Rights of Nature to the interconnected and highly complex network of zoning and planning laws in the United States would be a huge undertaking. However, since zoning is communal by definition and thus speaks to communal necessities, its underlying principles are consonant with Nature's Rights. Once Nature's Rights are placed in federal and state Constitutions, it will become easier to adjust zoning laws so that Rights of Nature become the substrate of all local zoning ordinances. Here are some guiding principles and management considerations for Nature's Rights with respect to urban and rurally developed private lands:

1. Change local zoning laws to conform to an overarching Rights of Nature framework, with Nature as the primary rights-holder. Zoning has traditionally been based solely on human desires, whether individual or communal. In the most recent layer of environmentally based zoning there is, for the first time, a recognition of specific, environmental requirements, such as the protection of riparian areas, wetlands, and the restricted development around lakes. Nevertheless, local zoning does not take Nature's Laws of Reciprocity into account, because it is so local and fragmented. Rights of Nature will, for the first time, provide a wide-ranging framework that will guide local zoning decisions to collaborate with, rather than largely ignore, the Laws of Reciprocity.
2. Encourage and incentivize via research grants the innovative thinking to provide additional solutions for implementing Rights of Nature through land use regulation and zoning. For example, to protect a river that passes through many local jurisdictions, parties could craft a pact between the local indigenous culture, for which the river is its homeland, and the state. Zoning implemented through a community-based Master Plan approved by the state would then ensure the designated protections. This kind of solution builds on the work of New Zealand and the Maori to protect the Whanganui River's integrity.[54]
3. Seek and codify the consensus in the local region around what developments can no longer be allowed under the Rights of Nature framework because of destructive effects on the local environment's ability to function. Zoning based on Rights of Nature is essential to all people's ability to live in a sustainably productive environment, which means that some kinds of development will no longer be acceptable because of the damage they cause.
4. Maintain and strengthen the *relationship of pattern and connectivity* in the local landscape through Rights of Nature zoning decisions. An urban landscape is a mosaic of interconnected patches of habitats.[55] Stability flows from the patterns of relationship among different species. Habitat fragmentation is the most serious threat to biological diversity.[56] Thus, connectivity protects species biodiversity

and creates ecological resilience in the landscape. Modifying connectivity among habitat patches strongly influences species abundance and use.[57]

5. Design Rights of Nature zoning paradigms to incorporate "smart growth" planning goals based on principles of connectivity, community, environmental protection and ecological integrity. This gives local zoning as wide a range of place in which to situate the local decisions that affect the surrounding landscapes and watersheds.
6. Ensure that Rights of Nature state laws and municipal zoning laws based on Rights of Nature principles curb urban sprawl beyond the existing urban footprint. Urban sprawl is a major culprit in loss of agricultural lands, forestlands, grasslands, and wetlands nationwide. There are many tools for curbing urban sprawl, such as the "urban growth boundaries" required around cities in Oregon. Increasing urban density is another important tool for limiting urban sprawl. Nature cannot flourish free of human interference if an ever-increasing portion of the landscape is urbanized.
7. Ensure that Rights of Nature laws, and state and local zoning laws based on Rights of Nature principles in rural areas, curb rural sprawl—the spread of rural residential housing onto agricultural lands and relatively intact ecosystems such as grasslands, forestlands, and river valleys. Rural sprawl accelerates habitat fragmentation, intensifies ecosystem degradation, increases building of transportation and other infrastructure, speeds the loss of agricultural lands and forestlands, and intensifies watershed fragility. It also commonly increases groundwater contamination from septic and rural sewer systems, and stretches scarce water supplies.
8. Ensure that Rights of Nature laws, and zoning laws based on Rights of Nature principles, limit the spread of rural resorts. These range from hunting lodges in wilderness areas to golf resorts on coastlines, as well as rural all-purpose recreation resorts that frequently include massive housing developments. While offering outdoor amenities ranging from fishing to golf to horseback riding, these resorts fragment habitat, increase the reach of urbanization, often harm surrounding communities, frequently wreak havoc on local water supplies, and unravel ecological integrity. Existing resorts must be inventoried to see if they can be redesigned and/or shrunk in size as necessary to minimize harm to the surrounding ecosystem's integrity *as it was before any development took place.* If so, they should continue operations to ensure people have a chance to participate in the natural world of which humans are a part. Any resort too destructive of local natural processes must be fully removed and the land restored to its natural state. Create programs of development transfer and/or financial remuneration and grants to aid in this restoration process.
9. Place regulation of the public health and safety—the bedrock of local zoning law—under Nature's own right to exist free of human interference as the primary rights-holder. This places traditional zoning in the only true context that can lead to health and safety in regulating land use—as part of Nature's Laws of Reciprocity.
10. Clarify in statute that the primary standard for "takings" lawsuits is whether the Rights of Nature, and the related community rights, are well-served by an owner's proposed use. If not, either a different use must be proposed, or some kind of compensation scheme designed, at least in the transitional period between traditional

zoning and Rights of Nature zoning. Placing zoning and land use in a Rights of Nature framework will greatly reduce the rash of "takings" lawsuits that continually crop up at both state and federal levels when a landowner thinks he has not been granted adequate "economic benefit" from his land.

The National Landscape: Forests, Grasslands, and Wetlands

Nothing better illustrates the problems with implementing a Rights of Nature concept in the United States than contemplating the millions of acres covered by forests and grasslands (including deserts, prairies, and shrub steppes). These lands, owned both privately and publicly, are now and have been for more than a century heavily manipulated to suit human uses with little or no consideration for ecological cycles, biophysical principles, or ecological resilience—neither the need for it nor what is required to sustain it. In addition, both inland freshwater and coastal saltwater/tidal wetlands—some of the most ecologically productive areas in the country—are scattered throughout the national landscape, where they have been drained, impounded, filled in, and generally treated as useless swampland.

It is imperative to understand how forests, grasslands, and wetlands fit into the Rights of Nature, as outlined in the Constitution and laws of Ecuador and Bolivia, as well as the vision of the Global Alliance of the Rights of Nature. Clearly under these Rights, the requirements of resilience, diversity, regeneration, and productive sustainability of the various ecosystems that combine to create our national landscape and its life cycles must come first within our national and state Constitutions and laws.

For example, the Bolivian "Law of Mother Earth" gives Mother Earth an inherent guarantee of regeneration: "The State … and society, in harmony with the common interest, must ensure the necessary conditions in order that the diverse living systems of Mother Earth may absorb damage, adapt to shocks, and regenerate without significantly altering their structural and functional characteristics…."[58] Chapter III, Article 7, gives Mother Earth the right to maintain the integrity of living systems, to the diversity of life, to water, clean air, and equilibrium of interrelationship and interdependence, as well as restoration.[59]

Similar provisions in the Ecuadorian Constitution are briefer, more restrained, and yet broader, as befitting a Constitution under which specific laws will be promulgated. But it too provides Nature with the right to "maintenance and regeneration of its life cycles, structures and functions."[60] In addition, it explicitly gives Nature the right to be restored. An especially important provision, given the discussion in this chapter of public lands, is Article 74, which provides, in part, "Persons, communities, peoples and nations shall have the right to benefit from the environment and the natural wealth…. Environmental services shall not be subject to appropriation…."[61]

The Universal Declaration of the Rights of Mother Earth, while not yet law in any nation, provides a good model of a statute.[62] Its list of Mother Earth's rights is similar to that from the Bolivian statute, but includes such additional provisions as the right to maintain its identity and integrity as a "distinct, self-regulating and interrelated being;" the right to be free from contamination and pollution; and the right maintain its genetic structure free from any modification or disruption "in such a manner that threatens its integrity or healthy functioning."[63]

Human responsibilities are laid out in detail, including: defense, protection, and conservation of Mother Earth; rectifying damages caused by violation of Earth's rights; establishing laws in defense of Mother Earth; and establishing "precautionary and restrictive measures" to prevent human activities from disrupting ecological cycles and ecosystems. Humans must also promote economic systems in harmony with the Earth and its recognized Rights.

Every ecosystem adapts to process and novelty of change, with or without the human hand. The challenge is that our lack of understanding and thus our heavy-handedness precludes our ability to guess, much less know, what kind of adaptations will emerge, as we alter the biophysical world in the course of our living. This being the case, it behooves us to pay particular attention to the "nuts and bolts" of diversity, as both creator and protector of ecological back-up systems—the life-saving legacy embodied in the Rights of Nature that can only be fulfilled by protecting extensive natural areas in which to learn the importance of:

1. *Biological diversity:* A critical part of maintaining the Rights of Nature, especially on public lands (see discussion below), is to maintain and/or restore quality habitat based on the connectivity of habitat patches through maintenance and/or restoration of both latitudinal and elevational corridors for the seasonal migration of species, as well as the dispersal of young animals. Because it is important to know what species one is dealing with and what their habitat requirements are, periodic "presence/absence" surveys must be conducted on both aquatic and terrestrial species. In addition to these standard surveys, a special survey might at times be needed, requiring the expertise of a herpetologist, entomologist, geomorphologist, mycologist, or some other discipline.[64]
2. *Genetic diversity:* The protection of genetic diversity could be accomplished through the maintenance of native plants and natural seeding in order to maintain and/or enhance the in-place gene pool and thus an ecosystem's potential adaptability to environmental change, like that of climate. Because global climate change, such as warming, may force plants and animals to migrate upward in elevation, it is critical to maintain the necessary habitat connectivity along potential migration routes.
3. *Functional diversity:* Functional diversity is an outcome of species composition and the physical structures that composition creates and maintains in space through time. Functional diversity under a Rights of Nature system would be determined in part by the aggregate of human manipulations across the landscape, but human use would emulate the mosaic of vegetation patterns throughout the ecosystem, be it a forest, grassland, or wetland.
4. *Disease and Pests:* It is important to understand the role of disease and animal pests (both invertebrate and vertebrate) in the creation, modification, and maintenance of ecological diversity. This understanding is critical because disease and animal pests are often seen as doing little more than damaging some desired commodity—"stealing" short-term economic gain from those who would profit.

   The flipside is that, in the long term, every living system is to some extent self-nurturing by reinvesting biological capital in itself. This means part of every living system is always in the process of dying, even as other parts are in the process of living. Within this cyclical nature of flow and ebb, give and take, life and death, each system tends to retain and recycle the elements of its dying and dead components to nurture those of its living and growing components.

For example, declining and dead trees account for much of the ongoing diversity of ecological processes within a forest, processes that are largely hidden from sight and continually changing; processes that often require decades or centuries to complete, such as the addition of vital organic material to the soil. In turn, declining and dead trees, both standing as snags and fallen, are important habitat over time for a changing clientele of plants, as well as invertebrate and vertebrate animals.

Further, declining and dead trees open gaps in the canopy that let more light into the forest and stimulate a diversity of light-loving plants to grow that, in turn, add to the habitat diversity of a particular place in time and space. Fallen trees also add vertical relief to the forest floor that arrests soil from creeping down slope with the pull of gravity and aid the infiltration of water into the soil, where it is purified and stored over long periods.[65]

5. *Ecological Backups:* Ecological backups create system resilience. In turn, biological diversity, genetic diversity, and functional diversity, as well as the diseases and pests that help keep an ecosystem in the cyclical motion of living and dying, are the bedrock of such backups. Hence, if diversity (biological, genetic, and functional) is accounted for in the patterns of vegetation and habitat across the landscape, ecological backup systems will take care of themselves.
6. *Natural-Area System:* Beyond the ecological diversity discussed above, there are in every landscape special features that add to the overall diversity of an ecosystem. These include such things as cliffs, caves, talus slopes, and edaphic (= soil) habitats with endemic plants (those limited to small areas in geographical distribution), to name a few. Because these areas have a singular distinctiveness within the landscape, they require special care and must be maintained within a system of protected "natural areas."[66]
7. *The Importance of Pattern:* From such repositories, in addition to monitoring human-caused changes and maintaining habitat for particular species, it is possible to learn how to maintain and/or restore biological processes in various portions of a given ecosystem. In Western thought, economic and ecological systems are perceived to operate on different time scales, meaning that the long-term, detrimental effects to the environment caused by decisions made in favor of short-term profits are ignored. For this reason, it is important to remember, when considering diversity and its stabilizing influence, that it is the relationship of *pattern*, rather than numbers, that confers stability on ecosystems.[67] To create a sustainable, culturally-oriented system, even a very diverse one, it is necessary to account for these co-evolved relationships, or the system has about as much chance of succeeding as has a sentence made out of so-called randomly selected words.[68]

Lands in Public Ownership

The United States is well placed to initiate the wide-ranging changes envisioned in existing Rights of Nature language because the federal government oversees vast tracts of the country owned by the public. For example, approximately 8.5 percent of the United States is National Forest land, in 155 National Forests adding up to approximately 190 million acres. The Bureau of Land Management administers another 245 million acres of public land, more than any other agency in the country—mostly desert, grassland, or shrub steppe, with some forestland in the Pacific Northwest. Most of the grassland west of the Rockies (approximately 332 million acres) is in federal, mainly in Bureau of Land Management, ownership. In addition, the U.S. Fish and Wildlife Service is in charge of the National Wildlife Refuge system and thus responsible for approximately 150 million acres of ecologically important public lands. Furthermore, the National Park Service manages around 80 million acres of land, National Parks, and many National Monuments. Finally, many states and local governments own state forests, grasslands, waterways, wetlands, and other public lands.

As a result, the United States has a significant portion of its landmass in public ownership, which gives it major opportunities to enhance and restore degraded ecological function according to the principles of the Rights of Nature. National Parks and Wildlife Refuges are explicitly managed to protect natural values and ecosystem sustainability, along with outdoor recreation.

However, the current statutory framework governing the majority of these public lands—those managed by the Bureau of Land Management and the Forest Service—is not conducive to a Rights of Nature vision. The current framework seeks to "balance" various, often completely incompatible, human uses with almost no focus on the biophysical requirements of the lands and waters themselves for maintenance of ecological sustainability.

For example, the fundamental legislation governing the sprawling acreage managed by the Bureau of Land Management is the Federal Lands Policy and Management Act of 1976.[69] The cornerstone of its vision is "multiple use," defined in the Act as "management of the public lands and their various resource values so that they are utilized in the combination that will best meet the present and future needs of the American people."[70] The Act contains many details, including giving the Bureau power to designate wilderness areas. But, fundamentally, the statute's vision is management for human use, and the management structure is oriented to allow extractive activities such as grazing, oil and gas leasing, and logging alongside non-extractive uses, such as wildlife conservation and wilderness areas.

The Forest Service statutes are similar, though more complex. The first statute to use the term "multiple use" and seek to balance uses on National Forest lands was the Multiple Use Sustained Yield Act of 1960,[71] which listed the purposes of National Forests as including outdoor recreation, range, timber, watershed, and fish and wildlife. The bedrock governing statute for the Forest Service is the National Forest Management Act of 1976, which has been amended and expanded several times.[72] This statute requires the Forest Service to develop a management program based on multiple use and implement resource-management plans for each Forest.

In addition, all the federal land-owning agencies are subject to the major environmental statutes. These include the National Environmental Policy Act of 1969,[73] requiring federal agencies to integrate environmental values into decision-making processes, and analyze environmental effects of proposed actions. Another important statute is the Endangered Species Act of 1973[74] that lays out the process of identifying threatened and endangered species, and provides protections for them. Federal statutes protecting clean water[75] and air[76] and providing for improvement in air and water quality, also apply on all federal public lands.

Perhaps most importantly for the Rights of Nature, federal public lands are subject to the Wilderness Act of 1964,[77] the statute creating the National Wilderness Preservation System.[78] The initial lands protected under the Act totaled 9.1 million acres, much of it administratively protected by executive order. Currently, the wilderness system protects about 109 million acres of federally-owned public lands in 44 states. Some National Parks and National Wildlife Refuges, but not all, are part of the wilderness system. The Act's definition of wilderness is framed in human terms, but specifies that Nature is the primary actor: "A wilderness, in contrast with those areas where man and his own works dominate the landscape, is hereby recognized as an area where the earth and its community of life are untrammeled by man, where man himself is a visitor who does not remain."[79]

To be designated as wilderness, a parcel of public land has to be at least 5,000 acres in size, and have minimal human imprint, including no, or minimal, roads. Logging, as well as oil and gas drilling, are prohibited in wilderness areas. As the Act states, "the earth and its community of life are untrammeled by man."[80] Motor vehicles and invasive research are also prohibited, but some extractive uses were grandfathered in, such as mining and grazing. The 1964 Act did not include the vast acreage managed by the Bureau of Land Management as possible for wilderness designation, but that changed in 1976, and Bureau lands can now be considered for wilderness designation.

Lands in Private Ownership

Forestlands

The largest swath of forestland in the United States is in private hands. The majority of privately owned forestland is in the eastern part of the country, though there are substantial private holdings in the West as well, in addition to the public forests. According to the 2011–2013 U.S. Forest Service survey, families and individuals own 61 percent of private forests, averaging approximately 66 acres of forest per owner. Corporations own 28 percent of the forests nationwide, averaging 775 acres of forest—including well-known companies like Weyerhaeuser or Plum Creek Timber that own thousands. The majority of corporate-owned forestland is in the South. Two-thirds of corporate ownerships own land for "investment purposes." Native American tribes, conservation organizations, or other groups own the final 6 percent of privately held forestland. A majority of these ownerships cluster in the Western states.[81]

Grasslands

The majority of grasslands are also privately owned, and their use tends to be split between intensive agriculture and grazing land. Pre-settlement grasslands constituted nearly half of the landmass of the 48 states, mostly west of the Mississippi, with more than half of those in the Great Plains east of the Rocky Mountains. Some 90 percent of those grasslands are in private hands today.

Of those grasslands west of the Rockies in private ownership, approximately 80 percent are pasture and rangeland in partnerships or family-held corporations. The vast majority of private grassland owners in the West possess 6,000 acres or more.

Grasslands have been heavily altered by human activities, especially agriculture and grazing, which have led to habitat fragmentation, problems with invasive species that reduce habitat diversity, and the severe habitat changes introduced by fire suppression. Moreover, grasslands are exceedingly important to the livestock industry, with nearly 95 percent of the grazing land needed for beef cattle lying in the Great Plains and Western part of the country.[82]

Wetlands

There continues to be a devastating loss of wetlands nationwide. In the early 1600s, when English and Dutch colonists first set foot in what was to become the coterminous United

States, there were between 221 and 224 million acres of wetlands—primarily in three regions: the Midwestern states (27 percent), the Southeastern states (24 percent), and the Delta and Gulf states (24 percent). The greatest loss of these wetlands occurred between colonial times and the early decades of the 20th century, with most occurring since 1885. Approximately 103 million acres remained as of the mid-1980s—six states having lost 85 percent or more of their original wetland acreage, whereas 22 states had lost 50 percent or more. By 1992, the estimated acreage of wetlands was down to 124 million acres in the contiguous 48 states (including an approximately 12 million acres on public lands). This decline represents a 55 percent loss since 1780, almost half (45 percent) of which is due to agricultural uses.[83]

The loss of wetlands is continuing for a variety of reasons, including agricultural conversion, diversion of water for irrigation, overgrazing, and urbanization. Some coastal wetlands have been dredged for shipping; others have been drained and filled for resort, industrial, and residential development.

Take Alabama, for example. Wetlands currently cover around 10 percent of the state and range in size from small areas of less than an acre to the 100,000-acre forested tract in the Mobile Tensaw River Delta. Most of the forested wetlands are bottomland forests in alluvial floodplains. (An alluvial floodplain is a largely flat area created by the deposition of water-borne sediment, called "alluvial soil," over a long period by one or more rivers coming from highland regions.) Coastal waters, on the other hand, support extensive salt marshes.

Wetland acreage in the area that is now Alabama has been reduced by around one-half in the last two centuries. Major causes for the loss of wetlands have been agricultural and forestry-related conversions in the interior; dredging on the coast; industrial, commercial, and residential development; erosion; and natural succession of vegetation.[84]

Forestland, Grassland, and Wetland Mismanagement

The management of forests, grasslands, and wetlands since European settlement has focused on human requirements and desires. Grazing, ranching, logging, mining, shipping, and similar uses have defined the management paradigm. This has begun to change, however, especially on federally owned lands, where statutes at least require "multiple use." Although multiple use mandates are still framed in terms of human needs, they do include goods such as wildlife, clean water, and Nature-based recreation, leading to management goals that—at least minimally—approximately parallel biophysical necessities that allow Nature to function as an ecosystem. Laws such as the Endangered Species Act increase the focus on holistic ecosystem management by federal agencies.

Private lands are another matter. Many, but not all, woodland owners maintain their woods primarily for recreation and wildlife purposes rather than commercial timber. But corporate forest management, which covers vast acreages, is extremely timber-oriented. It generally focuses on an agricultural style of management that features extensive clearcutting and replanting to a monoculture of fast-growing species—a "tree farm." These plantations are in turn managed on short rotations, generally less than 70 years, to produce timber. Forest values are generally ignored or minimally considered, as required by governing laws. In addition, urban conversion is a highly important, emerging threat to private forestlands nationwide.

Privately owned grasslands also suffer heavily from habitat degradation owing to intensive crop use and/or intensive grazing, which can include both simple overgrazing the native habitats and planting foreign species, such as crested wheatgrass, preferred by livestock.

In making decisions about patterns across the landscape, it is important to consider the consequences of these decisions in terms of future generations. Although the current trend toward homogenizing grassland with monocultures of crested wheatgrass or the forested landscape with tree farms may make sense with respect to maximizing short-term profits, it bodes ill for the long-term, biological sustainability and adaptability of the landscape. Moreover, intensive agriculture-style manipulation of grasslands and forestlands, as well as draining wetlands, are highly destructive, economic expedients that emulate nothing in Nature and strip Nature of any fundamental rights to flourish and maintain life processes. These practices must, therefore, be unequivocally eliminated from all land management.

Forestlands

The newest, and one of the gravest, threats to forest integrity in the United States is conversion to urban, suburban, or rural residential uses, thus not only accelerating habitat fragmentation but also permanently destroying the ecosystem services the forests can provide to all of Nature and to humans. These include clean drinking water, wildlife habitat, carbon sequestration, and scenic views, among others. These conversions also reduce an ecosystem's fundamental resilience and ability to recover after natural or human-made disasters.

For example, a 2009 study showed that people tend to migrate toward counties with more public forestland, resulting in conversion of adjacent, private forestland to residential parcels. The study reported, "If trends continue, substantial amounts of private forest land will shift from a low-density rural character to a more densely populated exurban categorization."[85] Conversion rates could be as high as 6 percent of fourth-order watersheds—each watershed approximately one million acres in size—having 10 percent to 20 percent of its forest area converted to converted urban use by 2030.[86]

Apart from urban and rural residential development on forestlands, intensive management (including public forests) for tree farming remains a menace to forest ecosystems nationwide. Unfortunately, it has been crystallized in American forestry by verbal sleight-of-hand in cornerstone statutes that hide crucial assumptions highly damaging to forest integrity. As mentioned above, the first major forestry law governing federal forests was the Multiple Use Sustained Yield Act of 1960. It enshrines an important misuse of language based on an economic assumption that is totally at odds with ecological reality. The erroneous assumption is that ecological processes in a forest ecosystem remain constant, even as we humans strive to maximize whatever product or amenity seems immediately desirable. The Act defines several terms related to timber-cutting volumes and calculations. For example, the term "non-declining, even flow" means a *sustained level of cut* based on the *existing* "inventory" of commercial-aged, standing trees (volume of wood)—even though that level of cut is not "biologically sustainable" over the long term, when the forest as a whole is taken into account. The discrepancy in the use of and interpretation of "sustained" versus "sustainable" is the fundamental flaw with the Multiple Use Sustained Yield Act.

To help focus on the important fact that all relationships are inseparable, dynamic, and eternally novel, we will consider: (1) the salmon's life cycle; (2) the symbiotic relationship among truffles, squirrels, and forest trees; (3) converting a forest to a tree farm; and (4) forestry mismanagement in Europe.

The Salmon's Life Cycle

To help the reader understand how salmon integrate the biophysical system between forest and sea, I (Chris) am going to tell a story about salmon in the Columbia River Basin of the Pacific Northwestern United States, which drains more than a quarter million square miles of land. Although the focus here is on salmon in the Pacific Northwest, the Atlantic salmon of the Eastern seaboard has the same general life cycle (Figure 10.5).

A flash of silver, a swirl of bright water, a female salmon flexes her tail against the swift current as she propels herself to a small gravel bar just under the surface in the headwaters of a Pacific Coast stream. Again, a flash of silver, then another, and another as other salmon press against the rush of clear, cold water, each seeking the exact spot to which its inner drive to spawn impels it.

Suddenly, from somewhere in the shadow of trees overhanging the tiny stream, there comes a large, magnificent male; he swims alongside the female with powerful undulations. They touch, and the female immediately turns on her side and fans the gravel with strong beats of her tail.

She continues spraying gravel into the current until she creates a shallow depression, after which she begins depositing hundreds of reddish-orange eggs, as the male squirts milky-white sperm into the water. The cloud of sperm, enveloping the eggs as the current carries it downstream, fertilizes them as they settle into the shallow "nest."

Now, she and her mate, having fulfilled the inner purpose of their lives, swim into deeper water, where they rest and die.

But in the gravelly stream bottom lies an orange, opaque egg inside of which a salmon is developing. In time, the baby salmon hatches and struggles out of the gravel into the open water of protected, hidden places in the stream. Here it grows until it is time to leave the stream of its origin and venture forth. It can go only one way—downstream to larger and larger streams and rivers until at last it reaches the ocean, all the way beset by increasing dangers to overcome.

On its way to the ocean, however, the young salmon depends on large driftwood (frequently termed "large-wood debris") that accumulates in the streams and rivers as

FIGURE 10.5
Adult male Chinook salmon. Scanned from plates in: Barton Warren Evermann and Edmund Lee Goldsborough. The Fishes of Alaska. United States Government Printing Office, Washington, DC (1907) (https://commons.wikimedia.org/wiki/File:Chinook_Salmon_Adult_Male.jpg).

instream habitat to protect it from both the swiftness of the current and predators. In addition, large, well-anchored pieces of wood also help stabilize the stream's channel, increasing the predictability of its configuration from year to year.

Only after some years at sea will the inner urge of individual salmon dictate their approaching time to spawn. Remembering the Earth's magnetic field where they entered the ocean years earlier, they use it to navigate the open waters, as they return to their home rivers, which they identify by the river's chemical signature. In so doing, salmon will differentiate into identifiable, freshwater populations that are reproductively isolated from one another, each with its affinity to a particular river. Once in the river, they will again separate into discrete subpopulations, each with its own affinity to a particular stream within the river system. A salmon's ability to find its spawning area depends on hidden genetic guidance that leads it back to its home waters when the time to spawn finally arrives.

As the dead salmon wash into the shallow water along the edge of the stream's banks, the elements of their bodies become concentrations of nutrients and energy that subsidize the forest that helped nourish them as fertilized eggs. This massive infusion of decomposing salmon in the forested stream promotes the growth of algae and bacteria that help sustain aquatic insects.

Juvenile salmon, steelhead, and cutthroat trout also poke around the expired, rotting bodies, eating the eggs left in the females and, eventually, picking off pieces of flesh. This huge addition of nutriments is critical for the young salmon because the rich banquet of dead fish enables youngsters to double their weight in about six weeks. The added body weight greatly increases the chances that a particular fish will survive to swim the gauntlet from the stream of its origin far out into the North Pacific Ocean and return again years later to spawn in the place it was hatched.

As a carcass decomposes underwater, its dissolved nitrogen and carbon are soaked up by algae and diatoms, which are one-celled plants that form a scum on the gravel and rocks, which in turn is grazed by aquatic insects that will become food for the salmon that will hatch the next spring. In addition, the birds and mammals that feast on the carcasses, such as golden eagles, Steller's jays, ravens, wrens, spotted skunks, river otters, raccoons, bears, foxes, mice, and shrews, deposit their droppings on the forest floor.

The upshot of this great infusion of nutrients is that the scum on the gravel and rocks and the plants along the stream's banks, including trees, suck up the nitrogen from the rotting salmon because nitrogen is an element in short supply in the soils of the Pacific Northwest. Sitka Spruce in southeast Alaska, for example, take only 86 years to reach a trunk thickness of 20 inches when fed by the decomposing carcasses of spawned-out salmon, as opposed to the normal 300 years to reach that trunk thickness without the benefit of the salmon carcasses. The forest plants then drop their leaves, needles, and twigs into the stream, providing more food for the aquatic insects and ultimately the young salmon, as well as shade for young fish in which to hide, eat, and grow.

Nature's feat of nourishing the plants and animals requires about one salmon carcass per every 3 square feet of stream edge. This can be roughly translated into approximately one dead salmon for the amount of water that would today fill a standard bathtub.[87]

There is today, however, a serious problem for salmon in forested areas (both public and private), namely, roads and culverts that impede their access to the spawning and rearing areas—a requirement for all wild salmon. Thousands of miles of spawning streams are unavailable throughout the Pacific Northwest because of improperly installed culverts at numerous points along the region's seemingly inexhaustible network of roads. These culverts were designed and their installation was supervised over many years by a variety of engineers who thought about roads—not salmon. In Oregon alone, more than 4,000 faulty culverts are preventing salmon and steelhead from accessing roughly 8,000 miles of streams with good spawning habitat, a situation that diminishes the overall populations.[88]

These culverts not only prevent the salmon from reaching their spawning streams but also deprive the forest and myriad animal life of the marine-derived nitrogen, phosphorus, and other nutrients that nourished the spawned-out salmon before their deaths. This nutrient deficit is one indication of an ecosystem in trouble, of a negative, cumulative effect. The cumulative effect of each improperly installed culvert progressively eliminates more and more spawning habitat, adversely affecting the salmon population's ability to maintain its biological viability—*a negative cumulative effect* that results from thinking strictly about the interaction of road construction and water.[89]

The Symbiotic Relationship among Truffles, Squirrels, and Forest Trees

Truffles are an important food of both Douglas squirrels[90] and northern flying squirrels that inhabit coniferous forests of the Pacific Northwest.[91] Truffles are the reproductive bodies of belowground-fruiting fungi, termed hypogeous fungi (from the Greek *hupo* ["under"] and *geo* ["earth"]). These fungi form a mutually beneficial symbiotic relationship with roots of certain plants, via the fungi's mycorrhizae (from the Greek *mykós* ["fungus"] and *riza* ["roots"]—literally "fungus-root"). Woody plants such as pine, fir, spruce, larch, Douglas fir, hemlock, oak, birch, and alder in particular depend on mycorrhiza-forming fungi for nutrient uptake. This phenomenon can be traced back some 400 million years to the earliest known fossils of plant rooting structures.

The host plant, say Douglas fir, provides simple sugars from photosynthesis and other metabolites to the mycorrhizal fungi, which lack chlorophyll. Fungal hyphae (the thread-like mold, from the Greek *hyphé*, ["web"]), in turn, penetrate the tiny, non-woody rootlets of the host plant to form a balanced, harmless mycorrhizal symbiotic relationship with the roots. The fungus absorbs minerals, other nutrients, and water from the soil and translocates them into the host. Further, nitrogen-fixing bacteria that occur inside the mycorrhiza use a fungal "extract" as food and in turn fix atmospheric nitrogen. (To "fix" nitrogen is to take gaseous, atmospheric nitrogen and alter it in such a way that it becomes available and useable by plants.) Nitrogen thus made available can be used both by the fungus and the host tree.

In effect, mycorrhiza-forming fungi serve as a highly efficient extension of the host root system. Many of the fungi also produce growth regulators that induce production of new root tips and increase their useful life span. At the same time, host plants prevent mycorrhizal fungi from damaging their roots. Mycorrhizal colonization enhances resistance to attack by pathogens. Some mycorrhizal fungi produce compounds that prevent pathogens from contacting the root system.[92]

Truffles are the initial link between belowground mycorrhizal fungi and the two kinds of squirrels, both of which nest and reproduce in the tree canopy and come to the ground, where they dig and eat truffles—the Douglas squirrel during the day and the flying squirrel at night. As a truffle matures, it produces a strong odor that attracts the foraging squirrel.[93]

When the squirrels eat truffles, they consume fungal tissue that contains nutrients, water, viable fungal spores, nitrogen-fixing bacteria, and yeast—all become important in the forest ecosystem. The fate of the squirrel's fecal pellets varies, depending on where they fall. In the forest canopy, the pellets might remain and disintegrate in the treetops, or a pellet could drop to a fallen, rotting tree and inoculate the wood. On the ground, a squirrel might defecate on a disturbed area of the forest floor, where a pellet could land near a conifer feeder rootlet that may become inoculated with the mycorrhizal fungus when spores germinate. If environmental conditions are suitable and root tips are available for colonization, a new fungal colony may be established.

Thus, both the Douglas squirrel and the northern flying squirrel exert a dynamic functionally diverse influence within the forest. The complex of effects ranges from the crown of the tree, down through the surface of the soil into its mantle where, through mycorrhizal fungi, nutrients are conducted through roots, into the trunk, and up to the crown of the tree, perhaps into the squirrel's own nest tree.[94]

Such relationships are by no means confined to the Douglas squirrel and northern flying squirrel. Many mammals, such as deer mice and white-footed mice,[95] red-backed voles,[96] jumping mice,[97] chipmunks,[98] mantled ground squirrels, western gray squirrels, and other rodents, depend more or less on hypogeous fungi for food.[99]

This is only a tiny glimpse of total complexity of a Pacific Northwest forest. The whole squirrel-fungal-forest feedback loop is complex beyond imagination. To put it simply, the squirrels eat the truffles, the truffles germinate and feed the trees in which the squirrels live, and the trees feed the fungi that feed them. In this sense, the squirrels take care of the trees and the trees take care of the squirrels—with the mycorrhizal fungi taking care of both, as the intermediary. Moreover, this small mammal, hypogeous fungal, and in tree interdependence has been documented in various small mammals of the high cloud forest of Argentina,[100] the coniferous and mixed conifer-hardwood forests of Europe,[101] and the eucalyptus forests of Australia.[102] Considering these intricate, complex relationships and the Rights of Nature, the question is: do trees and animals have rights?[103]

Converting a Forest into a Tree Farm

Native forests are being cut down at an accelerated rate, often through illegal logging, with exceedingly little understanding of how they function, especially in terms of environmental sustainability they influence over time. People connected to the timber industry, however, are proceeding as though they know what they are doing: moving from complex, diverse, native forests designed by Nature toward grossly simplified, uniform, tree plantations designed by humans solely to meet economic desires. The reasons given for these actions, such as jobs and community stability, do not alter the fact that forests are in jeopardy for lack of patience with Nature's design and time frame[104] (Figure 10.6).

According to the research of Suzanne Simard, forest ecology professor at the University of British Columbia, the largest trees in an old-growth forest provide the greatest amount of nutrients and soil stability, as well as the largest and most intricate mycorrhizal network, which provides seedlings with a much larger pool of nutrients than they could create on their own:

> In the shaded forest understory, conifer seedlings depend on carbon and nutrients from the old giants through these networks...the fungus, too, depends on interplant mychorrizal networks to secure a plant carbon source in new generations. As the young trees mature, they themselves become hubs for nurturing subsequent generations of seedlings. ... Loss of these hubs to high-grade logging thus ultimately affects recruitment of old-growth trees that provide habitat for "nest-webs" of cavity nesting birds and mammals, and thus dispersed seed and spores for future generations of trees.[105]

But, owing to the symptomatic thinking within the timber industry, a forest is thought of and "managed" for its conversion potential into monetary capital in a way that simulates

FIGURE 10.6
Two clearcut, first-order streams (headwater streams) that join shortly above the road to form a second-order stream in the Oregon Cascade Mountains. (USDA Forest Service photograph.)

modern agriculture. Tree farms are simplified and intensified with no knowledge of the long-term consequences, straining to have ever more of everything simultaneously, but at a terrible cost—loss of soil fertility; reliance on chemicals that pollute the land, air, and water; and vast acreages that will not again grow a forest in the foreseeable future.

As "forest management" intensifies, it comes closer and closer to merging with intensive agriculture, particularly in three respects: (1) an increasing attempt to purify and specialize the crop trees, (2) an increasing move toward monoculture, which decreases biophysical diversity and therefore ecological sustainability, and (3) an increasing view of plants and animals that exert any perceived negative effect on crop trees as pests.[106]

This view necessitates an increasing outlay of monetary capital, time, energy, and such materials as fertilizers, herbicides, pesticides, plastic tubing to protect seedlings, and so on. Intensified management ensures that innumerable biological processes—most not understood in the timber industry—will be viewed as management-created competition, which conflicts with production goals and will call for continued artificial simplification of the forest. The most ubiquitous and irreversible environmental problem society already faces worldwide is the loss of biological diversity through the growing extinction of species and their environmental services—usurping the Rights of Nature.[107]

The transition from Nature's forests to economic tree farms is marked by many changes and accompanied by grave ecological uncertainties. Not only do species of plants and animals but also "grandparent trees" become extinct with the conversion of the old-growth forests to tree farms. As young trees replace liquidated old trees in crop after crop, the ecological functions performed by the old trees, such as creation of the "pit-and-mound" topography on the floor of the forest, with its mixing of mineral soil and organic top soil, become extinct processes because there are no more grandparent trees to blow over.

The "pit" in pit-and-mound topography refers to the hole left when a tree's roots are pulled from the soil, and "mound" refers to the soil-laden mass of roots, called a rootwad, suddenly projected into the air above the floor of the forest. The tree-farm seedlings that replace the grandparent trees are much smaller and are different in structure. They cannot perform the same functions in the same ways.

Uprooted trees enrich the forest's topography by creating new habitats for vegetation. Falling trees create opportunities for new plants to become established in the bare mineral soil of the root pit and the mound. In time, a fallen tree itself provides a habitat that can be readily colonized by tree seedlings and other plants. Falling trees also open the canopy, which allows more light to reach the floor of the forest. In addition, pit-and-mound topography is a major factor in mixing the soil of the forest floor as the forest evolves.

In addition, water moves differently over and through the soil of a smooth forest floor that is devoid of large fallen trees to act as reservoirs, storing water throughout the heat of the summer and holding soil in place on steep slopes. The huge snags (standing dead trees)[108] and fallen trees that acted as habitats are gone, as are the stumps of the grandparent trees with their belowground "plumbing systems," which guided rain and melting snow deep into the soil, where it is retained during the hotter months.

This plumbing system of decomposing tree stumps and roots comes from the frequent formation of hollow, interconnected, surface-to-bedrock channels that drain water rapidly from heavy rains and melting snow. As roots rot completely away, the collapse and plugging of these channels force more water to drain through the soil matrix, which reduces soil cohesion and increases hydraulic pressure that, in turn, causes mass soil movement.[109]

Clear-cutting and subsequent tree farming is completely oblivious to all these subtle and intricate relationships that create the forest. This short-term, economically driven management scheme is also completely at odds with the forest ecosystem's ability to flourish and maintain its life cycles, the fundamental rights integral to a Rights of Nature system.

Forestry Mismanagement in Europe

Errors in the practice of forestry in Europe over the past couple of centuries illustrate well the results of ignoring ecological realities while attempting to maximize short-term profits, based on the "certainty" of economic assumptions. Coupled with the false certainty is the failure to recognize that a forest is ultimately controlled by Nature's Laws of Reciprocity principles, rather than by human desires, scientific knowledge, or technological ingenuity. This failure has resulted planting the "wrong" type of trees in Europe.

Prior to 1750, the mid- and low-elevation forests in central Europe were historically either deciduous hardwoods or mixture of deciduous hardwoods and coniferous softwoods.

These forests consisted of European beech and English oak, often occurred together in these forests with common hornbeam and European linden, as the main species of trees. Silver birch was also found in many areas in the north. Oak grew especially well on the more nutrient-rich soils, establishing excellent stands of trees where climatic conditions were favorable. The beech, with its more modest requirements of site and climate, covered large areas. And many different species of trees, such as European maple, grew along the river valleys.[110]

Europe's green forest canopy was dramatically thinned between 1750 and 1850, when the forested area diminished by 73,359 square miles. Thereafter, a greater use of fossil fuels, particularly coal, slowed the reliance on timber, and from 1850 to the present day, Europe's forests have grown by some 150,000 square miles and now cover 1 percent more land than before the industrial revolution. However, the forested areas of today differ dramatically from the original forest, because conifer forests have expanded by 244,000 square miles while broadleaved forests have shrunk by 168,000 square miles.

In the Europe of today, some 85 percent of the trees are managed by people who, over the past 150 years, have adopted a scientific approach to tree farms, where they plant such fast-growing, commercially valuable trees as Scots pine and Norway spruce.[111] Today's exploitative forestry practices in Europe are once again based on the precepts of tree farming. This concept holds everything in Nature as a constant value, except the age at which the trees can be harvested one can calculate length of a "rotation" (say 80 years from planting to harvest) that will give the highest rate of return on the economic capital invested.

The economic decision to select conifers over broadleaved varieties, such as hornbeam and birch, has had a significant impact on the forest's albedo effect—the amount of solar radiation a given surface reflects back into space. The lighter the surface, the more reflective it is—hence the trade-off. The exchange of broadleaved trees, such as beech and maple, which have lighter leaves that reflect more of the sun's radiation back into space, with such conifers as pine and spruce that are clothed in darker needles and thus absorb more of the sun's radiant heat, caused the continent's summer temperatures to increase by 0.2° Fahrenheit.

According to Kim Naudts, of the Max Planck Institute for Meteorology in Hamburg, Germany, "Two and a half centuries of forest management in Europe have not cooled the climate. The political imperative to mitigate climate change through afforestation and forest management therefore risks failure, unless it is recognized that not all forestry contributes to climate change mitigation."[112]

There has been some recognition of this failure in Germany, based on historical errors in the practice of forestry. Richard Plochmann, a professor at the University of Munich and District Chief of the Bavarian Forest Service, commenting on German Forestry in 1989, indicated that forestry would henceforth include mixtures of two or three species of trees, as opposed to the historic, single-species monocultures. In such mixes, at least one species must be indigenous.

Further, the ages of at which trees are cut will depend on their highest value: quality wood as opposed to the historic, inferior, fast-grown wood. This will require rotations of 120 to 140 years. Wherever possible, natural regeneration, as opposed to genetically selected or manipulated seedlings from nurseries, will be used. Therefore, clear-cutting will be replaced by shelter woods in which some of the mature trees are left to reseed the area, cutting trees in small groups, or selectively harvesting trees. In addition, herbicides will no longer be used, and insecticides and fertilizers will be used only rarely.

Finally, there will be no highly mechanized operations within the plantations as they are brought closer and closer to the physical structure and biological functions of a real forest.[113]

There are other recent indications that an understanding of the importance of forests is growing. In June 2016, Norway became the first country in the world to implement a policy of zero deforestation. This includes a pledge that the government's public procurement policy would not contribute to deforestation of the world's rainforests. Though individual corporations have pledged to free their supply chains of rainforest-damaging products, Norway is the first government to do so. This follows on the heels of a joint Norwegian declaration with Germany in 2014 to promote deforestation-free supply chains and seek sustainable sources for products, like beef, palm oil, and timber. Tropical countries often cut their forests to produce these products, leading to exceedingly high deforestation rates. Norway is a leader in funding both forest conservation programs worldwide and human rights programs for forest communities.[114]

Grasslands

America's grasslands have been heavily managed for profit, whether via conversion to agriculture, housing subdivisions, or use as rangelands for cattle or other herbivores.

Grasslands declined by about 260 million acres by 1950, most having been turned into cultivated cropland—frequently encouraged by federal agricultural policies. From 1950 to 1990, approximately 27 million more acres of grassland were lost, some to cropland, the rest to other uses, such as livestock grazing. Only a tiny sliver of this once-vast ecosystem remains intact; it is the most highly affected by human activity of all North American ecosystems.[115]

Though traditionally the greatest threats to grasslands have been intensive grazing and agriculture, today the pressure of housing subdivisions, "ranchettes," and consequent suburban growth, looms as another danger to ecological integrity of the remaining grassland habitat.[116] However, the largest single threat to privately owned grasslands remains conversion to cropland, responsible for about five times as much conversion as urban expansion or development. New uses, such as cropping for biofuel, have added pressure for cropland conversion. For example, between 2008 and 2012, in excess of 6 million acres of grassland were converted to cropland—an area the size of Maryland. Most new cropland tends to be on marginal land with severe limitations to cultivation. This means higher rates of erosion and soil loss, as well as lower yields.[117]

It is difficult to estimate the trends, since some rangelands have significant grassland characteristics, especially in the Western states, where many of the ecological grassland functions are partially or totally lost with the introduction of livestock grazing.

The effects of grazing on the arid systems of the world depend to a large extent on the evolutionary history of grazing animals. If large herbivores were present before the introduction of cattle, the effects of cattle grazing, generally speaking, seem to be smaller, allowing native grasslands a better chance of becoming self-sustaining.

The native grasslands of the arid West had been without large, grazing herbivores for 10,000 years before Europeans introduced cattle. The Great Plains grasslands, on the other hand, had been grazed by herds of bison, and have had a less traumatic experience from cattle grazing. Nevertheless, the impacts on grasslands have still been severe because cattle and bison use the ecosystem very differently—the bison having evolved with the grasslands[118] (Figure 10.7).

FIGURE 10.7
North American bison (buffalo) on the range in Crook County, Wyoming. Photograph by Ron Nichols for the U.S. Department of Agriculture (https://commons.wikimedia.org/wiki/File:Buffalo_USDA94c4147.jpg).

Grasslands, Riparian Zones, and Cattle

Riparian zones—the vegetation zones adjacent to streams and rivers—are highly productive biological areas throughout the arid West. Though they make up only half a percent to 2 percent of the landscape, their ecological value is incalculable. In the Great Basin and the Intermountain West, approximately 85 percent of the native wildlife depend on riparian zones for all or part of their life cycles. In Oregon and Washington, about 71 percent of the native species of wildlife depend on these rich riverine wetlands. Native fish are especially dependent on riparian zones to maintain the water quality necessary for their life cycles—temperature, nutrient composition, water free of sedimentation, production of complex stream habitats.[119]

Unfortunately, cattle favor riparian zones for the same reasons as wildlife—shade, high-quality food, and water: "Without controls on animal numbers, timing and duration of use, cattle can rapidly and severely degrade riparian areas through forage removal, soil compaction, stream bank trampling, and the introduction of exotics"[120] (Figure 10.8). This pattern of overuse especially affects streamside forests of cottonwood, aspen and willow, which have the highest densities of breeding songbirds in the West.[121] In 1990, the U.S. Environmental Protection Agency found that "extensive field observations suggest that riparian areas throughout much of the West are in their worst condition in history."[122]

In addition, cattle urine and manure in and near streams greatly elevate the levels of nitrogen and phosphorus above those needed by the stream for healthy functioning. This nutrient concentration, combined with the denuding of vegetation, depletes soils and disrupts aquatic life cycles.[123]

What is the best long-term solution to this cumulative degradation of riparian zones? Halting cattle grazing permanently, difficult as it may be politically. Simply ending cattle

FIGURE 10.8
Cattle grazing shortgrass rangeland in northern Colorado. Larimer County, Colorado. Photograph by Jeff Vanuga, USDA Natural Resources Conservation Service (https://photogallery.sc.egov.usda.gov/res/sites/netpub/server.np?find&catalog=catalog&template=view.np&field=itemid&op=matches&value=1605&site=PhotoGallery).

grazing on the publicly managed arid lands of the West would enormously boost ecological productivity on millions of acres. In many cases, the streams can recover their natural function on their own. Where restoration is needed, land managers, observing the functioning of sustainable riparian areas, can give a helping hand. The first step, after removing cattle, is usually the replanting of riparian vegetation, which improves water quality and slows channel degradation.

Native Grasses and Mycorrhizal Symbiosis

Most native arid and semiarid grassland plants form a mycorrhizal symbiosis with certain fungi through which the host plant absorbs water and nutrients from the soil. In a study of 575 mammals, 16 genera and 26 species (all from areas of native vegetation outside of the seedings, including the black-tailed jackrabbit, white-tailed jackrabbit, and mountain cottontail rabbit) were found to eat hypogeous, mycorrhizal-forming fungi, representing 15 genera, and thereby become a significant means of dispersing the fungal spores. These fungi form a mutually beneficial symbiotic relationship with roots of certain rangeland plants, where they serve as a highly effective extension of the host root system.

Many of these fungi also produce growth regulators that induce production of new root tips and increase their useful life span. At the same time, a host plant prevents its mycorrhizal fungus from damaging its roots. Mycorrhizal colonization enhances resistance to attack by pathogens. Indeed, some mycorrhizal fungi also produce compounds that prevent pathogens from contacting the root system. Moreover, these fungi are dispersed throughout prairie and other ecosystems by such organisms as earthworms and small mammals, which eat the belowground fruiting bodies and defecate the viable spore onto and within the soil as they move about.[124]

In addition to the likelihood of an exotic plant's being mycorrhizal-free, how closely related it is to the native species will largely determine not only how invasive it is but also how likely it is to undermine the repair of a relic piece of prairie. The more closely related an alien is to the indigenous plants, the less invasive it is likely to be, whereas the less related it is, the more invasive it will be.[125]

The Problem with Crested Wheatgrass

Crested wheatgrass was imported during the drought-stricken Dust Bowl years of the Great Depression to augment livestock forage because local prairie grasses were failing to provide sufficient fodder for the cattle. The invasion of wheatgrass was sealed when farmers discovered that it withstood drought and overgrazing, had a long growing season, and made good hay. Today, through the unflagging efforts of ranchers and government agencies such as the Bureau of Land Management, wheatgrass covers 25 million acres of prairie and shrub-steppe north of Mexico. It is not, however, the problem-free panacea for the livestock industry that it is trumpeted as being. The vast acreages of wheatgrass harbor subtle, long-term problems, some of which require decades to become apparent, by which time they are out of control.

An immediate, ecological difficulty lies in the propensity of wheatgrass to devote most of its energy above ground to the production of shoots, while maintaining only a meager root system. Indigenous grasses, in contrast, do not grow as tall, but they form prodigious networks of roots, which anchor soil in place and enrich it with nutrients and organic matter. As a result, soil in wheatgrass plantations contains significantly fewer nutrients and less organic matter than soil in native grasslands.[126]

In the prairie remnants in the Loess Hills of Iowa, grazing by domestic livestock promoted the greatest overall species richness, whereas grazing and burning resulted in the lowest cover by woody plants. Burning by itself, however, achieved the best overall increase in the cover and diversity of native species while simultaneously reducing exotic forbs and grasses, the latter being predominantly cool-season in habit, such as cheatgrass.[127]

In contrast, livestock grazing appears to be an exotic, ecological force in grama grasslands of southeastern Arizona, destructive to certain components of native flora and fauna. The destructiveness of livestock grazing there may result from the absence of extensive grazing by indigenous ungulates in the Southwest since the Pleistocene. The tolerance of particular grasslands to their use by domestic livestock may depend on their historic association with native grazing animals, combined with the pervasiveness of certain exotics like cheatgrass.[128]

Unlike the indigenous bison of the Great Plains, cattle are an exotic species that was introduced into North America more than two centuries ago. Whereas habitat and food items are partitioned among coexisting native herbivores (like bison, pronghorn antelope, and elk) in a pristine prairie ecosystem, domestic cattle are much more generalist in their use of habitat and their foraging.[129]

The cover of perennial grass has declined in many arid types of grassland over the past two centuries, while shrub density has increased. These changes, which are characteristic of desertification, are thought to have occurred most often after prolonged periods of intense grazing by domestic livestock. At many such sites, however, the subsequent removal of livestock for up to 20 years did not increase the cover of grasses.

To understand the time required for grasses to recover in historically arid grasslands dominated by shrubs, vegetation was examined at two desertified sites that differed in the length of time they had been free of livestock. There was little noticeable difference between vegetation at the site from which livestock had been fenced out for 20 years and the shrub-dominated vegetation just outside the exclusion fence. Nevertheless, there was a significantly higher cover of perennial grasses in the area from which livestock had been removed 39 years earlier, and all the increase had occurred within the last 20 years.[130]

It thus seems that perennial grasses in historic grassland ecosystems dominated by shrubs require a period of 20 or more years to recover from grazing once the domestic livestock have been eliminated. The grasses' requirement of two decades or more for their recovery from livestock grazing is a beautiful illustration of the dynamic shared by all ecosystems: cumulative effects, lag periods, and thresholds. It also demonstrates our limited powers of spontaneous observation, which can be thought of as the snapshot effect.

Wildlife and Keystone Species

As grasslands are converted into cropland or intensive grazing land, their value for wildlife and biodiversity—with the associated ecosystem services ranging from clean water to healthy bird populations—plummets. Keystone species, whose presence has wide-ranging effects in shaping the ecosystem, are especially vulnerable in intensive management regimes. However, with innovative partnerships and courageous leadership, these relationships can be repaired and remain sustainable.

The black-footed ferret is a perfect example. These ferrets lived in prairie dog colonies across the Great Plains and Intermountain West from Canada to Mexico, in populations tens of thousands strong. But the ferret was brought to the uttermost red line of extinction by habitat destruction and disease, until in 1986 only 18 known animals remained in the wild. Black-footed ferrets eat prairie dogs, and require large prairie dog colonies across thousands of acres to retain viable populations. Thus, sustainable ferret populations also signal healthy prairie dog populations and prairie ecosystems. The ferrets are now being bred and released back to the wild, numbering a few hundred and living in 27 locations within their former range in Canada, Mexico, and eight U.S. states. The population is slowly increasing.[131]

The plight of the greater sage grouse is another instructive example. Livestock grazing is probably the principal threat to sage grouse survival, except for the permanent removal of all habitat value via conversion to agriculture or housing.[132] Sage grouse are especially iconic because of their fascinating mating ritual, in which males congregate at leks (ancestral mating grounds) and engage in elaborate dancing and calling displays to attract hens.

Sage grouse need sagebrush to survive, which they use for food and cover. But they also require tall grasses as cover from predators, and seek different habitats, depending on the stage of the life cycle. Nesting habitat, for example, requires a sagebrush overstory but a thick understory of grasses. Wet meadows and riparian areas are important as the chicks grow. Consequently, livestock grazing affects sage grouse in multiple ways: trampling of nests and eggs, grazing of the grasses and forbs preferred by sage grouse chicks, increasing the vulnerability for the takeover of invasive plants, and creation of an intensive ranching infrastructure of fences, power lines, watering facilities, dams, and winter pasturage.[133]

In addition to livestock grazing, the sagebrush steppe continues to be subjected to many uses that threaten the sage grouse, including mining, hydraulic fracturing, suburbanization, and off-road vehicle use. Sage grouse breeding populations plummeted to about 140,000 individuals in 11 Western states as of 2002. In 2010, the U.S. Fish and Wildlife

Service designated the sage grouse as a candidate for listing under the Endangered Species Act, which galvanized collaborative efforts to aid the sage grouse and stem the accelerating habitat loss and fragmentation across ranching communities.[134]

The Natural Resources Conservation Service launched the Sage Grouse Initiative in 2010, using federal funds to aid conservation covering 78 million acres in 11 Western states. Private lands are the primary focus, with private and federal partners collaborating in a voluntary, incentive-based model to both conserve wildlife habitat and use ranchlands in less damaging ways.[135]

The results: More than 1,100 ranches in the 11 Western states are conserving 4.4 million acres of land, including reducing habitat fragmentation with more than 450,000 acres of conservation easements to limit development. Since 40 percent of the sage grouse range is on privately-owned grazing lands, the Initiative's work to improve grazing systems on 2.4 million acres since 2010 is crucially important. The Sage Grouse Initiative has also prioritized removing invasive conifers and managing invasive cheatgrass, which can trigger monster wildfires, as it is very flammable and reseeds extensively after burns.

The Service aims to spend $211 million on the Initiative through 2018, having already spent nearly $300 million since 2010. Conservation partners and landowners have contributed $128 million to the effort as well. The goal is to conserve 8 million acres of sage grouse habitat by the end of 2018, using the many federal conservation programs in the Farm Bill combined with scientific research to prioritize actions needed to reduce threats. Ultimately, the Service and its partners will create a cohesive, range-wide plan to guide conservation efforts across the range of sage grouse habitat.[136]

As a result of the Sage Grouse Initiative, the Fish and Wildlife Service decided in 2015 not to list the grouse under the Endangered Species Act at this time, since so much positive, wide-ranging, and collaborative habitat restoration in the sagebrush sea was underway. Substantial increases in sage grouse numbers will tell the many Initiative partners whether their efforts are bearing fruit—Nature's Laws of Reciprocity always provide clear feedback on the effect of human actions—or whether listing is required, as some partners believe is already the case. The U.S. Fish and Wildlife Service plans to revisit its sage grouse decision in another 5 years.[137]

Wetlands

Although it might not seem necessary to ask what defines a wetland, it is a good question, because they are not necessarily wet all the time. Instead, they comprise areas that are covered by water for varying periods, including during the growing season. They are transition zones between terrestrial and aquatic ecosystems, where the water table is either near the surface or the land is inundated by shallow water. Consequently, an area's saturation determines not only how the soil develops but also the types of plant and animal communities living there, both aquatic and terrestrial. In areas where the presence of water is prolonged, conditions favor the growth of specially adapted plants ("hydrophytes" from the Greek *hydro* ["water"] combined with *phyton* ["plant"]) and promote the development of characteristic, so-called "hydric soils" (from the Greek *hydro* ["water"] combined with *ikos* ["pertaining to"]).

Wetlands range in size from less than an acre to thousands of acres and can take many forms, from the tundra to the tropics on every continent except Antarctica. Moreover, they vary widely throughout their global distribution due to differences in regional and local soils, topography, climate, hydrology, water chemistry, vegetation, and other factors, including human disturbance.

Although unevenly distributed, they occur in every state and U.S. territory, where they are important to the nation's environment in multiple ways. Nevertheless, the current 124 million acres of wetlands in the continental United States is only approximately 55 percent of their original extent as a result of wetland draining, livestock grazing, and other destructive uses.

Interior Freshwater Wetlands

When the soil is saturated in typical freshwater wetlands, the oxygen used by the microbes and other decomposers in the water is slowly replaced by oxygen in the air. Because oxygen moves roughly 10,000 times slower through water than through air, all wetlands have one trait in common—oxygen-poor soils. As a result, plants that live in wetlands—the aforementioned "hydrophytes"—have genetic adaptations that either allow them to survive temporarily without oxygen in their roots or make it possible for them to transfer oxygen from the leaves or stem to the roots. This anaerobic condition (from the Greek *an* ["without"] + *aer* ["air"] = oxygen + *bios* ["life"]) causes waterlogged soils to have the sulfurous odor of rotten eggs.

Interior wetlands are most common on floodplains along rivers and streams, along the margins of lakes and ponds, and other low-lying areas. They also occur in isolated depressions, such as playas (which are the flat bottom of an undrained desert basin that, at times, and with sufficient precipitation, becomes a shallow lake). In addition, a wetland occurs where groundwater intercepts the soil surface or where precipitation sufficiently saturates the soil to create vernal pools. (Vernal pools are a type of ephemeral wetland that is usually associated with forest settings that seasonally flood and other times are dry, and are generally isolated from stream systems.) Inland wetlands also include marshes and wet meadows dominated by herbaceous plants, swamps dominated by shrubs, and wooded swamps dominated by trees, which not only are unique habitats in and of themselves but also are important stopovers for migrating waterfowl.[138]

Many of these wetlands are dry one or more seasons annually and, particularly in the arid and semiarid West, may be wet only periodically. The quantity of water present and the timing of its presence partly determine a wetland's ecological function within its setting. Even those appearing dry—at times for significant parts of the year, such as vernal pools—can provide critical habitat for wildlife adapted exclusively to breeding in these areas.[139]

Coastal Saltwater Wetlands

Tidal wetlands (also called tidal flats or salt marshes) are closely linked to our nation's estuaries where seawater mixes with freshwater to form an environment of varying salinities. The tidal action combines the saltwater and fluctuating water levels to create a rather difficult environment for most plants. Consequently, many shallow coastal areas are mudflats or sand flats devoid of vegetation. Nevertheless, certain grasses and grass-like plants have successfully adapted to these saline conditions and form the tidal salt marshes found along the Atlantic, Gulf, and Pacific coasts. In addition, mangrove swamps, with salt-loving shrubs or trees, are common in tropical climates, such as those found in southern Florida and Puerto Rico.[140]

Salt marshes of the Pacific Northwestern United States, for example, are densely vegetated at elevations within the annual, vertical range of regular tidal fluctuations. Plants of the salt-marsh community, although of terrestrial origin, are capable of growing in saturated estuarine sediments and of withstanding the stresses of both salinity and tidal inundation. Salt marshes have high annual rates of vegetative production, a significant portion of which is exported to the estuary as detritus (Figure 10.9).

The requisite conditions for salt marshes are lowered salinity, extensive areas of soft sediments at the level of high tide, and low wave energy. Such conditions are virtually restricted to estuaries in the Pacific Northwest, where there are no true, open coastal salt marshes.

Salt marshes are, nevertheless, important components of estuaries in the Pacific Northwest. Despite their relatively high elevation at the upper limits of tidal influence, salt marshes function as buffers, muting the effects of floods and storms by allowing turbulent waters to dissipate much of their energy over the marsh's vast expanse. Further, the tidal creeks draining salt marshes are extensively used at high tide by migratory waterfowl and juvenile anadromous fishes, especially salmon and steelhead trout. (An anadromous fish is one that spawns in freshwater, migrates to the ocean to mature, and then returns to freshwater to spawn and complete its life cycle—in other words, they must acclimate to saltwater before entering the ocean, and acclimate to freshwater on their return to rivers and streams to spawn.)

In addition, large driftwood, namely, trees, is also scattered throughout the marsh and remains in place for long periods, which allows the general level of the marsh around them to increase through the deposition of silt and the accumulation of organic matter. When

FIGURE 10.9
South Carolina intracoastal waters. Photograph by Gentry George, U.S. Fish and Wildlife Service (https://commons.wikimedia.org/wiki/File:A_view_of_the_south_Carolina_intracoastal_waters.jpg).

these trees are refloated during unusually high tides, floods, or coastal storms, shallow depressions remain in the marsh's sediment. These depressions increase habitat diversity and hold water at low tide in summer, which harbors juvenile fishes.[141]

The State of America's Wetlands

Wetlands are important to the nation's environment because they store floodwater, trap nutrients and sediment, help recharge groundwater, provide habitat for fish and wildlife, such as migrating waterfowl, and buffer shorelines from damaging waves.[142]

To glean some idea of what is currently happening to the wetlands of the U.S., we will look at a brief review of how they are faring in five states:

1. Alaska has about 170 million acres of wetlands—more than the other 49 states combined. And, more than 70,000 swans, 1 million geese, 12 million ducks, and 100 million shorebirds depend on them for resting, feeding, or nesting. The freshwater wetlands include: bogs, fens, tundra, marshes, and meadows. The saltwater wetlands include tidal flats, beaches, rocky shores, and salt marshes. Most of the freshwater wetlands are peatlands covering as many as 110 million acres. ("Peat" is a heterogeneous mixture of more-or-less decomposed plant material that has accumulated in a water-saturated wetland in the absence of oxygen. Its structure ranges from the relatively decomposed remains of plants to a fine, shapeless mass of minute particles.) Alaska's coastal wetlands are cooperatively protected and managed by local governments, rural regions, and the state.
2. California's wetlands have significant economic and environmental value, providing such benefits as the maintenance of water quality, attenuating floods and erosion, preventing saltwater from intruding into belowground freshwater aquifers along the coastal lowlands, and providing wildlife habitat. The Sacramento–San Joaquin Delta regularly harbors up to 15 percent of the waterfowl on the Pacific Flyway. California has lost as much as 91 percent of its original wetlands, primarily to agriculture. Flooded rice fields, which are converted wetlands, covered roughly 658,600 acres in the mid-1980s. Rice farmers, state and university researchers, and private organizations are cooperatively studying the feasibility of managing rice fields for migratory-waterfowl habitat. Wetland protection is identified as a goal of The California Environmental Quality Act of 1970.[143] As of 2008, the acres of rice harvested in California had dropped to 517,000 acres.[144]
3. New Mexico's wetlands cover around 482,000 acres (0.6 percent) of the state, mostly in the eastern and northern areas, where they include: forested wetlands, bottomland shrublands, marshes, fens, alpine snow glades, wet and salt meadows, shallow ponds, and playas. (Fens are a kind of mire usually fed by mineral-rich surface water or groundwater.) Riparian wetlands and playa lakes are especially valuable to migratory waterfowl and wading birds. New Mexico has lost about one-third of its wetlands, mostly attributed to agricultural diversion of water for irrigation, overgrazing, and urbanization. Other causes of loss or degradation have been mining, clear-cut logging, road construction, regulation of streamflows, and invasion by nonnative plants.
4. As of the mid-1980s, Oregon's statewide wetlands covered nearly 1.4 million acres—a decline of more than one-third over the previous 200 years. Most of the losses are attributed to agriculture, primarily in the Willamette River Valley and

Upper Klamath Basin. To improve the effectiveness and efficiency of Oregon's efforts to conserve, restore, and protect wetlands, the state has developed the Wetland Conservation Strategy,[145] which is based on the recommendations of advisory committees representing federal, state and local agencies and interest groups.

5. Washington's wetlands cover only approximately 2 percent (939,000 acres) of the state, but their benefit is both ecological and economic. Wetlands are nursery and feeding areas for anadromous fish, such as salmon and steelhead trout. Around 75 percent of the state's wetlands contain freshwater, which includes forested and shrub swamps, bogs, fens, marshes, wet prairies and meadows, vernal pools, and playas. Approximately 25 percent of the wetlands are estuarine or marine, which includes marshes, tidal flats, beaches, and rocky shores. Estimates of wetland loss in Washington range from 20 percent to 50 percent, owing to degradation from agricultural, urban expansion, placement of shipping ports and industries, logging, and invasion of nonnative plants and animals.[146]

Since European settlement, which began about 1792 and was burgeoning by 1800, estuarine marshlands of the Pacific Northwest (Oregon and Washington) have changed greatly. On the one hand, salt marshes not only have been put in a straightjacket of dikes along the rivers to disconnect them from the estuaries but also have been simultaneously ditched to drain them quickly of water. This extensive separation of salt marshes from their respective estuaries has been to gain land for agriculture and pastures for dairy cattle. Additional salt marshes have been destroyed by road building and through infilling to gain higher ground for construction of housing developments and shopping centers.

On the other hand, the rapid seaward growth of some remaining salt marshes is the result of such human activities as logging, road construction, and cultivation of the land, all of which disrupt the stability of the watersheds. Such disruption increases erosion and thus the amount of sediment carried down streams and rivers, which in turn increase the rate with which sediments accumulate in the estuaries.[147]

The Future of Wetlands: A Legal Framework

Wetlands have been greatly misunderstood in the United States, and for at least 250 years have been the target of "development" aiming to fill and put them to "productive" uses ranging from agriculture to subdivisions. The United States still loses tens of thousands of acres a year.

However, they are now understood to be regions of great biological richness, with many values ranging from flood control to water filtering and tremendous biological productivity that sustains vast legions of migrating birds. A system based on the Rights of Nature, with a focus on allowing the Earth and natural processes to flourish, must ensure that natural wetlands are protected and restored nationwide. Existing laws are helpful, but insufficient.

Federal legislation has slowly begun to turn around the problem of wetland loss, beginning with the Migratory Bird Conservation Act of 1929.[148] Especially important have been provisions, beginning in 1985 with the Federal Food Security Act,[149] that reduced or eliminated subsidies to farmers who convert wetlands to agriculture, and eliminated tax incentives for clearing land. Most recently, the North American Wetlands Conservation Act of 1989[150] provides matching grants to entities developing partnerships to undertake wetland conservation projects in North America and Mexico to the benefit of wetland-dependent migratory birds. In addition, many National Wildlife Refuges contain significant wetland acreage.

Five federal agencies play leading roles in the effort to conserve wetlands. Two of the most important are the Environmental Protection Agency and the Army Corps of Engineers, which have joint roles in wetlands protection through the Clean Water Act, Section 404.[151] This crucial portion of Clean Water Act regulates dredging and filling "waters of the United States" through artificial means, including wetlands. Another portion of the Clean Water Act, Section 401,[152] requires states and Tribal authorities to promulgate water-quality standards, under which they have the power to deny all federal permits or licenses that could harm state or Tribal waters, including wetlands. This has turned into a powerful tool for states to stop federally permitted projects a state may find unacceptably harmful environmentally.

Protecting Wetlands under Current Law

There usually are four main tools used in combination to protect wetlands: land purchase, conservation easements (to protect wetlands from being developed while remaining in private ownership), government funding (which ranges from subsidies to grants and loans), and regulations prohibiting the loss of a wetland and/or providing for mitigation. Many wetland-protection programs are innovative government and private partnerships, such as the National Estuary Program, which provides federal funding for local education and collaborative estuary protection programs around the country.[153]

But what about the surrounding lands, which are an integral part of the wetland's function? These lands are all interdependent, often in invisible ways. Protecting the wetland itself is usually not enough, although such protection has been—and continues to be—the main focus of regulatory programs. Wetlands are sensitive to their surroundings, and are influenced by many land uses that affect the quality of their waters, often from surprisingly far away. These include erosion from poor land use practices, stormwater runoff from impervious surfaces, flood storage, the application of pesticides/herbicides, overuse or pollution of groundwater, and the degradation of water quality in streams from development, road-building, agriculture, and logging. Merely protecting a wetland itself often does little to maintain its biological integrity—even though it is the crucial first step.

State and local governments usually tackle these problems with tools ranging from better land use planning and better site design of developments to control sediment, stormwater runoff programs, aquatic buffers, and restrictions on discharging pollution into waters affecting the wetland.[154] In addition, The Natural Resources Conservation Service oversees many voluntary programs encouraging, and providing incentives for, environmental stewardship of land and waters among agricultural producers.[155]

Restoration of Wetlands: What Does It Mean in Practice?

Since so many wetlands are lost annually to development of one kind or another, the United States has developed a large and complex wetland-regulatory system that uses every option to protect wetlands; yet, wetlands still succumb to the pressure of economic development. The over-arching goal of the regulatory framework is to create a "no-net-loss" of wetlands. This is an important concept because, under the Clean Water Act, real wetlands may be destroyed—but must be compensated for by the restoration or enhancement of the existing wetland or the creation of a new one.

What exactly is meant by the "restoration" of a wetland? According to the Environmental Protection Agency, "The objective is to emulate a natural self-regulating system that is integrated ecologically with the landscape in which it occurs."[156] This may include reintroducing native plants and animals, or allowing a former wetland to re-establish itself by removing dikes that drained wetlands from a pasture. Wetland restoration is supposed to be holistic, though wetland enhancement projects frequently emphasize one aspect of a wetland over others, resulting in at least some change in how the wetland functions.

However, two of the main methods of "restoration," although widely used, have much less effect on protecting wetlands than may at first be apparent. Creation of wetlands is one of these: constructing a wetland in an area that historically was not a wetland, does not have a wetland soil profile, and is not adjacent to an existing wetland. "Created" wetlands do not have a high success rate in becoming actual wetlands.[157]

"Mitigation" refers to restoration, enhancement, or creation of other wetlands to compensate for wetlands legally destroyed under the existing regulations. In other words, mitigation allows destruction of highly productive wetlands, and allows the developers to compensate by providing some wetland function elsewhere. Usually, but not always, the mitigation project should be near the destroyed wetland and provide similar functions. These programs vary widely in effectiveness, but often lead to the loss of truly productive wetlands in exchange for less viable wetlands in locations less desirable for development.[158]

Still Wetlands Disappear and Are Degraded: What Can Be Done?

It is clear that the existing regulatory framework for the protection of wetlands is a *partial* failure. It is not a complete failure, however, because wetland loss in the United States has slowed, and the many overlapping wetland protection programs have resulted in significant protection and restoration of biologically productive lands and waters.

Nevertheless, they are not enough to truly protect ecological functions of wetlands, for at least these reasons:

1. The entire wetland protection framework is a regulatory structure based, first and foremost, on human needs for wetlands. We need wetlands for agriculture, highways, land for subdivisions, water sources, fish refugia, and many other purposes. But the regulatory framework does not begin with the needs of Nature for its wetlands, and the needs of the wetlands themselves for full functioning.
2. Programs aim to protect or restore wetlands, but even purchasing a wetland does not protect it. All wetlands are affected by pollution or overuse of groundwater and streams, stormwater discharges, and land use decisions from far away. No wetland protection program can encompass these many subtle forms of degradation.
3. Mitigation programs frequently turn into shell games that lead to wetland losses without appearing to do so: a wetland is, for example, filled for development, and another non-wetland area is made into a "created" wetland, which has little or no true wetland function. A policy of "no net loss" of wetlands is meaningless if the exchange is from true wetlands to artificially created ones, or even true wetlands enhanced for human purposes.

The only way to truly protect wetlands is to adopt a Rights of Nature framework, in which the requirements of the ecosystem in order to flourish are the standard by which human desires for alterations of wetlands is measured. In a Rights of Nature system, mitigation is unlikely to have much place, because it is not focused on the necessities of the ecosystem itself. Wetland restoration and enhancement, guided by the standard of Nature's Rights, and a legal framework to enforce it, will have a role.

There are many wetland projects nationwide that have been dramatically successful in restoring degraded or lost wetland function, and many more crying out for help. We choose three here to showcase what can be done—and show the limitations of the efforts—under a regulatory framework that does not prioritize Nature's own need to flourish.

I (Cameron) have been involved in a large restoration project on the Oregon coast that aims to restore about 525 acres of coastal wetland in Tillamook County. The reason: The town of Tillamook, surrounded by diked pasturelands and managed former wetlands, has flooded regularly and seriously for years in the wet climate. Finally, a large group of stakeholders, including farmers, began collaborating on how best to restore wetland function to mitigate flooding. The project took 10 years of hard work by all parties. The result was the Southern Flow Corridor Project, funded largely by the federal government, to remove dikes, restore wetland function, improve water quality, and provide flood control by restoring some of the wetlands bordering Tillamook Bay.

Although this project looks to be an unqualified success, once again it merely restores the minimum amount of wetlands necessary to meet an immediate human goal, of flood control in this instance. The ecosystem's requirements are not primary, and no legal framework exists to change that. The following three examples illustrate similar problems on a larger scale: (1) the Everglades, (2) the Klamath Basin: the "Everglades of the West," and (3) the Limberlost.

1. The Everglades

The Everglades is an enormous, sprawling, watershed-wetland complex in South Florida covering (before drainage and infilling) some 4,000 square miles, a third of the entire Florida peninsula. The main actor is the Okeechobee River, which not only drains into the lake of the same name but also regularly spreads over the land, progressing in wide, shallow, concentrated flow, termed a "sheet flow," to Florida Bay.

Unfortunately, the Everglades became the focus of agricultural development beginning in the early 1800s. Efforts to drain portions of the Everglades were begun in 1882 by Hamilton Disston, a Pennsylvania land developer.[159] In the 1920s, many canals were constructed to drain the Everglades, and create "reclaimed" wetlands to plant vegetables and sugarcane. The Army Corps of Engineers constructed a dike around Lake Okeechobee in the 1920s after two devastating hurricanes, followed by creation of the Central and Southern Florida Flood Control Project in 1948[160] (Figure 10.10). The Project built a staggering 1,400 miles of canals and levees to control water in the Everglades, along with hundreds of pumping stations. The Everglades Agricultural Area[161] it created encompassed 700,000 acres or 27 percent of the Everglades, and the use of fertilizers by the agricultural interests created massive growth of invasive species in that part of the Everglades still in its natural state.

Urban development in South Florida proceeded apace; only about 25 percent of the Everglades—protected in Everglades National Park—remains in its natural state. The turning point came in 1969, with the battle over building a massive jetport in Big Cypress Swamp. Politics began to swing around to land protection: the federal government created Big Cypress National Preserve in 1974,[162] and Floridians approved $240 million in bonds to purchase environmentally important lands.

In the new 1970s political climate, programs of the Central and Southern Florida Flood Control Project came under increasing fire, especially its last project: to straighten the meandering, 90-mile Kissimmee River into a 52-mile channel, supplanting 45,000 acres of marshland with retention ponds and dams. Wildlife plummeted. Agriculture claimed the land and its fertilizers and pesticides washed into Lake Okeechobee. A huge algal bloom blanketed a fifth of the lake in 1986.

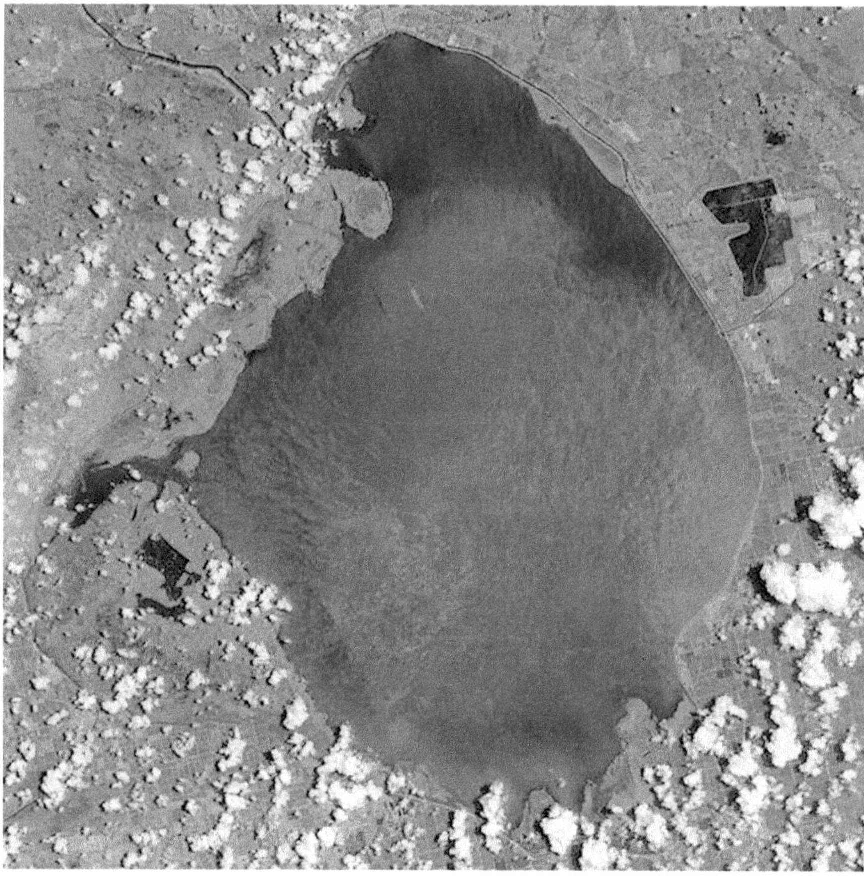

FIGURE 10.10
Lake Okeechobee, Florida. Photograph credit: NASA Earth Observatory images by Joshua Stevens, using Landsat data from the U.S. Geological Survey. Caption by Kathryn Hansen (https://commons.wikimedia.org/wiki/File:Cropped_lake_okeechobee_oli_2016184_lrg.jpg).

The degradation from this project tipped the balance politically, and in 1983, the Save Our Everglades campaign was launched. The canal channeling the Kissimmee began to be removed or backfilled, and Congress approved the Kissimmee River Restoration Project via the Water Resources Development Act of 1992.[163] The environmental damage and restoration were costly: to undo only 22 miles of the canal was projected to cost $578 million.

But it was clear by 2000 that existing restoration measures could not go far enough to protect the Everglades (Figure 10.11). Consequently, President Clinton signed the bipartisan Comprehensive Everglades Restoration Plan in 2000.[164] The new law provided $1.3 billion for immediate implementation of the total $7.8 billion cost. The state of Florida has since purchased more than 210,000 acres, a little more than half of the land needed for restoration, and spent more than $2 billion under the plan. But the Comprehensive Everglades Restoration Plan has been dogged with trouble:

1. The sugar industry managed to get the Legislature to approve an increase in acceptable phosphorus levels in 2003, and stretch out the mandatory deadlines.
2. Congress stalled in providing funding, having spent only $400 million of a legislated $7.8 billion by 2008.

FIGURE 10.11
Everglades restoration near Miccosukee Indian Village on U.S. Route 41 (facing west). National Park Service photograph (https://commons.wikimedia.org/wiki/File:Just_E_of_SV_Hwy41_Miccosukee,_NPSPhoto,_2008_(9250152394).jpg).

3. Everglades National Park is still being starved of water.
4. Rampant urban development means the state of Florida often cannot even bid on lands that should be protected for restoration.
5. Restoration projects, such as removing the C-111 canal built for farmland irrigation, have been proceeding slowly and with inadequate funding as of 2010.

As is common with wetlands, merely purchasing and restoring the wetlands themselves cannot solve the problems because wetlands are so interrelated to their surroundings. Serious ecological ills continue to plague the Everglades despite partially successful restoration efforts:

1. Severe water-quality problems continue, notably with such chemicals as phosphorus resulting from agriculture, especially the sugarcane industry. "Big Sugar," as it is called, has fought fiercely to avoid responsibility for maintaining water quality standards in Everglades National Park and surrounding areas.
2. Mercury was discovered in Everglade's fish in the 1980s. Waste incinerators and fossil fuel power plants expel mercury into the atmosphere, which returns to the Everglades in the rain or dust. Although tighter air quality controls have reduced the mercury level by 60 percent to 70 percent, it is still of concern.
3. Urban encroachment continues to eat away at the Everglades' integrity, especially in such cities as Naples and Ft. Myers.
4. Florida panther numbers continue to drop; the American crocodile, native to Florida, has been listed as endangered since 1995; wading birds such as the roseate spoonbill, great white egret, and reddish egrets have seen populations spiral downward as much as 90 percent since the 1970s.
5. Profoundly disruptive, invasive melaleuca trees, also known as paperbark tea trees, from Australia thrive in the Everglades.[165] Invasive reptiles of all kinds, ranging from Burmese pythons, which grow up to 20 feet long, to green iguana, which reproduce rapidly in wilderness habitats, also flourish in the Everglades.[166]

Even if all the projects under the Comprehensive Everglades Restoration Plan are funded and successfully completed, what will the Everglades be? The so-called Central and South Florida Project Review Study, submitted to Congress in 1999 as a preliminary to the Comprehensive Everglades Restoration Plan, admitted that the restored ecosystem "will not completely match the pre-drainage system. This is not possible, in light of the irreversible physical changes…. It will be an Everglades that is smaller…. But it will be a successfully restored Everglades, because it will have recovered those hydrological and biological patterns, which defined the original Everglades…."[167]

2. Klamath Basin: The "Everglades of the West"

The Klamath Basin stretches for more than 12,000 square miles (approximately the size of the state of Maryland) in Oregon and northern California (Figure 10.12). There are three large freshwater lakes, and a sprawling network of more than 350,000 acres of marshes, wet meadows, and rivers that once were host to massive bird migrations of the Pacific Flyway, and prolific runs of salmon.

The federal Klamath Irrigation Project, begun in 1905, sought to bring farming to the high desert country. This devastating project resulted in about 80 percent of the Basin's wetlands being drained—only around 75,000 acres remain. Most of these are protected in the six National Wildlife Refuges in the Basin. In 1908, President Roosevelt set aside 81,000 acres of marsh and open water in Lower Klamath Lake—one of the first refuges for wild birds. But the Bureau of Reclamation drained the lake in 1917, which devastated the fledgling wildlife refuge. Clear Lake Refuge, created in 1911, was gutted by a Bureau of Reclamation dam and then subsequent dams, which radically altered the water flowing to Tule Lake. The 1928 creation of Tule Lake National Wildlife Refuge

FIGURE 10.12
Aerial view of Upper Klamath Lake, near Klamath Falls, Oregon. U.S. Bureau of Reclamation photograph (https://commons.wikimedia.org/wiki/File:Upperklamathlake.jpg).

sought to protect the remnants of its once-magnificent marshes, roughly 37,000 of the original 100,000 acres.[168]

The Bureau of Reclamation's successful efforts to drain the "worthless" wetlands for agriculture emptied Lower Klamath Lake completely for two decades; it was partly revived in the 1940s, but its normal size of 100,000 acres shrank to a 13,000-acre pond with no ties to the Klamath River. The Bureau of Reclamation's dam re-plumbed the Upper Klamath Lake to provide water for the crops grown in the former Lower Klamath lakebed, so Upper Klamath's water level depends entirely on human decisions about water allocation from the dam. Upper Klamath Refuge, created in 1928, is sometimes drained dry, all 14,000 acres of its marsh, by the demands placed on it by agriculture. This has devastating consequences for wildlife and birds.

Unfortunately, after fierce debate in the 1940s and 1950s, Congress approved the Kuchel Act in 1964. This ended homesteading in the Refuges. But it also ratified leasing 22,000 acres of land in Tule Lake and Lower Klamath Refuges for commercial agriculture—as long as it remained compatible with refuge purposes.[169] Over the years, it has become crystal clear that the two purposes have nothing in common, as agriculture has further degraded the remaining marshes via massive water use, pesticides, and fertilizers in addition to being responsible for draining thousands of acres of marshland in the Refuges for potatoes, onions, and alfalfa.[170]

The wetlands of Tule Lake and Lower Klamath National Wildlife Refuges are especially important to migrating waterfowl. Bear Valley National Wildlife Refuge contains the largest wintertime population of bald eagles in the lower 48 states—often close to a thousand birds. The wetlands also provide high concentrations of waterfowl and nesting areas for various species of birds.

The Basin has been the site of a decades-long, fierce, and drawn-out battle for a dwindling water supply, with competitive demands from fish advocates, indigenous tribes, and farmers/ranchers—all of whom depend on the Basin's fragile and insufficient water supply. Irrigators are sometimes left without water, and crops and cattle die. At other times, the National Wildlife Refuges and rivers are without water, and massive salmon die-offs occur. As a result of these die-offs, commercial fishers cannot make a living due to low salmon runs. The most nightmarish salmon die-off was in 2002, when 34,000 adult salmon died before spawning.

Efforts at settlement of water differences began in the 1990s, and there has been a bewildering parade of agreements, partial agreements, breakdowns in negotiation, and urgent conclaves for many years. The conflicts became even more desperate when the Klamath Tribes were granted senior water rights in the Upper Basin in 2013—resulting in water shutoff to irrigators to protect tribal fishing rights, which led to 100,000 cattle without water.

In 2014, the parties (including the United States and the state of Oregon) settled on a draft agreement, the Upper Klamath Basin Comprehensive Agreement.[171] This built on two earlier important agreements. Removal of the four Pacificorp dams on the Lower Klamath River had been the subject of a separate earlier agreement, the Klamath Hydroelectric Settlement Agreement,[172] which laid out the financial structure for the dam removals. The parties have signed an amendment that, if approved by the regulating federal agency, will guarantee removal of the four dams by 2020, rather than requiring Congressional authorization before action could occur. Funding for the dam removals will remain split between California voters and Pacificorp customers. A third Klamath Basin Restoration Agreement[173] provided resolution of the water wars between ranchers/farmers and downstream fish-dependent communities.

The new, overarching Upper Klamath Basin Comprehensive Agreement details the retirement of water rights from willing sellers in the Upper Basin to provide more water for Upper Klamath Lake, and lays out habitat restoration requirements. Although battles could continue over tribal water rights, the Agreement seeks to settle that issue once and for all. All three agreements are interrelated, and the hope is that, by meeting the needs of all parties, the agreement can be finalized. If successful, it would represent a blueprint for river restoration on a grand scale, similar to that attempted for the Everglades. However, Congressional action is necessary to finalize the Upper Klamath Basin Comprehensive Agreement, and this approval has not yet been secured.

This complex of interrelated agreements is relevant to wetland protection because the water rights provisions affect the National Wildlife Refuges that cover a portion of the Basin. Under these agreements, the Refuges—especially Tule Lake and Lower Klamath Lake—will be able to count on assured water deliveries for waterfowl conservation.[174]

Whatever the political winds, the main outlines of current Klamath restoration are clear. They include: (1) removal of the dams; (2) phasing out leased agriculture in the Lower Klamath and Tule Lake refuges; (3) purchasing water rights of those willing to sell; (4) restoring marshes on lakebeds reclaimed from agriculture, including the Upper Basin wetlands; and (5) maintaining adequate water to Upper Klamath National Wildlife Refuge.

Will it happen? Like the Everglades, the wetlands that cry out for restoration have no way to flourish and replenish themselves. The ecosystems are forced to depend on human consciousness and goodwill. Any restoration plan—even with the best intentions—will ultimately fall far short of what the Klamath Basin needs to flourish and maintain its ecological resilience. Only a Rights of Nature legal framework can change that.

3. The Limberlost Swamp

Limberlost Swamp was a sprawling, forested wetland of 13,000 acres in east central Indiana. But early settlers were interested in farmland, not swamps. Between 1888 and 1910, steam-powered dredges drained the Loblolly Marsh area at its heart, and loggers cut the huge, old trees for lumber.

As it began to disappear, writer and naturalist Gene Stratton-Porter (born Geneva Grace Stratton) made the Limberlost famous in literature. Porter, born in 1863 in Wabash County, Indiana, was an author, naturalist and photographer who wrote best-selling novels and Nature books. Her photographs focused on the birds and moths of the Limberlost. Stratton-Porter and her husband and daughter built a large cabin in Geneva, Indiana, near the Limberlost, so she could explore it close to hand. In 1912, when developers had drained the swamp, Stratton-Porter moved away, but her books about the vanished swamp made it famous even once it was gone, especially *Girl of the Limberlost*. Stratton-Porter's writings were the catalyst for the later restoration of the Limberlost—a famous example of literature catalyzing environmental protection.[175]

Even though the Limberlost became agricultural land, and was cropped starting in 1910, it kept trying to revert to wetland. Water soaking the land ruined crops. Drainage tiles clogged with silt, drainage ditches flooded. Farmers could not grow many of the crops they planted. Finally, in 1991, a former dairy farmer, Ken Brunswick, with other volunteers, created a restoration project, Limberlost Swamp Remembered.

By 2002, the project had purchased more than 1,000 acres of farmland for restoration. Loblolly Marsh, a central part of the Limberlost and the restoration project, began to thrive within a few years of the restoration, as seeds from wetland plants, latent in the soil, awakened and began to recolonize the restored areas. Much of Loblolly is floodplains that connect to the Wabash River. By 2013, the restored Limberlost totaled 1,600 acres. Restoration work continues on the once-magnificent Limberlost Swamp.[176]

Management Considerations for a Rights of Nature System

The boiled-down suggestions below are meant to be a high-level set of goals for which detailed management plans may be developed, always following the dynamics of Nature's Laws of Reciprocity, as required in a Rights of Nature System. Powerful and innovative ideas frequently emerge from people with divergent experiences working out a design for each goal, such as cattle ranchers and Native Americans; small farmers dependent on tidal dikes and conservation groups; flood control district managers, urban residents, and wetland biologists; small woodlot owners, carpenters, and forest scientists. We encourage all these groups—and many more—to begin thinking about ways to create and maintain a Rights of Nature framework to protect these three critically important ecosystems nationwide.

Forests and Grasslands

1. Pass comprehensive laws under the Rights of Nature Constitutional frameworks that provide that forests and grasslands will be used by humans only in the amount and to the extent that allows the natural ecological systems to recover and flourish. These laws must be oriented first and foremost toward Nature's own requirements, rather than humans' economic desires.
2. Set up methodologies for researching and compiling baseline data that allow agencies and communities to estimate the acreage and ecosystem sub-types needed for forests and grasslands to regain their true functional processes, rather than struggling with a meager remnant of highly vulnerable function. This will involve using tools developed through interactive mapping and ecological risk assessment, both of which are now sophisticated technologies.
3. Write safeguards into the legal system requiring adoption after mandatory public process of the Rights of Nature–based management recommendations of government or semi-government researchers who compile baseline data. This safeguards the integrity of those who make decisions based on placing Nature first. A general tax or surcharge, rather than fees from extractive resource users, must fund the government offices performing this work.
4. Certify qualified nonprofit organizations to monitor and enforce in court both the data-gathering and subsequent implementation of forest and grassland protection and restoration programs, to ensure an external layer of trusteeship for these critically important conservation measures.
5. Draw initial forests and grasslands needed for ecological function from the public lands, which belong to the people of the United States and are frequently in a more resilient ecological state than private lands. If lagging ecological function shows private lands are also needed, compensation and personal retraining programs can be designed and implemented to smooth the transitions.
6. Provide federal funding and expertise to allow communities to purchase cutover industrial forestlands that provide their drinking water, recreation opportunities, flourishing wildlife, carbon sequestration, and other local ecosystem services. Pass statutes requiring that these community forests be managed under a Rights of Nature framework to ensure that trees are allowed to grow in natural succession into older forests of species biodiversity and canopy complexity in order that forest ecosystems flourish across the American landscape once again.
7. Prohibit clear-cut logging in all native forests on both public and private lands. Thinning or selective tree removal to open up the canopy in existing tree farms may be necessary to aid the forest in regaining its understory cover and undercanopy complexity. "Understory" refers to the shrubs and plants growing beneath the main canopy of a forest.
8. Prohibit selective high-grading—removal of the largest trees—in all native forests on both public and private lands. The largest trees provide the most intensive root and nutrient networks for large regions of the forest.
9. Prohibit all pesticides, herbicides, and fertilizers in native forests or recovering tree farms on both public and private lands. Though these may appear benign or quickly broken down, they have long-term, worldwide effects on ecosystems,

including groundwater and oceans, as research in many different biomes is now showing. (See *Interactions of Land, Ocean and Humans: A Global Perspective.*[177])

10. Fund and encourage research, both at the federal level and locally, on the kinds of forest uses humans can undertake in different native forest ecosystems of the United States to fulfill their necessities without degrading the forests' ecological sustainability. This research must be collaborative and cross-cultural—for example, exploring ways of partnering with indigenous communities and investigating collaboration with ecologically critical species and their habitats to enrich and strengthen the ecological resilience and sustainability of a region.
11. Decommission all but the most necessary main roads in publicly owned forests nationwide, including removal of all but essential culverts, to reduce habitat fragmentation and increase the passage of native fish into their traditional spawning habitat.
12. Fund and encourage research on best and least invasive methods of aiding restoration of native grasslands, riparian areas, and wetlands degraded by the grazing of cattle and other domesticated livestock. This includes restoration of endangered species, such as the black-footed ferret and sage grouse. Design programs to accomplish these goals as innovative partnerships in which government and private landowners collaborate under a Rights of Nature framework to restore sustainable ecosystem processes and their function to American prairies, high desert, and shrub steppe landscapes.
13. End all cattle grazing on federal and state public lands throughout all the grassland/steppe biomes in the United States, unless objective research shows the ecosystem can sustain, or benefit from, modest amounts of grazing that aids in strengthening the ecosystem's flourishing. In general, cattle grazing is too damaging to riparian areas, soils, native plants, streams, and ecosystem processes to continue on the public lands, which are repositories of the biodiversity necessary to strengthen and maintain resilience.
14. Fund and encourage research on organically oriented, sustainable carrying capacity-based cattle management on private lands to determine the best methods of aligning cattle management with the Rights of Nature. This research must focus locally, as grassland ecosystems vary greatly in their resilience and capacity to handle cattle (among other types of livestock), even under the best of circumstances. This must also embrace native alternatives, such as raising domesticated or, as appropriate, semi-wild bison in their native ecosystems, which would benefit both the ecosystem and a fundamental species of the Great Plains.
15. Expand and broaden federal programs run through the U.S. Department of Agriculture that encourage farmers to restore native grasslands. Currently, these often run for a few years and provide payments, but should be shifted to create permanent protections and funding for restoration if needed.
16. Determine via interactive mapping and ecological risk assessment (as detailed above) which areas in grassland biomes currently planted in agriculture crops are critical for Nature to flourish without human interference, and return these areas to natural grasslands.
17. Develop compensation and retraining programs for ranchers, farmers, loggers, and forestland owners who can no longer manage ecosystems for profit, as currently allowed under the relatively unfettered industrial system that prevails

today. This could include learning techniques of less intensive use, forest and grassland restoration skills, or the creation of alternatives.

18. Create more wilderness areas on public lands nationwide wherever possible to increase the resilience of native ecosystems and water quality. These must not be focused solely on "ecologically important" areas, but need to be broad-based, including ecosystems of every kind, at every elevation and size. Focus primarily on wilderness designation for grasslands, as none have so far been designated in the United States. The proposed 48,000 Buffalo Gap Wilderness, formed out of the National Grassland of the same name, would be the first if Congress passed the legislation introduced in 2010 to designate it. Many wilderness areas in the West already include high desert or shrub steppe landscapes, such as the 170,000-acre Steens Mountain Wilderness in southeastern Oregon, part of a larger 428,000-acre Steens Mountain Cooperative Management and Protection Area.[178] Wilderness has extreme value as wildlife refugia, native plant repositories, and areas for human education with respect to caring for the land and bringing the Rights of Nature to fruition.
19. Create a network of research and restoration offices at federal and state levels, funded by general revenues, to spearhead restoration projects, especially large-scale ones, such as those emanating from restoring Great Plains agricultural lands and massive tree farms in the Southeast and Northwest parts of the country. Ensure in the legal framework authorizing these projects that they will have both policy and funding priority under the Rights of Nature framework so they do not get bogged down or blocked part way through to completion.

Wetlands

1. Pass laws under a Rights of Nature Constitutional framework to ensure that wetland integrity and restoration has the highest priority.
2. Create regional and state Wetland Ombudsman offices with planning, restoration, and enforcement authority to coordinate research on wetland needs, and to oversee restoration of large and very large wetland complexes, as well as provide assistance to the state, local governments, and private owners in restoring smaller wetlands. Funding must be a dedicated general fund source to avoid problems of corruption in the agency.
3. Certify qualified nonprofits to oversee the Wetland Ombudsman offices and enforce the mandates if there are problems or insufficient political will to complete major wetland restorations in any state or across state boundaries.
4. Inventory larger wetland complexes (such as the Mississippi River delta, Canaan Valley, West Virginia and Maine's Scarborough Marsh, to name three of many nationwide) to determine the original size of the wetland and then work to restore its full ecological sustainability through purchase, easements, wetland re-plantings, dike removal, and similar activities.
5. Create legal prioritization mandates through the Ombudsman offices (see above) to oversee and complete critically needed large-scale wetland restoration projects of dramatically important complexes, such as the Everglades, the Klamath Basin, and the Limberlost. In order to prevent political sabotage of such important programs, create retraining, agriculture relocation, and/or alternatives projects for those displaced by large-scale restoration programs.

6. Map and research tidal and estuarine wetlands nationwide in order to restore them to fully sustainable productivity. Begin with a comprehensive legal framework detailing the harmful activities that must stop prior restoration work, such as dredge-and-fill and building marinas, industrial complexes, and other infrastructure. The program must also create and enforce strict protocols to limit stormwater runoff, sewer overflows, septic leakages, pollution flows, and similar problems. Include innovative buyout and community collaboration programs to foster understanding of estuarine habitats, along the lines of the existing National Estuary Program.
7. End wetland mitigation programs. These are designed to cause "no net loss" in wetlands, but in fact frequently create "new" wetlands in areas that have no natural ecological history as wetlands, and which do not provide much or any wetland function. They merely lead to increased loss of actual wetlands. Instead, pass laws that prioritize wetland protection, and create innovative programs such as development transfers and easements to protect them from development under a Rights of Nature system.
8. Expand and broaden federal programs run through the U.S. Department of Agriculture that financially aid farmers and ranchers in protecting wetlands on their property, as well as helping to pay for wetland and riparian restoration projects on agricultural lands. Payments, which often expire after a few years, need to be permanent. These programs, as the active arm of a vigorous wetland protection policy, would be able to dramatically increase the acreage of protected wetlands on private lands.

Endnotes

1. The foregoing two paragraphs are based on: Rebecca Adamson. People who are indigenous to the Earth. *YES! A Journal of Positive Futures*, Winter (1997):26–27.
2. Usufruct Law & Legal Definition. US Legal Definitions. http://definitions.uslegal.com/u/usufruct/ (accessed December 25, 2015).
3. The following discussion of the Ecuadorian Constitution is based on: Constitution, full language: http://pdba.georgetown.edu/Constitutions/Ecuador/english08.html (Title II, Chapter 7, Article 71) (accessed August 10, 2015).
4. *Ibid.* (Title II, Chapter 7, Article 72).
5. Language of the Bolivian law (compete text): http://www.worldfuturefund.org/Projects/Indicators/motherearthbolivia.html (Article 2, sec. 3) (accessed August 12, 2015).
6. *Ibid.*
7. Universal Declaration of Rights of Mother Earth: Global Alliance for the Rights of Nature. https://therightsofnature.org/universal-declaration/Article 2 sec. (1) (a–c) (accessed August 16, 2015).
8. This paragraph is based on: John R. Nolon. Historical overview of the American land use system: A diagnostic approach to evaluating governmental land use control. *Pace Environmental Law Review*, 23 (2006):821–854.
9. William Blackstone. Commentaries on the Laws of England: Volume 2, The Rights of Things. George T. Bisel Company, Philadelphia, PA. (1922) 569 pp.
10. Rutherford H. Platt. *Land Use and Society: Geography, Law, and Public Policy* (revised ed.). Island Press, Washington, DC. (1996) 504 pp.

11. The preceding two paragraphs are based on: Jean Manco. History of Building Regulations. http://www.buildinghistory.org/regulations.shtml (accessed 27 January 2016).
12. John R. Nolon. Historical Overview of the American Land Use System: A Diagnostic Approach to Evaluating Governmental Land Use Control. *op. cit.*
13. VILLAGE OF EUCLID, OHIO v. AMBLER REALTY CO. (1926). http://caselaw.findlaw.com/us-supreme-court/272/365.html (accessed January 27, 2016).
14. Berman v. Parker, 348 U.S. 26 (1954). https://www.law.cornell.edu/supremecourt/text/348/26 (accessed January 27, 2016).
15. Kelo v. City of New London 545 U.S. 469 (2005). https://www.law.cornell.edu/supct/html/04-108.ZS.html (accessed January 27, 2016).
16. Kelo v. City of New London. https://en.wikipedia.org/wiki/Kelo_v._City_of_New_London (accessed January 27, 2016).
17. John R. Nolon. Historical Overview of the American Land Use System: A Diagnostic Approach to Evaluating Governmental Land Use.
18. State Environmental Policy Act 1971. http://www.ecy.wa.gov/programs/sea/sepa/overview.html (accessed January 29, 2016).
19. John R. Nolon. Historical Overview of the American Land Use System: A Diagnostic Approach to Evaluating Governmental Land Use Control. *op. cit.* 46. The Constitution of the United States. http://www.usconstitution.net/const.pdf (accessed January 28, 2016).
20. *Ibid.*
21. *Ibid.*
22. The National Environmental Policy Act of 1969. http://www.fws.gov/r9esnepa/RelatedLegislativeAuthorities/nepa1969.PDF (accessed January 28, 2016).
23. History of the Clean Water Act. http://www.epa.gov/laws-regulations/history-clean-water-act (accessed January 28, 2016).
24. Coastal Zone Management Act of 1972. https://coast.noaa.gov/czm/act/sections/#305 (accessed January 28, 2016).
25. Endangered Species Act of 1973. http://www.nmfs.noaa.gov/pr/pdfs/laws/esa.pdf (accessed January 28, 2016).
26. David M. Bearden. Comprehensive Environmental Response, Compensation, and Liability Act: A Summary of Superfund Cleanup Authorities and Related Provisions of the Act. https://www.fas.org/sgp/crs/misc/R41039.pdf (accessed January 28, 2016).
27. Robert P. Hartwig and Claire Wilkinson. The National Flood Insurance Program (NFIP). http://www.iii.org/sites/default/files/FloodWhitePaper1.pdf (accessed January 28, 2016).
28. John R. Nolon. Historical Overview of the American Land Use System: A Diagnostic Approach to Evaluating Governmental Land Use Control. *op. cit.*
29. Oregon Coastal Management Program. http://www.oregon.gov/lcd/ocmp/pages/index.aspx (accessed January 28, 2016).
30. Encyclical Letter **Laudato Si** of The Holy Father **Francis** on Care for Our Common Home [in English]. (Sections 152 & 156). http://w2.vatican.va/content/francesco/en/encyclicals/documents/papa-francesco_20150524_enciclica-laudato-si.html (accessed August 19, 2015).
31. *Ibid.* (Section 159).
32. Jay Griffiths. The speed craze. http://www.consumer.org.my/index.php/development/cultural/327-the-speed-craze (accessed January 2, 2016).
33. Chris Maser. Social–Environmental Planning: The Design Interface between Everyforest and Everycity. *op. cit.*
34. The preceding discussion of architecture and history is based on: James Howard Kunstler. Home from nowhere. *The Atlantic Monthly*, 278 (1996):43–66.
35. *Ibid.*
36. Jane Silberstein and Chris Maser. *Land-Use Planning for Sustainable Development, Second Edition.* CRC Press, Boca Raton, FL. (2014) 296 pp.

37. The preceding two paragraphs are based on: (1) Andrew V. Suarez, Karin S. Pfennig, and Scott K. Robinso. Nesting success of a disturbance-dependent songbird on different kinds of edges. *Conservation Biology,* 11 (1997):928–935; (2) Denis A. Saunders, Richard J. Hobbs, and Chris R. Margules. Biological consequences of ecosystem fragmentation: A review. *Conservation Biology,* 5 (1991):18–32; and (3) Robert G. Lee, Richard Flamm, Monica G. Turner, and others. Integrating sustainable development and environmental vitality: A landscape ecology approach. pp. 499–521. *In*: *Watershed Management.* Robert J. Naiman, editor. Springer-Verlag, New York, NY. 1992.
38. Jane Silberstein and Chris Maser. *Land-Use Planning for Sustainable Development, Second Edition. op. cit.*
39. Matthew R. Falcy and Cristián F. Estades. Effectiveness of corridors relative to enlargement of habitat patches. *Conservation Biology,* 21 (2007):1341–1346.
40. James O. Wilson and James Howard Kunstler. The War on Cars. Slate. (1998). http://www.slate.com/id/3670/entry/24044/ (accessed January 22, 2009)
41. Bill Bishop. To reduce congestion, don't build more roads—close'em. *Corvallis Gazette-Times,* Corvallis, OR (May 20, 1998).
42. Peter Headicar. Traffic in towns. *Resurgence,* 197 (1999):22–23.
43. The discussion of riparian zones is based on: (1) Jack Ward Thomas, Chris Maser, and Jon E. Rodiek. Riparian zones. pp. 40–47. *In*: *Wildlife Habitats in Managed Forests—The Blue Mountains of Oregon and Washington.* USDA Forest Service, Agricultural Handbook No. 553 (Jack W. Thomas, Technical Editor). U.S. Government Printing Office, Washington, DC (1979); (2) Jack Ward Thomas, Chris Maser, and Jon E. Rodiek. Riparian zones. *In*: *Wildlife Habitats in Managed Rangelands—The Great Basin of Southeastern Oregon.* USDA Forest Service/USADI Bureau of Land Management. General Technical Report PMW 80, Portland, OR. (1979) 17 pp.; and (3) Chris Maser. Framing objectives for managing wildlife in the riparian zone on eastside federal lands. *In*: Managing Oregon's riparian zone for timber, fish, and wildlife. National Council of the Paper Industry for Air and Stream Improvement (NCASI). *Technical Bulletin,* 514 (1987):13–16.
44. Chris Maser and James R. Sedell. From the Forest to the Sea: The Ecology of Wood in Streams, Rivers, Estuaries, and Oceans. *op. cit.*
45. *Ibid.*
46. The preceding discussion of floodplains is based on: (1) What Is a Floodplain? http://www.friendsoftheriver.org/fotr/BeyondFloodControl/no5.html (accessed December 31, 2015); (2) Healthy floodplains yield multiple benefits. http://www.nature.org/ourinitiatives/habitats/riverslakes/benefits-of-healthy-floodplains.xml (accessed December 31, 2015); and (3) Chris Maser and James R. Sedell. From the Forest to the Sea: The Ecology of Wood in Streams, Rivers, Estuaries, and Oceans. *op. cit.*
47. The preceding discussion of wetlands is based on: (1) Basic Facts about Wetlands. Defenders of Wildlife. http://www.defenders.org/wetlands/basic-facts (accessed December 31, 2015); (2) What Is a Wetland? http://www.epa.gov/wetlands/what-wetland (accessed December 31, 2015); (3) What Are Wetland Functions? http://www.epa.gov/wetlands/what-are-wetland-functions (accessed December 31, 2015); (4) Why Are Wetlands Important? http://www.epa.gov/wetlands/why-are-wetlands-important (accessed December 31, 2015); and (5) Wetlands. http://www.worldwildlife.org/habitats/wetlands (accessed December 31, 2015).
48. (1) Sandi Doughton. New analysis shows Oso landslide was no fluke. http://www.seattletimes.com/seattle-news/science/new-analysis-shows-oso-landslide-was-no-fluke/ (accessed August 3, 2016); and (2) 2014 Oso mudslide. https://en.wikipedia.org/wiki/2014_Oso_mudslide (accessed August 3, 2016).
49. (1) Benjamin Earl. What Exactly Is a Geohazard? Sustainable Development Information. http://www.sustainabledevelopmentinfo.com/what-exactly-is-a-geohazard/ (accessed January 1, 2016); (2) Christian Jaedicke, Kalle Kronholm, Anders Solheim, and others. Hazards, Climate

Change and Extreme Weather Events. http://dc.engconfintl.org/geohazards/3/ (accessed January 1, 2016); and (3) Geohazard. http://www.sgs.org.sa/English/NaturalHazards/Pages/Geohazard.aspx (accessed January 1, 2016).
50. Joseph H. Chadbourne and Mary Chadbourne. *Common Groundwork: A Practical Guide to Protecting Rural and Urban Land: A Handbook for Making Land-Use Decisions.* Institute for Environmental Education. American Planning Association Planners Book Service, Chicago, IL (1993).
51. Vision and Mission. The Nature Conservancy. http://www.nature.org/about-us/vision-mission/ (accessed January 3, 2016).
52. (1) Benefits for Landowners. http://www.landtrustalliance.org/what-you-can-do/conserve-your-land/benefits-landowners (accessed January 3, 2016); and (2) Conservation Easements. http://www.nature.org/about-us/private-lands-conservation/conservation-easements/what-are-conservation-easements.xml (accessed January 3, 2016).
53. Purchase of Development Rights. https://www.uwsp.edu/cnr-ap/clue/Documents/PlanImplementation/Purchase_of_Development_Rights.pdf (accessed January 3, 2016).
54. Kate Shuttleworth. Agreement entitles Whanganui River to legal identity. http://www.nzherald.co.nz/nz/news/article.cfm?c_id=1&objectid=10830586 (accessed August 18, 2015).
55. (1) Chris Maser. Adaptable landscapes are the key to sustainable forests. *Journal of Sustainable Forestry*, 1 (1993):49–59; and (2) Chris Maser. Biodiversity is the "bottom line" for sustainable forestry. *In Good Tilth*, 4 (1993):14–16.
56. Chris Maser, Robert F. Tarrant, James M. Trappe, and Jerry F. Franklin (Technical Editors). From the forest to the sea: A story of fallen trees. USDA Forest Service General Technical Report PNW-229. Pacific Northwest Research Station, Portland, OR. (1988) 153 pp.
57. David H. Johnson and Tomas A. O'Neill (Managing Directors). Wildlife-Habitat Relationships in Oregon and Washington. Oregon State University Press, Corvallis, OR. (2001) 736 pp.
58. Language of the Bolivian law (compete text): http://www.worldfuturefund.org/Projects/Indicators/motherearthbolivia.html (accessed August 12, 2015) (Chapter 1, Article 2 (3)).
59. *Ibid.* (Chapter III, Article 7).
60. [Ecuadorian] Constitution, full language: http://pdba.georgetown.edu/Constitutions/Ecuador/english08.html (accessed August 10, 2015) (Chapter 7, Article 71).
61. [Ecuadorian] Constitution, full language: http://pdba.georgetown.edu/Constitutions/Ecuador/english08.html (accessed August 10, 2015) (Chapter 7, Article 74).
62. Universal Declaration of Rights of Mother Earth: Global Alliance for the Rights of Nature. https://therightsofnature.org/universal-declaration/ (accessed August 16, 2015).
63. *Ibid.* (Article 2).
64. Chris Maser. Biodiversity is the "bottom line" for sustainable forestry. *op. cit.*
65. The preceding two paragraphs are based on: Chris Maser, James M. Trappe, Steven P. Cline, and others. The Seen and Unseen World of the Fallen Tree. *USDA Forest Service General Technical Report PNW-164* (Chris Maser and James M. Trappe, Technical Editors.) Pacific Northwest Forest and Range Experiment Station, Portland, OR. (1984) 56 pp.
66. Chris Maser. Why protect "natural" areas within our dynamic, cultural landscape? pp. 25–29. *In: Science and the Management of Protected Areas.* Willison, J.H.M., S. Bondrup-Nielsen, C. Drysdale, and others (eds.). Elsevier Science Publishers B.V., Amsterdam, The Netherlands (1992).
67. David A. Perry. Landscape pattern and forest pests. *Northwest Environmental Journal*, 4 (1988):213–228.
68. Chris Maser. Patterns across the landscape. 1995. *Forum*, 10 (1995):103–106.
69. The Federal Lands Policy and Management Act of 1976. The Federal Lands Policy and Management Act of 1976 (accessed April 29, 2016).
70. *Ibid.*
71. Multiple-Use Sustained-Yield Act of 1960. http://www.fwspubs.org/doi/suppl/10.3996/042013-JFWM-031/suppl_file/042013-jfwm-031r-s03.pdf (accessed April 29, 2016).
72. National Forest Management Act of 1976. http://www.fs.fed.us/emc/nfma/includes/NFMA1976.pdf (accessed April 29, 2016).

73. National Environmental Policy Act of 1969. http://www.epw.senate.gov/nepa69.pdf (accessed April 29, 2016).
74. The Endangered Species Act of 1973. http://www.google.com/search?hl=en&source=hp&biw=&bih=&q=endangered+species+act+of+1973&gbv=2&oq=Endangered+Species+Act&gs_l=heirloom-hp.1.1.0i131j0l9.183134.186268.2.188606.1.1.0.0.0.0.107.107.0j1.1.0....0...1ac.2.34.heirloom-hp..0.3.288.L-_Uz3rZEGQ (accessed April 29, 2016).
75. Summary of the Clean Water Act. https://www.epa.gov/laws-regulations/summary-clean-water-act (accessed April 29, 2016).
76. Summary of the Clean Air Act. https://www.epa.gov/laws-regulations/summary-clean-air-act (accessed April 29, 2016).
77. (1) The Wilderness Act of 1964. http://wilderness.nps.gov/document/wildernessAct.pdf (accessed April 29, 2016).
78. Wilderness. http://wilderness.nps.gov/faqnew.cfm (accessed April 29, 2016).
79. The Wilderness Act of 1964. *op. cit.*
80. *Ibid.*
81. Who owns America's trees, woods, and forests? Results from the U.S. Forest Service 2011–2013 national woodland owner survey. http://www.nrs.fs.fed.us/pubs/48027 (accessed April 27, 2016).
82. The foregoing discussion of grasslands is based on: Richard Conner, Andrew Seidl, Larry Van Tassell, and Neal Wilkins. United States Grasslands and Related Resources: An Economic and Biological Trends Assessment. http://irnr.tamu.edu/media/252770/us_grasslands.pdf (accessed April 28, 2016).
83. (1) Thomas E. Dahl and Gregory J. Allord. History of wetlands in the conterminous United States. *United States Geological Survey Water-Supply Paper* 2425:19–26 (accessed April 27, 2016); (2) Thomas E. Dahl. Wetlands-losses in the United States, 1780's to 1980's: *Washington, D.C., U.S. Fish and Wildlife Service Report to Congress*, 1990:1–13 (accessed April 27, 2016); and (3) George A. Pavelis. Farm drainage in the United States: History, Status, and Prospects. U.S. Department of Agriculture, Economic Research Service, Washington, D.C. Volume No. 1455. (1987) 192 pp.
84. The preceding discussion of wetland is based on: (1) Thomas E. Dahl and C.E., Johnson. Wetlands—Status and trends in the conterminous United States, mid-1970's to mid-1980's. *Washington, D.C., U.S. Fish and Wildlife Service*, 1991:1–22 (accessed April 27, 2016); and (2) National water summary on wetland resources. *United States Geological Survey Water-Supply Paper* 2425. https://water.usgs.gov/nwsum/WSP2425/state_highlights_summary.html (accessed April 27, 2016).
85. Robert A. Smail and David J. Lewis. Forest Land Conversion Ecosystem Services, and Economic Issues for Policy: A Review. http://www.fs.fed.us/pnw/pubs/pnw_gtr797.pdf (p. 6) (accessed August 23, 2016).
86. Robert A. Smail and David J. Lewis. Forest Land Conversion Ecosystem Services, and Economic Issues for Policy: A Review. *Ibid.*
87. The foregoing story of salmon is based on: (1) Chris Maser and James R. Sedell. *From the Forest to the Sea: The Ecology of Wood in Streams, Rivers, Estuaries, and Oceans.* St. Lucie Press, Delray Beach, FL. (1994) 200 pp; (2) C. Jeff Cederholm, David H. Johnson, Robert Bilby, and others. Pacific salmon and wildlife—Ecological contexts, relationships, and implications for management. pp. 628–684. *In*: David H. Johnson and Thomas A. O'Neil, Managing Directors. *Wildlife-Habitat Relationships in Oregon and Washington.* Oregon State University Press, Corvallis, OR (2001); (3) James R. Sedell, Joseph E. Yuska, and Robert W. Speaker. Study of Westside Fisheries in Olympic National Park, Washington. U.S. Department of the Interior, National Park Service, Final Report CX-9000-0-E 081. (1983) 74 pp; (4) J.M. Helfield and R.J. Naiman. Effects of salmon-derived nitrogen on riparian forest growth and implications for stream productivity. *Ecology,* 82 (2001):2403–2409; (5) Ellen Morris Bishop. Years of adapting separate steelhead from hatchery cousins. *Corvallis Gazette-Times*, Corvallis, OR (March 5, 1998); (6) Timothy J. Beechie, George Pess, Paul Kennard, Robert E. Bilby, and Susan Bolton. Modeling recovery rates and

pathways for woody debris recruitment in northwestern Washington streams. *North American Journal of Fisheries Management*, 20 (2000):436–452; (7) James M. Helfield and Robert J. Naiman. Effects of salmon-derived nitrogen on riparian forest growth and implications for stream productivity. *Ecology*, 82 (2001):2403–2409; (8) Ted Gresh, Jim Lichatowich, and Peter Schoonmaker. An estimation of historic and current levels of salmon production in the northeast Pacific ecosystem. *Fisheries*, 25 (2000):15–21; (9) Bruce P. Finney, Irene Gregory-Eaves, M.S.V. Douglas, and J.P. Smol. Fisheries productivity in the northeastern Pacific Ocean over the past 2,200 years. *Nature*, 416 (2002):729–733; and (10) Nathan F. Putman, Kenneth J. Lohmann, Emily M. Putman, and others. Evidence for geomagnetic imprinting as a homing mechanism in Pacific salmon. *Current Biology* (2013) http://www.cell.com/current-biology/abstract/S0960-9822(13)00003-1 (accessed February 7, 2013).

88. Based on: The Associated Press. Culvert fixes hamper salmon recovery. *Corvallis Gazette-Times*, Corvallis, OR (February 13, 1999).
89. *Ibid.*
90. (1) Christopher C. Smith. The adaptive nature of social organization in the genus of tree squirrels, *Tamiasciurus*. *Ecological Monographs*, 38 (1968):31–64; and (2) Chris Maser. *Mammals of the Pacific Northwest: From the Coast to the High Cascade Mountains*. Oregon State University Press, Corvallis, OR. (1998) 406 pp.
91. (1) Chris Maser. *Mammals of the Pacific Northwest: From the Coast to the High Cascade Mountains. op. cit.*; (2) Chris Maser and Zane Maser. Interactions among squirrels, mycorrhizal fungi, and coniferous forests in Oregon. *Great Basin Naturalist*, 48 (1988):358–369; (3) Zane Maser, Chris Maser, and James M. Trappe. Food habits of the northern flying squirrel (*Glaucomys sabrinus*) in Oregon. *Canadian Journal of Zoology*, 63 (1985):1084–1088; (4) Chris Maser, Zane Maser, Joseph W. Witt, and Gary Hunt. The northern flying squirrel: A mycophagist in Southwestern Oregon. *Canadian Journal of Zoology*, 64 (1986):2086–2089.
92. The preceding three paragraphs are based, in part, on: (1) Chris Maser, James M. Trappe, and Ronald A. Nussbaum. 1978. Fungal–small mammal interrelationships with emphasis on Oregon coniferous forests. *Ecology*, 59 (1978):799–809; (2) Chris Maser, James M. Trappe, and Douglas Ure. Implications of small mammal mycophagy to the management of western coniferous forests. *Transactions of the 43rd North American Wildlife and Natural Resources Conference*, 43 (1978):78–88; (3) Wes Colgan III, Andrew B. Carey, James M. Trappe, and others. Diversity and productivity of hypogeous fungal sporocarps in a variably thinned Douglas-fir forest. *Canadian Journal of Forest Research*, 29 (1999):1259–268; and (4) Suzanne W. Simard, Kevin J. Beiler, Marcus A. Bingham, and others. Mycorrhizal networks: Mechanisms, ecology and modeling. *Fungal Biology Reviews*, 26 (2012):39–60.
93. S. Pyare and W. S. Longland. Mechanisms of truffle detection by northern flying squirrels. *Canadian Journal of Zoology*, 79 (2001):1007–1015.
94. The foregoing two paragraphs are based on: (1) C.Y. Li, Chris Maser, and Harlan Fay. Initial survey of acetylene reduction and selected microorganisms in the feces of 19 species of mammals. *Great Basin Naturalist*, 46 (1986):646–650; and (2) C.Y. Li, Chris Maser, Zane Maser, and Burce Caldwell. Role of three rodents in forest nitrogen fixation in Western Oregon: Another aspect of mammal–mycorrhizal fungus–tree mutualism. *Great Basin Naturalist*, 46 (1986):411–414.
95. Chris Maser and Zane Maser. Notes on mycophagy in four species of mice in the genus *Peromyscus*. *Great Basin Naturalist*, 47 (1987):308–313.
96. (1) Douglas C. Ure and Chris Maser. Mycophagy of red-backed voles in Oregon and Washington. *Canadian Journal of Zoology*, 60 (1982):3307–3315; and (2) Chris Maser and Zane Maser. Mycophagy of red-backed voles, *Clethrionomys californicus* and *C. gapperi*. *Great Basin Naturalist*, 48 (1988):269–173.
97. G. S. Jones, John O. Whitaker, Jr., and Chris Maser. Food habits of jumping mice (*Zapus trinotatus* and *Z. princeps*) in Western North America. *Northwest Science*, 52 (1978):57–60.
98. Zane Maser and Chris Maser. Notes on mycophagy of the yellow-pine chipmunk (*Eutamias amoenus*) in Northeastern Oregon. *Murrelet*, 68 (1987):24–27.

99. Michael A. Castellano, James M. Trappe, Zane Maser, and Chris Maser. *Synoptic Spore Key to Genera of Hypogeous Fungi in Northern Temperate Forests with Special Reference to Animal Mycophagy*. Mad River Press, Inc., Eureka, CA. (1989) 186 pp.
100. Javier G. Perez Calvo, Zane Maser, and Chris Maser. A note on fungi in small mammals from the *Nothofagus* Forest in Argentina. *Great Basin Naturalist*, 4 (1989):618–620.
101. Susanne Schickmann, Alexander Urban, Katharine Kräutler, and others. The interrelationship of mycophagous small mammals and ectomycorrhizal fungi in primeval, disturbed and managed Central European mountainous forests. *Oecologia*, 170 (2012):395–409.
102. Chris Maser, Andrew W. Claridge, and James M. Trappe. Trees, truffles, and beasts: How forests function. *op. cit.*
103. (1) Chris Maser. A tree is the quintessential plant. *International Journal of Ecoforestry*, 12 (1997):271–278, 289–294; and (2) Chris Maser. Do animals have rights? *Trumpeter*, 10 (1993):104–106.
104. Chris Maser. The trilogy of extinction. *Trumpeter*, 9 (1992):135–138.
105. Suzanne Simard and Kathy Martin. *Friends of Clayoquot Sound* (Summer 2012):6.
106. (1) David A. Perry. Landscape pattern and forest pests. *Northwest 104*. (2) Chris Maser. Economically sustained yield versus ecologically sustainable forests. pp. 1–13. *In*: *Restoration of Old-Growth Forests in the Interior Highlands of Arkansas and Oklahoma*. Proc. of the Conf. Sept. 19–20, 1990 (D. Henderson and L.D. Hedrick, eds.) USDA. Forest Service, Ouachita National Forest and Winrock International Institute for Agricultural Development, Morrilton, AR (1991); (3) Chris Maser. *Sustainable Forestry: Philosophy, Science, and Economics*. St. Lucie Press, Delray Beach, FL. (1994) 373 pp.; and (4) Larry D. Harris, Arthur McKee, and Chris Maser. Patterns of old-growth harvest and implications for cascades wildlife. *Proceedings of the North American Wildlife and Natural Resources Conference*, 47 (1982):374–392.
107. (1) Chris Maser. Economically sustained yield versus ecologically sustainable forests. pp. 1–13. *In*: *Restoration of Old-Growth Forests in the Interior Highlands of Arkansas and Oklahoma*. Proc. of the Conf. Sept. 19–20, 1990 (D. Henderson and L.D. Hedrick, eds.) USDA. Forest Service, Ouachita National Forest and Winrock International Institute for Agricultural Development, Morrilton, AR. (1991); (2) Chris Maser. *Sustainable Forestry: Philosophy, Science, and Economics*. St. Lucie Press, Delray Beach, FL. (1994) 373 pp.; and (3) Larry D. Harris, Arthur McKee, and Chris Maser. Patterns of old-growth harvest and implications for cascades wildlife. *Proceedings of the North American Wildlife and Natural Resources Conference*, 47 (1982):374–392.
108. Jack W. Thomas, Ralph G. Anderson, Chris Maser, and Evelyn L. Bull. Snags. pp. 60–77. *In*: *Wildlife Habitats in Managed Forests—The Blue Mountains of Oregon and Washington*. USDA Forest Service, Agricultural Handbook No. 553 (Jack W. Thomas, Technical Editor). U.S. Government Printing Office, Washington, DC (1979).
109. The foregoing discussion of fallen trees is based on: (1) Chris Maser and James M. Trappe, Technical Editors. The Seen and Unseen World of the Fallen Tree. *USDA Forest Service General Technical Report PNW-164*. Pacific Northwest Forest and Range Experiment Station, Portland, OR. (1984) 56 pp.; (2) Chris Maser, James M. Trappe, and C.Y. Li. Large woody debris and long-term forest productivity. *Proceedings of the Pacific Northwest Bioenergy System: Policies and Applications*. Bonneville Power Administration, May 10 and 11, Portland, OR. (1984) 6 pp.; (3) Chris Maser and James M. Trappe. The fallen tree—A source of diversity. pp. 335–339. *In*: Forests for a changing world. *Proceedings of the Society of American Foresters 1983 National Conference* (1984); (4) Chris Maser, Ralph G. Anderson, Kermit Cromack, Jr., and others. Dead and down woody materials. pp. 78–95. *In*: *Wildlife Habitats in Managed Forests—The Blue Mountains of Oregon and Washington*. USDA Forest Service, Agricultural Handbook No. 553 (Jack W. Thomas, Technical Editor). U.S. Government Printing Office, Washington, DC (1979); and (5) Chris Maser, Steven P. Cline, Kermit Cromack, Jr., and others. What we know about large trees that fall to the forest floor. pp 25–45. *In*: From the Forest to the Sea—A Story of Fallen Trees (Chris Maser, Robert F. Tarrant, James M. Trappe, and Jerry F. Franklin, Technical Editors). *USDA For. Serv. Gen. Tech. Rep. PNW-229*. Pacific Northwest Research Station, Portland, OR. (1988) 153 pp.
110. Chris Maser. *Sustainable Forestry: Philosophy, Science, and Economics*. St. Lucie Press, Delray Beach, FL. (1994) 373 pp.

111. The preceding two paragraphs are based on: (1) Matt McGrath. 'Wrong type of trees' in Europe increased global warming. http://www.bbc.com/news/science-environment-35496350 (accessed February 5, 2016); (2) Kim Naudts, Yiying Chen, Matthew J. McGrath, and others. Europe's forest management did not mitigate climate warming. *Science*, 351 (2016):597–600; and (3) Michael Holy. Why planting some trees could accelerate global warming. http://www.csmonitor.com/Environment/2016/0205/Why-planting-some-trees-could-accelerate-global-warming (accessed February 5, 2016).
112. E. Kim Naudts, Yiying Chen, Matthew J. McGrath, and others. Europe's forest management did not mitigate climate warming. *op. cit.*
113. The preceding two paragraphs are based on: Richard Plochmann. *The Forests of Central Europe: A Changing View*. pp. 1–9. *In*: *Oregon's Forestry Outlook: An Uncertain Future*. The Starker Lectures. Forestry Research Laboratory, College of Forestry, Oregon State University, Corvallis, OR (1989).
114. Norway Becomes First Country in the World to Commit to Zero Deforestation. http://www.independent.co.uk/news/world/europe/norway-becomes-first-country-in-the-world-to-commit-to-zero-deforestation-a7064056.html?platform=hootsuite (accessed August 8, 2016).
115. Aviva Glaser, Editor. America's Grasslands: Status, Threats, and Opportunities. https://www.nwf.org/pdf/Policy-Solutions/Americas%20Grasslands%20Conference%20Proceedings061312.pdf (accessed August 21, 2016).
116. Richard Conner, Andrew Seidl, Larry Van Tassell, and Neal Wilkins. United States Grasslands and Related Resources: An Economic and Biological Trends Assessment. *op. cit.*
117. Lekha Knuffman, Editor. America's Grasslands Conference: Partnerships for Grassland Conservation. Proceedings of the 3rd September 29–October 1, 2015, Fort Collins, CO. Washington, DC: National Wildlife Federation.
118. The preceding two paragraphs are based on: Allison Jones. Surveying the West: A summary of research on livestock impacts. pp. 171–173. *In*: *Welfare Ranching: The Subsidized Destruction of the American West* (George Wuerthner and Mollie Matteson, eds.) Island Press, Washington, DC (2002) 368 pp.
119. (1) Jack W. Thomas, Chris Maser, and Jon E. Rodiek. Riparian zones. *In*: *Wildlife Habitats in Managed Rangelands—The Great Basin of Southeastern Oregon* (Jack W. Thomas and Chris Maser, eds.) USDA Forest Service General Technical Report PNW-80. Pacific Northwest Forest and Range Experiment Station, Portland, OR. (1979) 18 pp; (2) Jack W. Thomas, Jon E. Rodiek, and Chris Maser. Riparian zones in managed rangelands—Their importance to wildlife. pp. 21–31. *In*: *Proceedings of the Forum—Grazing and Riparian/Stream Ecosystems* (Oliver B. Cope, ed.). Trout Unlimited, Inc., Washington, DC (1979); and (3) J. Boone Kauffman. Lifeblood of the West: Riparian zones, biodiversity and degradation by livestock. pp. 175–178. *In*: *Welfare Ranching: The Subsidized Destruction of the American West* (George Wuerthner and Mollie Matteson, eds.) Island Press, Washington, DC (2002).
120. J. Boone Kauffman. Lifeblood of the West: Riparian zones, biodiversity and degradation by livestock. pp. 176. *op. cit.*
121. *Ibid.*
122. Joy Belsky, Andrea Matzke, and Shauna Uselman. What the river once was: Livestock destruction of waters and wetlands. p. 179. *In*: *Welfare Ranching: The Subsidized Destruction of the American West* (George Wuerthner and Mollie Matteson, eds.) Island Press, Washington, DC (2002).
123. John Carter. Stink water: Declining water quality due to livestock production. pp. 189–195. *In*: *Welfare Ranching: The Subsidized Destruction of the American West* (George Wuerthner and Mollie Matteson, eds.). Island Press, Washington, DC (2002).
124. The discussion of mycorrhizae is based on: (1) Chris Maser, Zane Maser, and Randy Molina. Small-mammal mycophagy in rangelands of Central and Southeastern Oregon. *Journal of Range Management*, 41 (1988):309–312; (2) David Read. The ties that bind. *Nature*, 388 (1997):517–518; (3) David Read. Plants on the web. *Nature*, 396 (1998): 22–23; (4) Marcel G. A. van der Heijden, John N. Klironomos. Plant biodiversity, ecosystem variability and productivity. *Nature* 396

(1998): 69–72; and (5) M.C. Wicklow-Howard. Mycorrhizal ecology of shrub-steppe habitat. *In: Proceedings—Ecology and Management of Annual Rangelands* (S.B. Monsen and S.G. Kitchen, eds.). Intermountain Research Station, USDA Forest Service, Fort Collins, CO. (1994) pp. 207–210.

125. Sharon Y. Strauss, Campbell O. Webb, and Nicolas Salamin. Exotic taxa less related to native species are more invasive. *Proceedings of the National Academy of Sciences*, 103 (2006):5841–5845.
126. The preceding discussion of crested wheatgrass is based in part on: (1) Janice M. Christian and Scott D. Wilson. Long-term ecosystem impacts of an introduced grass in the Northern Great Plains. *Ecology*, 80 (1999):2397–2407; and (2) C.M. D'Antonio and P.M. Vitousek. Biological invasions by exotic grasses, the grass/fire cycle, and global change. *Annual Review of Ecology and Systematics*, 23 (1992):63–87.
127. Lars A. Brudvig, Catherine M. Mabry, James R. Miller, and Tracy A. Walker. Evaluation of Central North American prairie management based on species diversity, life form, and individual species metrics. *Conservation Biology*, 21 (2007):864–874.
128. Carl E. Bock and Jane H. Bock. Cover of perennial grasses in Southeastern Arizona in relation to livestock grazing. *Conservation Biology*, 7 (1993):371–377; (2) W.D. Billings. Ecological impacts of cheatgrass and resultant fire on ecosystems. *In*: The Western Great Basin. pp. 22–30. *Proceedings—Ecology and Management of Annual Rangelands* (S.B. Monsen and S.G. Kitchen, eds.) Intermountain Research Station, USDA Forest Service, Fort Collins, CO. (1994); and (3) S.B. Monsen. The competitive influences of cheatgrass (*Bromus tectorum*) on site restoration. pp. 43–50. *In: Proceedings—Ecology and Management of Annual Rangelands* (S. B. Monsen and S.G. Kitchen, eds.). Intermountain Research Station, USDA Forest Service, Fort Collins, CO (1994).
129. Brent E. Johnson and J. Hall Cushman. Influence of a large herbivore reintroduction on plant invasions and community composition in a California grassland. *Conservation Biology*, 21 (2007):515–526.
130. (1) Thomas J. Valone, Marc Meyer, James H. Brown, and Robert M. Chew. Timescale of perennial grass recovery in desertified arid grasslands following livestock removal. *Conservation Biology*, 16 (2002):995–1002; and (2) M. Lisa Floyd, Thomas L. Fleischner, David Hanna, and Paul Whitefield. Effects of historic livestock grazing on vegetation at Chaco Culture National Historic Park, New Mexico. *Conservation Biology*, 17 (2003):1703–1711.
131. Basic Facts about Black-Footed Ferrets. http://www.defenders.org/black-footed-ferret/basic-facts (accessed August 9, 2016).
132. Randy Webb and Mark Salvo. Sage grouse: Imperiled icon of the Sagebrush Sea. pp. 237–39. *In*: *Welfare Ranching: The Subsidized Destruction of the American West* (George Wuerther and Mollie Matteson, eds.). Island Press, Washington. DC (2002).
133. *Ibid.*
134. The Natural Resources Conservation Service. Sage Grouse Initiative. 2.0 Investment Strategy, FY 2015–2018. http://www.sagegrouseinitiative.com/wp-content/uploads/2015/08/SGI2.0_Final_Report.pdf (accessed August 21, 2016).
135. Sage Grouse Initiative, Wildlife Conservation through Sustainable Ranching: New Paradigm. http://www.sagegrouseinitiative.com/about/new-paradigm/ (accessed August 21, 2016).
136. The Natural Resources Conservation Service. Sage Grouse Initiative. *op. cit.*
137. Darryl Fears. Decision not to list sage grouse as endangered is called life saver by some, death knell by others. https://www.washingtonpost.com/news/energy-environment/wp/2015/09/22/fewer-than-500000-sage-grouse-are-left-the-obama-administration-says-they-dont-merit-federal-protection/?utm_term=.f23ae9420681 (accessed August 21, 2016).
138. The preceding three paragraphs are based on: (1) What Is a Wetland? https://www.epa.gov/wetlands/what-wetland (accessed April 28, 2016); and (2) Water: No Longer Taken for Granted. Chapter 7: What Are Wetlands? 2008. http://www.encyclopedia.com/topic/wetlands.aspx (accessed April 28, 2016).
139. The preceding three paragraphs are based on: *Ibid.*
140. What Is a Wetland? *op. cit.*
141. The discussion of salt marshes is based on: Chris Maser and James R. Sedell. From the Forest to the Sea: The Ecology of Wood in Streams, Rivers, Estuaries, and Oceans. *op. cit.*

142. Agricultural Resources and Environmental Indicators, Chapter 6.5:1–28. http://www.ers.usda.gov/media/873717/wetlands.pdf (accessed April 27, 2016).
143. Planning and Conservation League. The California Environmental Quality Act "CEQA" Frequently Asked Questions. http://www.pcl.org/projects/ceqafaq.html (accessed April 30, 2016).
144. California Rice Statistics and related National and International Data. http://calrice.org/pdf/2009-Statistical-Report-FINAL.pdf (accessed August 11, 2016).
145. Oregon's Wetland Conservation Strategy. https://www.oregon.gov/dsl/WETLAND/docs/wet_cons_strat.pdf (accessed April 30, 2016).
146. The summary of the six states is based on: National Water Summary on Wetland Resource: State Summary Highlights. https://water.usgs.gov/nwsum/WSP2425/state_highlights_summary.html (accessed April 28, 2016).
147. Discussion of the Pacific Coast estuarine marshlands is based on: Chris Maser and James R. Sedell. From the Forest to the Sea: The Ecology of Wood in Streams, Rivers, Estuaries, and Oceans. *op. cit.*
148. Migratory Bird Conservation Act [of 1929]. Migratory Bird Conservation Act of 1929 (accessed May 9, 2016).
149. 99-198-Food Security Act of 1985. http://www.fns.usda.gov/sites/default/files/99-198%20-%20Food%20Security%20Act%20Of%201985.pdf (accessed May 9, 2016).
150. North American Wetlands Conservation Act. http://www.epw.senate.gov/envlaws/wetlands.pdf (accessed May 9, 2016).
151. Section 404 of the Clean Water Act. http://www.usace.army.mil/Portals/2/docs/civilworks/regulatory/materials/cwa_sec404doc.pdf (accessed May 9, 2016).
152. Section 401 Certification Best Practices in Dredge and Fill Permit Programs. http://www.aswm.org/pdf_lib/401_best_practices_summary.pdf (accessed May 9, 2016).
153. National Estuary Program (NEP). https://www.epa.gov/nep (accessed May 11, 2016).
154. The foregoing discussion of wetlands is based on: Adapting Watershed Tools to Protect Wetlands. http://wetlandprotection.org/images/stories/PDFs/5_wetlandsarticle3.pdf (accessed May 11, 2016).
155. Natural Resources Conservation Service (NRCS) Overview. http://www.usda.gov/wps/portal/usda/usdahome?contentidonly=true&contentid=NRCS_Agency_Splash.xml (accessed May 11, 2016).
156. Wetlands Restoration Definitions and Distinctions. https://www.epa.gov/wetlands/wetlands-restoration-definitions-and-distinctions (accessed May 12, 2016).
157. (1) R.J. Hunt, D.P. Krabbenhoft, and M.P. Anderson. Groundwater inflow measurements in wetland systems. *Water Resources Research*, 32 (1996):495–507; (2) Randy J. Hunt, David P. Krabbenhoft, and Mary P. Anderson. Assessing hydrogeochemical heterogeneity in natural and constructed wetlands. *Biogeochemistry*, 39 (1997) 271–293; (3) Randall J. Hunt, Thomas D. Bullen, David P. Krabbenhoft, and Carol Kendall. Using stable isotopes of water and strontium to investigate the hydrology of a natural and constructed wetland. *Ground Water*, 36 (1998):434–443; and (4) A. Clay, C. Bradley, A.J. Gerrard, and M.J. Leng. Using stable isotopes of water to infer wetland hydrological dynamics. *Hydrology and Earth System Sciences*, 8 (2004):1164–1173.
158. Wetlands Restoration Definitions and Distinctions. *op. cit.*
159. Hamilton Disston. https://en.wikipedia.org/wiki/Hamilton_Disston (accessed May 12, 2016).
160. Development of the Central & South Florida (C&SF) Project. http://141.232.10.32/about/restudy_csf_devel.aspx (accessed May 12, 2016).
161. Vision for a Sustainable Everglades Agricultural Area. https://nicholas.duke.edu/wetland/eaa.htm (accessed May 12, 2016).
162. Big Cypress National Preserve—One Land, Many Uses. https://www.nps.gov/bicy/planyourvisit/upload/Park_Preserve.pdf (accessed May 12, 2016).
163. Water Resources Development Act of 1992. https://www.gpo.gov/fdsys/pkg/STATUTE-106/pdf/STATUTE-106-Pg4797.pdf (accessed May 13, 2016).

164. Comprehensive Everglades Restoration Plan. https://www.nps.gov/ever/learn/nature/cerp.htm (accessed May 13, 2016).
165. BJ Jarvis. Melaleuca: An Invasive Tree of Florida. http://pasco.ifas.ufl.edu/gardening/melaleuca.shtml (accessed May 14, 2016).
166. The overall discussion of the Everglades is based on: (1) Restoration of the Everglades. https://en.wikipedia.org/wiki/Restoration_of_the_Everglades (accessed May 12, 2016); and (2) Restoring America's Everglades. http://www.evergladesrestoration.gov/ (accessed May 12, 2016).
167. Central and South Florida Project Review Study. http://141.232.10.32/pm/projects/project_docs/pdp_asr_combined/052808_asr_report/052808_asr_ch1_restudy_feas_rpt_prog_eis.pdf (accessed May 14, 2016) (page xii).
168. (1) Stephen Most. Klamath Basin Project (1906). http://www.oregonencyclopedia.org/articles/klamath_basin_project_1906_/#.VzdtrBxm3TM (accessed May 14, 2016); and (2) Refuges in Peril: Fish, Wildlife, and the Klamath Water Crisis. http://oregonwild.org/sites/default/files/pdf-files/Klamath_Report.pdf (accessed May 14, 2016).
169. (1) Farm/Wetland Rotational Management—A Habitat Management Alternative for Tule Lake National Wildlife Refuge (accessed May 14, 2016); and (2) Brief History of Leaselands. http://www.usbr.gov/mp/kbao/programs/land-lease/1_Bidding%20Program/2008/Brief%20History.pdf (accessed May 14, 2016).
170. Stephen Most. Klamath Basin Project (1906). *op. cit.*
171. Upper Klamath Basin Comprehensive Agreement. http://klamathtribes.org/wp-content/uploads/2014/08/2014-4-18-UPPER-KLAMATH-BASIN-COMPREHENSIVE-AGREEMENT.pdf (accessed May 14, 2016).
172. Klamath Hydroelectric Settlement Agreement. http://www.klamathrestoration.org/index.php/klamath-hydroelectric-settlement-agreement (accessed May 14, 2016).
173. Summary of the Klamath Basin Settlement Agreements. http://www.edsheets.com/Klamath/Summary%20of%20Klamath%20Settlement%20Agreements%204-5-10.pdf (accessed May 14, 2016).
174. The preceding six paragraphs are based on: Upper Klamath Basin Agreement Reached. http://klamathriverrestoration.org/index.php/77-main-menu-items/118-upper-basin-agreement-for-front-page (accessed May 14, 2016).
175. Gene Stratton-Porter. https://en.wikipedia.org/wiki/Gene_Stratton-Porter (accessed May 14, 2016).
176. The preceding two paragraphs are based on: (1) Limberlost Restoration. http://landandlit.iweb.bsu.edu/about_us/limberlost.html (accessed May 14, 2016); and (2) Limberlost (Loblolly) Marsh. https://tippecanoeaudubon.wordpress.com/2013/08/11/limberlost-loblolly-marsh/ (accessed May 14, 2016).
177. Chris Maser. Interactions of Land, Ocean and Humans: A Global Perspective. CRC Press, Boca Raton, FL. (2014) 308 pp.
178. Steens Mountain. http://www.blm.gov/or/districts/burns/recreation/steens-mtn.php (accessed August 23, 2016).

11

Water

That water is essential to human life, and all other life, is a truism so obvious it should not need repeating. However, given the massive assault on the world's water cycle, and the pollution of individual water sources in a staggering number of ways, restating the obvious is the best way to begin.

The need for a complete overhaul of human legal systems to place the Rights of Nature at the center is nowhere more desperate than in dealing with water at every level of the ecosystem. All of the nascent legal structures on Nature's Rights recognize the high importance of water to ecosystems, and for the functioning of Nature's Laws of Reciprocity. They stress the necessity of restoring life cycles as needed, of which water is a crucial aspect—without which life would not exist.

Water: The Rights of Nature

The Ecuadorian Constitution states in Chapter 2: "Article 12. The human right to water is essential and cannot be waived. Water constitutes a national strategic asset for use by the public and it is unalienable, not subject to a statute of limitations, immune from seizure and essential for life."

Now let us revisit the critical sections of the new Constitution. Ecuador states in Chapter 7, Article 71: "Nature, or Pacha Mama, where life is reproduced and occurs, has the right to integral respect for its existence and for the maintenance and regeneration of its life cycles, structure, functions and evolutionary processes."[1] This of course, speaks to water, the primary ingredient of life. Critically important to wrestling with how best to take care of the earth's water systems, Ecuador also states, "Nature has the right to be restored." Interestingly, this right is separate from any obligation of the state or persons to "compensate individuals and communities that depend on affected natural systems."[2]

Turning now to Bolivia's new Law of the Rights of Mother Earth, recall that the initial definition describes Mother Earth as "a dynamic living system comprising an indivisible community of all living systems and living organisms, interrelated, interdependent and complimentary, which share a common destiny."[3] The new law guarantees Mother Earth many rights: to life, the diversity of life, and clean air, among others. But the primary focus on water is the language of Chapter III, Article 7, I (3), where Mother Earth is held to have: "The right to preserve the functionality of the water cycle, its existence in the quantity and quality needed to sustain living systems, and its protection from pollution for the reproduction of the life of Mother Earth and all its components."[4]

Since water comprises both the veins and the life-sustaining blood of the Earth and ties together all ecosystems of land and sea, Bolivia's provision of a right to equilibrium is also very important in thinking about water: "The right to maintenance or restoration of the interrelationship, interdependence, complementarity and functionality of the components of Mother Earth in a balanced way for the continuation of their cycles and reproduction of their vital processes."[5] Like Ecuador, the Bolivian law gives Mother Earth the right to "timely and effective restoration of living systems affected by human activities directly or indirectly."[6]

Building on the Ecuadorian and Bolivian frameworks, the Universal Declaration of the Rights of Mother Earth (UDRME) lists the fundamental rights of Mother Earth, including the right to life and to exist, and the right to be respected.[7] One of the rights flowing from these basic ones is "The right to water as a source of life."[8] The way humans must act to protect water and its critical life cycles include in Article 3:

(d) ensure that the pursuit of human wellbeing contributes to the wellbeing of Mother Earth, now and in the future;
(e) establish and apply effective norms and laws for the defense, protection and conservation of the rights of Mother Earth;
(f) respect, protect, conserve and where necessary, restore the integrity, of the vital ecological cycles, processes and balances of Mother Earth;
(g) guarantee that the damages caused by human violations of the inherent rights recognized in this Declaration are rectified and that those responsible are held accountable for restoring the integrity and health of Mother Earth.[9]

Water Is the Basis of Human Life and All Life on Earth

A brief overview of water's role will illustrate how essential water is. It is the basis for all life, and how much or little of it there is in a given region determines the nature of the ecosystem and the human communities it will sustain.

Water source and capacity for storage are finite in any given landscape. Fresh, usable water, once thought by non-indigenous peoples in the United States to be inexhaustible in supply, is now becoming scarce in many parts of the world. In the western United States, for example, water pumped from deep underground aquifers is today such a valuable commodity that it is often referred to as "sandstone champagne."[10]

The availability of water throughout the year will ultimately determine both the quality of life in a community and thereby the value of real estate. Consequently, every nation's supply of quality water is precious beyond comparison. Water is the *most valuable* commodity from our (U.S.) nation's forests and those of the world.

Seventy-one percent of the Earth's surface is covered with water. Of the freshwater, over 68 percent was—until now—frozen in glaciers and the polar ice caps, and another 30 percent is below ground, leaving only 1 percent of all water available in usable form for life. More than 97 percent of the water on Earth is saltwater that makes up the oceans.[11]

Streams are Nature's arterial system on land, and the stream-order continuum operates on a simple premise. Streams form a spectrum of physical environments, with associated aquatic and terrestrial plant and animal communities, in which downstream processes are linked to upstream processes.

As organic material floats downhill from its source to the sea, it becomes smaller and smaller in size, while the volume of water carrying it becomes larger. Thus, small streams feed larger streams and larger streams feed rivers with partially processed organic matter, such as wood, the amount of which becomes progressively smaller the farther down the continuum of the river system it goes.

A first-order stream is the smallest undivided waterway or *headwaters*. A first-order stream is the only aquatic entity with ecological integrity because it is not influenced by the condition of any other stream. Where two first-order streams join, they enlarge as a second-order stream. Where two second-order streams come together, they enlarge as a third-order stream, and so on. The concept of stream order is based on the size of the stream—the cumulative volume of water.[12]

Fresh potable water ultimately comes from the world's oceans in the form of precipitation. Most water used by communities initially falls in the form of snow and ice either at high elevations or in northern latitudes, where it is stored as snowpack until it melts and subsequently feeds the streams and rivers that eventually reach distant communities and cities, where it is available for human use. Human water use is generally taken from rivers, such as the Columbia in United States, the Amazon in South America, the Ganges in India, the Yangtze River in China, the Volga in Russia, and the Rhine in Europe.

The Greater Himalayas hold the largest mass of ice outside the Polar Regions and are the source of the 10 largest rivers in Asia. But the glaciers in the southern Himalayas are melting rapidly. The cascading effects of rising temperatures and dwindling amounts of ice and snow in that region of the Himalayas are affecting such things as availability of water, species biodiversity, shifts in the boundaries of biophysical systems (upward movement in tree line and other changes in high-elevation biophysical systems), and global feedback loops (shifts in the monsoons and loss of soil carbon). Climate change will also increase the uncertainty in water supplies and likely reduce agricultural production for human populations across Asia[13] (Figure 11.1).

FIGURE 11.1
Langtang Peak, 23,000 feet high. Taken from 12,500 feet in the Himalayas in the Nuwakot District, Nepal. Photograph by Chris Maser.

Aside from glacial ice, the annual accumulation of snow (snowpack) can, under good conditions, last as snowbanks late into the summer or even early autumn. How much water the annual snowpack has and how long the snowpack lasts depends on six things: (1) the timing, duration, and persistence of the snowfall in any given year; (2) how much snow accumulates during a given winter; (3) the moisture content of the snow (wet snow holds more moisture than dry snow); (4) where the snowfall accumulates in relation to shade and cool temperatures in spring and summer (e.g., under the cover of trees and on north-facing slopes vs. in the open and south-facing slopes with no protective shade); (5) when the snow begins melting and the speed at which it melts (the later in the year it begins melting and the slower it melts, the longer into the summer its moisture is stored above- and belowground); and (6) the health of the overall water catchment.

Although the first five points seem self-evident, the last one requires some explanation. In dealing with the health of water catchments, one must consider those of both high and low elevation. How we treat our high-elevation forests (and those at more northerly latitudes) is how we treat a major portion of the most important sources of our supply of potentially available water—the purity and longevity of the extant snow and ice.

Snow disappears in two ways: sublimation and melting. *Sublimation* means that snow, accumulating in such places as the upper surfaces of coniferous boughs above the ground, evaporates and recrystallizes without melting into water. When snow sublimates, it bypasses any role in our supply of available water. Melting snow, on the other hand, is a different story.

With the advent of late spring and early summer, snow begins to melt and gradually infiltrates the soil as water until every minute nook and cranny is filled, which all the while is obeying the unrelenting dictates of gravity as it journeys toward the streams and rivers of the land on its way to the sea from whence it came.

Thus, the amount and quality of water available for human use is largely the result of climate, topography, and the ecological integrity of the water catchments. In turn, water is stored in four ways: (1) in the form of snowpack and glaciers aboveground; (2) in the form of water penetrating deep into the soil, where it flows slowly belowground; (3) in belowground aquifers and lakes; and (4) in aboveground lakes, wetlands, and reservoirs.

Cities built in arid environments such as deserts (e.g., Las Vegas, Nevada, in the United States; Jerusalem, Israel; and Lima, Peru) have additional problems. Cities like Lima, which obtain their water strictly from distant glaciers, are particularly prone to encountering significant problems with their long-term supply of water because alternatives are either severely limited or nonexistent. The availability of pure water throughout the year determines the fundamental quality of life in a community. Water catchments in the local area under local control provide the essential lifeline. There are also those owned by an absentee owner who has no vested interest in the community's supply of water. Such absentee ownership could be a person, corporation, government body, or agency beyond local jurisdiction.

Israel is the world leader in solving problems of water availability for urban and agricultural use. As a small country in the desert with essentially no snowpack to supply natural water storage, Israel has devoted decades of research and innovative technology to solving water supply problems. Israel pioneered the desalinization of seawater, reuse of municipal waste for agriculture, drip irrigation to maximize irrigation efficiency, use of brackish water to meet agricultural needs, development of drought-resistant crop strains, water-safety technology to discover and fix urban water infrastructure leaks before they become large and troublesome, and market pricing to curb domestic water use, among other innovative ideas. Though the country uses nearly a billion cubic meters (35,314, 666, 721 cubic feet) a year more water than is available naturally, Israel is able to be self-sustaining in water supply. Approximately half of its water comes from reused, treated wastewater, brackish water, or desalinated water. Israelis also export both their technology and expertise to other regions, where the need for water is desperate, such as sub-Saharan Africa.[14]

Existing Legal Systems and Governance of Water

In the European settlement of the United States, two distinct and very different water doctrines appeared—ironically, based on the amount of water available in the bioregion. Broadly speaking, the eastern part of the country, with higher and steadier rainfall and abundance of lakes and rivers, bases its water use on the doctrine of Riparian Rights. The western half of the country, much of which (except for the Pacific coast) is drier grassland or desert landscapes, developed the doctrine of Prior Appropriation. In addition, the federal government has so-called reserved rights. In the last half-century, a fourth area of water use and water law appeared as a result of massive water projects to support burgeoning populations: the law of Water Infrastructure. There are many secondary legal issues dealing with water as well, ranging from interstate water compacts to navigability issues and environmental regulation of water.

Let us examine these four systems and their implications for a Nature's Rights of water: (1) riparian rights, (2) prior appropriation, (3) reserved rights, and (4) water infrastructure.

Riparian Rights

The "riparian" area is the land by a stream or river, and thus "riparian rights" accrued to the adjacent landowner who had the right to use the water, usually below the normal high-water mark—but not the one associated with floodwaters. States vary with respect to what rights are available to the riparian landowner in a given situation, such as diverting water to sell, mining the land under a watercourse, constructing docks and marinas, or for industrial purposes.

Generally speaking, the riparian owner may not deny the riparian rights of downstream owners along the waterway, which prohibits diverting or channeling the water away from its natural watercourse. Each user's right must be "reasonable" with respect to other users. The same is true for a landowner who sits over an underground stream or the underground flow of a surface stream. In the riparian doctrine, the water right exists because the land and the water are contiguous; an earlier riparian rights holder has no preference over a later one in terms of his right to the water.[15]

Prior Appropriation

Rules for the prior appropriation water use are commonplace in the Western states of the United States. It is not based on a landowner's proximity to the water; it simply gives first preference to the first user. As a rule of thumb, this doctrine may be described as "first in time—first in right." Holders of earlier water rights have preference over later water rights holders, if there is not enough water for them all.

A water right is usually based on both its physical control and its beneficial use, the latter being dependent on the water's continual availability. The person seeking to appropriate water has to show the intent to appropriate the water, then actually divert it and apply it to beneficial use. Historically, that meant a physical diversion of water, but now includes other methods. However, a "point of diversion" is still an essential part of a water right.[16]

What constitutes "beneficial use" of water—an essential aspect of the Prior Appropriation doctrine—continues to be contentious, because each state has developed its own body of water law to determine the nature of beneficial use. The underlying idea was always to prevent the waste of water, which was and remains a scarce resource in many regions.

The water right can be sold or transferred, and both are commonly done. Under this system, the water right is a valuable property right of its own, as many parcels of land in the arid West would be useless without a corresponding water right. Those with earlier priority dates are more valuable because they will provide water longer in times of drought. However, legal cases and state laws have limited some aspects of the system. For example, a senior water rights holder cannot change a component of his water right if it will harm a junior water rights holder, but the senior holder can require the government agency that manages the water to shut off the water to junior holder when there is insufficient water to go around.[17]

This doctrine of Prior Appropriation had many antecedents, but a formative cause was the discovery of gold in California in the mid-19th century. With many miners working claims, a rough and ready system of prior appropriation for both land and water was worked out in practice, and variations on it became commonplace in the legal systems of the Western states.[18]

Reserved Rights

It is important to note that the federal government has so-called "reserved rights" in water to ensure adequate water for Indian reservations and federal public lands. Federal reserved rights go back at least to the time when the lands were set aside. This doctrine grew out of the case of *Winters v. United States* (1908), concerning the use of Montana's Milk River by non-native settlers and the Fort Belknap Indian reservation. The U.S. government filed a suit on the reservation's behalf, and the Supreme Court held that, when Congress set aside land for Fort Belknap, the water necessary to it was implicitly set aside. The Winters Doctrine, as it has come to be called, applies also to federal public lands, such as national parks, wildlife refuges, national forests, military bases, and other public uses. Federal reserved water rights may go unused for many years, unlike other appropriative rights, and they are generally immune from state water laws.

However, federal court decisions have limited federal reserved rights in various ways. They are limited to the minimum use necessary for the purposes of the set-aside lands, which has to be determined by a state-based adjudication or negotiated agreement. In addition, they must be solely for the *primary* purpose of the set-aside lands, not secondary purposes. Thus, quantifying the federal right is not only very important but also the amount of water claimed by the federal government is often exceedingly controversial. For example, federal reserved water rights can be used in assessing oil shale resources on public lands, but not the water necessary to develop the oil shale—for that developers must apply for water rights through the state's legal system.[19]

Water Infrastructure

A somewhat shadowy but vitally important area of the water picture in the United States is water infrastructure. This ranges from the water pipes for a town of 500 people to vast federal water projects, such as the dam system on the Colorado River.

Local Water Infrastructure

Local water infrastructure nationwide is what ensures that water comes out when you turn the tap on. But the American water infrastructure is aging; in 2013, the American Society of Civil Engineers rated the nation's infrastructure as a "D+," recognizing that there are 240,000 water main breaks *per year* in the United States. Not only that, but aging pipes leak; as much as seven billion gallons of water leak from U.S. infrastructure every year. A 2012 report by the American Water Works Association estimated that something on the order of a million miles of underground pipes need replacing.[20]

The problem is that local governments spend their infrastructure money on operation and maintenance, not repair and replacement, as the facilities are expensive just to operate. Congress's decision to switch to loans for water infrastructure through state revolving funds has meant communities must shoulder the full costs of their water infrastructure, without federal funds to pay for it.

In 1977, at its peak, more than 70 percent of all capital spending for drinking water and sewage infrastructure came from the federal government. But the grant program was phased out in 1990, and federal funding for water infrastructure dropped to one-sixth of the total. Eighty-three percent of the funding now comes from state and local sources. Though federal loans help, the burden basically falls to local communities, where raising water rates to pay for infrastructure is politically unpopular.[21]

The solutions vary, ranging from better preventative maintenance for infrastructure to developing public–private partnerships for water delivery. Although most water delivery entities were private during the 19th century, by 1970, 80 percent were public. Now, about a quarter of U.S. consumers' water is at least partly privatized. Privatization of water—a public resource—can lead to deteriorating water quality and public inability to participate in the protection and conservation of their own drinking water.[22]

Not investing sufficiently in water infrastructure has drastic consequences, as exemplified by Superstorm Sandy, which hit the east coast in late October 2012. It caused 11 billion gallons of sewage to spill into waterways when the storm knocked out coastal sewage plants. Similarly, in July 2014, a 93-year-old water main ruptured near the University of California Los Angeles campus. This caused the loss of 20 million gallons of water during the state's worst drought in recorded history.[23]

Interstate Water Infrastructure

All the major rivers in the United States flow through more than one state and provide multiple uses to millions of people through large-scale water infrastructure projects, ranging from dams and reservoirs to intricate irrigation networks. Each river basin is enormous and diverse, and though each has multiple federal statutes applying to it, there is no single federal statute for solving interstate disputes. Though the U.S. Army Corps of Engineers and Bureau of Reclamation manage and/or operate the interstate infrastructure, they do not provide an overall structure for interstate water management.

Water planning is splintered between local, state, and the federal government, as well as private parties. Contributing to the fragmentation is the general lack of a comprehensive, long-term strategy governing activities in the interstate river systems. Thus the "state of the river" is impossible to even gauge, especially since the federal agencies tend to operate via annual plans, which have the force of law. Multiple, diverse interstate water compacts divide up the use of river water. There are also international boundary commissions to wrestle with water-use problems in rivers that flow between the United States and Canada or Mexico. Adding to the confusion, federal courts often face complex cases of interstate or between-user (such as Native peoples and a state) water allocation disputes, or cross-boundary environmental problems.[24]

The first case allowing federal courts to adjudicate interstate water allocation came before the U.S. Supreme Court in 1902. Kansas sued Colorado for diverting so much water from the Arkansas River that it ran dry in the summer. Colorado argued it could justifiably use all the water for beneficial purposes within the state's boundaries, including all the waters of the Arkansas River, even if that meant all residents of Kansas were deprived of any water from the river. The Supreme Court disagreed. Even so, the Court ruled that Kansas had not provided enough information for the Court to rule on the merits. Kansas tried again in 1907, but the Court ruled again that Kansas had not made its case. Nevertheless, the Court held a state could not withdraw waters in its borders to the "substantial detriment" of the downstream state. Each state had an "equitable" right to the river.[25]

Congress has only twice allocated interstate river water: with the Boulder Canyon Project Act of 1928, and in 1990 with the Truckee and Carson Rivers between Arizona and Nevada. The most common method of river allocation is through interstate compacts. Interstate water compacts differ greatly; some focus on water allocation, others on water operations, while some are merely planning documents. It is important to remember that an interstate compact is a binding, legal contract. There are 38 interstate compacts in the United States, most of them in the Western states, and they deal with subjects as diverse as water allocation, flood prevention, water planning, and pollution control.

A major purpose of federal involvement has been the building and maintaining of large dams. The majority of large dams on U.S. rivers have been built by the Army Corps of Engineers (75 dams) and the Bureau of Reclamation (58 dams). The regional Tennessee Valley Authority is responsible for another 39 dams. Large storage reservoirs are also usually owned and operated by the same two federal agencies; this includes reservoirs such as "Lake" Mead and "Lake" Powell.

The Corps of Engineers is the older agency, tracing its origins to the Continental Congress's establishment of a "chief of Engineers in the Continental Army" during the American Revolution. The Corps builds dams mainly for navigation, flood control, and hydropower, but not water supply and irrigation. The Bureau of Reclamation, on the other hand, was created in 1902 specifically to build irrigation and water storage projects in the arid West, and now has about 180 projects in 17 Western states, where its activities must conform to state water rights law (Figure 11.2).

As for private dams, they are usually constructed for hydropower, flood control, or irrigation. They not only are more numerous and much smaller than federal dams but also are most frequently located on within-state rivers. The Federal Energy Regulatory Commission licenses private dams over a certain size.[26]

FIGURE 11.2
Glen Canyon Dam on the Colorado River forms Lake Powell in northern Arizona. Photograph by Nikater (https://commons.wikimedia.org/wiki/File:Glen_Canyon_Dam02.jpg).

Pollution and the Hydrological Cycle

Chemical pollutants, such as pesticides, move through air, soil, and water—ultimately concentrating in the oceans. The nature of the pesticide and how it is transported determines where it will move, where it will collect, at what speed, and how long it will remain in the environment. Pesticides can build up anywhere that air, soil, and water transport them. In addition, they are transported over a wide range of environments in the tissues of such organisms as plants (crop residue, crop plants shipped to market, and driftwood from tree farms), insects, fish, birds, and mammals, including humans.

Agriculture and Pollution: Same Story, Different Place

As author Donald Worster says, "Obviously, agriculture involves the rearranging of Nature to bring it more in line with human desires, but it does not require exploiting, mining, or destroying the natural world."[27] This is an important concept, because the ocean ultimately waters all agricultural crops worldwide, with the aid of the winds that carry the risen moisture, which falls as rain, is stored as snow, or as ancient water pumped from below ground though an irrigation system. We, on the other hand, poison the very ocean that nurtures us by polluting the waters on their return to the ocean from which they came—the ocean, where all pollutants accumulate worldwide (Figure 11.3).

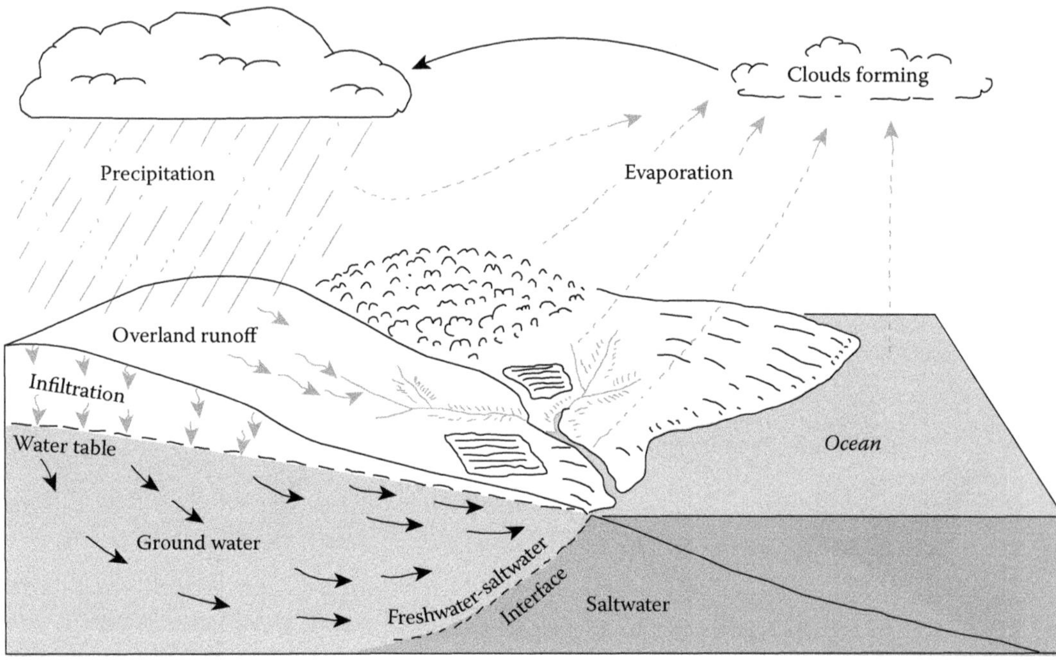

FIGURE 11.3
Hydrological cycle. Illustration by Ralph C. Heath, U.S. Geological Survey (https://commons.wikimedia.org/wiki/File:Hydrologic_cycle.png).

The majority of today's industrial farmers are not only addicted to synthetic chemicals but also addicted to the chemical-based method of farming. Beyond the application of synthetic fertilizers, the natural gas used in their production accounts for 90 percent of the cost of the ammonia, which is the basis for the nitrogen fertilizer applied to such crops as corn. The pesticides and herbicides required to produce these vast monocultures are also gas-based petrochemicals. Then there is the substantial amount of diesel fuel needed to operate the farm machinery. And this says nothing of the enormous quantity of water such exceedingly thirsty crops as corn require, which in turn leaches agricultural non-point source pollutants into ditches, streams, and rivers, as well as groundwater.

How Aboveground Water-Borne Pollutants Get into the Ocean

Probably the most wasteful water user worldwide is agriculture, which uses about 70 percent of the Earth's accessible freshwater—not only more than twice the 23 percent used by industry but also dwarfing the 8 percent used by municipalities. The primary, unsustainable, wasteful uses of water in agriculture are: (1) leaky irrigation systems, including open canals to transport water from place to place; (2) wasteful field methods of application, such as irrigating during the heat of summer days; and (3) cultivation of thirsty crops not suited to the environment. Besides, the problem is made worse by misdirected subsidies, low public and political awareness of the crisis, and weak environmental legislation and laws.[28]

Soil, the main terrestrial vessel, receives, collects, and passes to the water all airborne, human-caused pollutants. In addition, such pollutants as sewage, excess chemical fertilizers, pesticides, and oil are added directly to the soil—and through the soil to the water, which, bearing tons of toxic effluents, is transported into ditches, streams, and rivers to the point where they enter the ultimate vessel, the combined oceans of the world. Water is a captive of gravity, so all the pollutants it accumulates on its downhill journey eventually end up in oceans worldwide.[29]

Water pollution is usually invisible, with the effects appearing distant in time and place from where the pollution occurred. To illustrate, ditches along roads—such as those adjoining agricultural fields, golf courses, and so on—form a continuum or spectrum of physical environments (the same as streams and rivers), a longitudinally connected part of the biophysical system in which *downstream* processes are linked to and influenced by *upstream* processes. The ditch continuum begins with the smallest ditch and ends at the ocean. So, little ditches feed bigger ditches, and bigger ditches eventually feed streams and rivers that ultimately feed the ocean.

Here the question is: What happens to the continuum concept when a ditch is polluted? To pollute a ditch means to contaminate it with chemical-laden runoff from such areas of heavy chemical uses as agricultural fields and golf courses, or by dumping human garbage into it. Ditches are also contaminated by discharging noxious substances, such as oil and hydraulic fluid from vehicles and farm equipment, both of which disrupt biological processes, often by corrupting the integrity of their chemical interactions.

Although Nature's organic matter (food energy) from the forest is continually diluted the further down the stream continuum it goes, pollution (especially chemical pollution) is continually *concentrated* the further down the ditch continuum it goes because it gathers its potency from the discharge of every contaminated ditch that adds its waters to the passing flow. Hence, with every ditch that is polluted, the purity of the stream and river accepting its fouled discharge is to that extent compromised. The amount of pollution that is discharged from municipal water-treatment facilities and/or that bypasses them to be dumped directly into the estuaries and oceans of the world through the stream/ditch continuum is staggering.[30]

Consider, for example, just one small, local creek, Dixon Creek, flowing through the heart of north Corvallis, Oregon, my (Chris') hometown. The creek drains 4.8 square miles or 3,041 acres, which includes upstream use that is 18 percent agriculture, 20 percent forest, and 62 percent residential.

According to a 1998 study of water quality in the Willamette River basin of western Oregon conducted by the U.S. Geological Survey, Dixon Creek carried traces of nine pesticides in its waters, as well as fecal coliform bacteria. The creek, only a few miles long and almost entirely within the city limits, flows past churches, parks, homes, and even through the campus of Corvallis High School before emptying into the Willamette River, which in turn empties into the Columbia River, which empties into the Pacific Ocean.

The following pesticides found in Dixon Creek are listed in descending order of concentration:

1. *Dichlobenil*, an herbicide
2. *Tebuthurion*, an herbicide not known to be hazardous to aquatic organisms
3. *Diazinon*, an insecticide toxic to fish
4. *Carbaryl*, an insecticide moderately toxic to aquatic organisms
5. *Prometon*, an herbicide
6. *Metolachlor*, an herbicide moderately toxic to both cold- and warm-water fish
7. *Atrazine*, an herbicide slightly toxic to fish and other pond life

8. *Desethylatrazine*, an herbicide slightly toxic to fish and other pond life, which breaks down in the environment
9. *Simazine*, an herbicide of low toxicity to aquatic species

Now consider that these poisons add to the diversity of the environment not only through their different chemical makeup but also in how they kill, what they kill, in what concentrations they can kill, how they move through the food chain killing as they go, and the deadly synergism of their chemical interactions when they come in contact with one another. Consider also how they alter the ecosystem in which their respective effects become manifest.[31]

Where might such effects become visible? Ask the Mississippi River.

As the water of the Mississippi River flows toward the Gulf of Mexico, collecting runoff from the Appalachian Mountains to the Rocky Mountains and everywhere in between, it passes through 10 states, through massive agricultural fields and by numerous towns and cities. It not only gathers fertilizers and pesticides from the Corn Belt along its journey but also leached sewage from the urban areas. By the time the Mississippi enters the Gulf, its current has been transformed into a conduit for chemical nutrients, and this enriched current stimulates massive blooms of phytoplankton. Consequently, it forms a dead zone the size of Massachusetts (7,900 square miles) every summer, which has existed since the 1970s and supports almost no life beyond phytoplankton and bacteria.[32]

A dead zone is an aquatic area lacking sufficient dissolved oxygen to support life. Today, the world's oceans are being increasingly plagued by human-caused dead zones (analogous to an "oxygen desert"), of which 530 occur near inhabited coastlines worldwide, where aquatic life is most concentrated. A dead zone is caused by a process known as "eutrophication" triggered by an excess of plant nutrients from fertilizers, livestock manure, human sewage, and other pollutants. However, the use of chemical fertilizers is considered the major human-related cause of dead zones around the world. ("Eutrophic" come from the Greek *eutrophia* ["healthy, adequate nutrition"] and the Greek *eu* ["good"] plus *trophe* ["food" or "feeding"] which taken together means "well fed." Thus, "eutrophication" is the process of being well fed.)

The size of dead zones, which fluctuate seasonally, is driven largely by climate and such weather patterns as wind, precipitation, temperature, and the inflow of rivers. To illustrate, the dead zone in the Gulf of Mexico in 2012, which develops every spring and summer, was the fourth smallest since measurements of the zones began in 1985. The zone measured 2,889 square miles, according to the report released by the Louisiana Universities Marine Consortium on July 27, 2012. In 2011, it measured about 6,765 square miles. It shrank due to a record drought across the country, which meant fewer nutrients were washed into the Mississippi River and thus into the Gulf.

With respect to the Gulf of Mexico in particular, much of the water entering the Mississippi comes from massive fields of corn, which is grown in soil with tile drains that allow more nitrogen to seep into the river faster and in greater quantities than from crops without drainage tiles. Therefore, making ethanol from corn not only could cause more cornfields to be planted but also could exacerbate the dead zone in the Gulf of Mexico—perhaps beyond repair.[33]

Moreover, some pesticides break down into, or combine with, compounds that are even more toxic than the original chemical. Thus, under certain conditions, tiny amounts can accrue high toxicity. In addition, the only portions of pesticides that are tested for toxicity are the active ingredients, which usually form a minute portion of the solution. This means that the larger, untested portion of most pesticides, which contain other, so-called "inert" chemicals, can be even more toxic than the active ingredients. It is critical to understand, however, that "inert" is an industrial euphemism whereby the illusion of chemical inactivity is suggested. *But a truly inert substance is a biophysical impossibility in an interactive system* because there is no such thing as an independent variable.

While moving water can dilute chemicals put into it, they concentrate when moving water enters "still" bodies of water, such as lakes, which are separate entities, and oceans, which have common connections. The frequent flippant comment that "the solution to pollution is dilution" is not only untrue but dangerously misleading.

There is yet another profound difference between lakes and oceans, however. Most lakes have outlets, which allow inflowing water to flush them of pollutants to a greater or lesser degree, depending on the lake. But oceans have no outlets whereby these pollutants can be flushed, so they are continually concentrated through the inflow of contaminated streams and rivers. As the airborne moisture condenses into drops of rain, it collects pollutants on its journey back to the ocean, where they can only become part of the endless, self-reinforcing feedback loop of toxic chemical compounds—thereby affecting such animals as sharks and dolphins, which store pollutants in their body fat, and polar bears, which suffer from pollutant-induced shrinking of their gonads.[34]

How Belowground Water-Borne Pollutants Get into the Ocean

Water, in the form of rain or snow, not only washes and scrubs chemical pollutants from the air but also leaches them from the soil as it obeys the call of gravity. Not all pollutants are carried in trickle, ditch, stream, and river to be concentrated in the oceans of the world. Some are concentrated in groundwater, including subterranean lakes. Moreover, it's extremely difficult to stop the pollution of groundwater, especially from synthetic fertilizers like those used to produce corn, which include nitrogen.

Clearly, pollution of groundwater comes from many sources, especially the use of agricultural pesticides and fertilizer, combined with massive overuse of underground aquifers. Hydraulic fracturing, and specifically the injection of wastewater from the process, is a new and highly damaging form of groundwater pollution. "Once polluted," counsels ecologist Eugene Odum, "groundwater is difficult, if not impossible, to clean up, since it contains few decomposing microbes and is not exposed to sunlight, strong water flow, or any of the other natural purification processes that cleanse surface water."[35]

Unknown to most people, there is a *belowground* analog to the aboveground journey of water in subterranean seeps, trickles, and rivulets, which coalesce into streams and rivers that flow from the mountains to the ocean entirely below ground. On reaching the oceans, they enter the marine environment through porous soils along beaches, just below the salty surface, or erupt as freshwater springs on the ocean floor of the continental shelf near many of the world's shores, where the freshwater influences the dynamics of the marine biophysical system. Around 480 cubic miles of freshwater enter the world's oceans each year as submarine groundwater, although some coastlines provide considerably more than others.[36]

The pressure of ocean water does not control the groundwater discharge. In fact, a submarine spring can flow equally well whether in shallow coastal waters or at the bottom of a deep ocean trench. The volume of submarine-groundwater discharge to the oceans of the world represents an important vehicle for the delivery of myriad human-made, toxic, and carcinogenic chemical compounds from agricultural fields, tree farms, urban settings, industrial complexes, and from drilling for oil and natural gas.[37]

There is little doubt that human activities since the onset of the Anthropocene have distributed chemical pollutants worldwide through the air and both aboveground and belowground waterways. Moreover, these pollutants are primarily responsible for the increased concentrations of numerous chemicals in biological tissues worldwide, such as Arctic foxes that feed on oceanic prey. For example, the levels of mercury have not only doubled in the world's oceans over the past 100 years but also increased 10-fold in the Arctic's top predators.

It is possible that, during a rapidly changing climate, controlling the emission of some contaminants may be followed by long delays before a reduction in levels of contamination is measurable in the food web—particularly for those chemicals prone to biomagnification (= growing concentrations) in food webs, like mercury. This delay would be augmented for those chemicals that become archived in large quantities over long periods in reservoirs, such as global soils and oceans.[38]

Although some settle into the deep ocean, where a portion is buried in sediments, some are destroyed by sunlight in the atmosphere. Both of these are slow processes, however, because they are continually re-entering the atmosphere from the ocean, before being dissolved again in a recurring feedback loop.[39]

Climate Change and the Hydrological Cycle

Climate change is also affecting the water cycle and water supply, making it far more difficult for rivers to sustain themselves and still provide for human needs. For example, a team of Peruvian and international scientists estimated that Peru and Bolivia together account for more than 90 percent of the world's tropical glaciers. However, the glaciers have lost about a third of their surface area between the 1970s and 2006. And this loss includes glaciers on Huascaran, Peru's largest mountain, which reaches an elevation of 22,200 feet. Moreover, the changing climate is melting the glaciers faster than in decades past and making the flow of rivers increasingly irregular, which is leading to more droughts and floods.

On the economic side, the dwindling supply of water affects every Peruvian household, to say nothing of the 80 percent of the country's power, which is hydroelectric, as well as its agricultural exports and mining, both of which absorb huge volumes of this precious liquid.[40]

Another example is Spain's Ebro River, which at 578 miles long is the third longest river emptying into the Mediterranean. Beginning on Spain's north coast, it flows through many regions before fanning out into a 79,000-acre delta in the Cantabrian region. The delta is a national park, internationally recognized wildlife and bird area as well as a major agricultural region. But the Ebro is threatened by loss of its sediments, which protect the integrity of the wetlands. Development, irrigation, and dams have destroyed 90 percent of the sediments. These same factors threaten the river's flow and its ability to nourish the delta with regular flooding. Spain has been accused of contravening European Union law that requires setting a minimum environmental flow for rivers before parceling out excess water—if there is any. Spain has denied the charges. Decades of political and regional conflicts have prevented planning to grapple successfully with the Ebro's biophysical integrity.[41]

Environmental Laws

Fundamentally, the effort to protect the nation's waters is a regulatory system, beginning with federal laws and descending through state requirements to local laws. The Federal Water Pollution Control Act of 1948 was the first major U.S. law focusing on water pollution.[42]

But concern over pollution accelerated as bad news piled up in the following decades: pollution in Chesapeake Bay caused $3 million losses annually to the fishing industry in 1968; bacteria levels in the Hudson River in 1969 were reported to be 170 times the safe limit; 1969 also saw record numbers of fish kills nationwide, involving more than 41 million fish. This included the largest single fish kill ever, of 26 million fish in Lake Thonotosassa in Florida from discharges of four food-processing plants. Most frightening of all, a floating oil slick on the Cuyahoga River in Ohio burst into flames in June 1969. All these events and many more led to wide-ranging amendments in 1972, and the new law became known as the Clean Water Act.

Clean Water Act

The Clean Water Act created the basic structure for regulating polluting discharges into the "waters of the United States." Essentially, the law established a national commitment to restore and maintain the biophysical integrity of American waters. The Environmental Protection Agency (EPA) was given authority to create and implement pollution-control programs for industries. The Clean Water Act made it illegal for anyone to discharge a pollutant from a "point source" into navigable waters, unless the individual or entity obtained a Clean Water Act permit. (Point source pollution is that which comes from known, discrete source, such as a factory pipe.) The Clean Water Act also funded the construction of sewage-treatment plants in many communities nationwide. Unfortunately, the construction grants program was phased out beginning in the late 1980s.

A principal reason for the Clean Water Act's many successes is the provision in the law allowing private citizens with "standing" the ability to enforce the Clean Water Act laws when the government could not or would not undertake the task. ("Standing" means someone who can prove they have an interest that may be "adversely affected" by a polluter's actions.) Thus, citizens could sue violators in federal court.

With this provision available, local environmental groups nationwide took up the challenge, and there have been many astounding successes under these provisions. First, however, the local residents had to collect information on the suspected violator, including (where possible) water quality samples. This has led to a nationwide network of sophisticated groups who are familiar with water-sampling techniques and mechanisms whereby to detect pollution.

Yet even these provisions of citizen enforcement have not been enough. Why?

Despite the many successes through the Clean Water Act, improvements to water quality have declined; water quality is deteriorating nationwide, according to state and local monitoring reports. One reason is that the Clean Water Act's jurisdiction over small streams and wetlands, which are critical to downstream water bodies, is very unclear. The reason for this is inconsistent interpretation by the courts. The Environmental Protection Agency proposed a new Clean Water Rule to clarify their jurisdiction over these small bodies of water. But developers, and especially large-scale agriculture operations, which fear regulation of fertilizing and ditch-building, continue to oppose it.[43]

At the same time, it is clear that non-point sources of pollution—diffuse pollutants such as those from agriculture's use of fertilizers and pesticides, logging, and urban stormwater runoff—are fueling the aforementioned dead zones, such as that in the Gulf of Mexico.

Toxic algal blooms caused by non–point source pollution in the Lake Erie watershed recently poisoned drinking water for half a million people in Toledo, Ohio.[44] Non–point source pollution is increasingly the target of lawsuits at every level of the court system. The Clean Water Act does not regulate nonpoint-source pollution very well, and increasingly the nation's waterways show it.[45]

Safe Drinking Water Act

The other major federal law protecting the nation's waters is the 1974 Safe Drinking Water Act. This Act gives the Environmental Protection Agency authority to protect drinking water and its sources by setting national, health-based standards to protect against natural and human-caused contaminants. Drinking water standards vary by type and size, but the Safe Drinking Water Act applies to every public water system in the United States, of which there are more than 170,000 nationwide. The Safe Drinking Water Act sets the national standards, which determine maximum contaminant levels allowable in drinking water. State partners cooperate with the Environmental Protection Agency to enforce these standards in communities nationwide. The tools to limit drinking-water pollution are many, ranging from the protection of water sources, water treatment, integrity of distribution systems, and public information.[46]

Nevertheless, the system has many nationwide failures. I (Cameron) work with many communities on the Oregon coast, such as Yachats and Rockaway Beach, that are in a state of crisis because their source watersheds are partly or wholly owned by private timber companies. These companies, limited only by the very weak provisions of the Oregon Forest Practices Act, practice clear-cut logging. The result is vast swaths of clear-cut watersheds, webbed with logging roads, and thus subject to massive landslides and stream sedimentation. Once the lands are replanted to tree farms, the companies spray large quantities of herbicides to discourage unwanted vegetation from competing with their crop trees. Consequently, I frequently deal with frantic communities exhausted by efforts not only to learn when herbicides will be sprayed (state law has very lax and vague notification standards) but also to get the Department of Environmental Quality to test the water for contamination as a result of the spraying. Efforts to get the Oregon Legislature to provide protection of source watersheds have failed thus far, to the great distress of coastal residents.

Finding the Waters Again

Where, in this vast jumble of individual and federal water rights; local infrastructure; and massive, national water infrastructure are the waterways themselves, and their rights to exist and flourish? No existing law, federal or otherwise, considers this question.

In *Laudato Si*, Pope Francis devoted a good deal of thought to the problem of water, and committed an entire section of the encyclical to the looming problems: "Fresh drinking water is an issue of primary importance, since it is indispensable for human life and for supporting terrestrial and aquatic ecosystems...now in many places demand exceeds the sustainable supply, with dramatic consequences in the short and long term... *access to safe drinkable water is a basic and universal human right, since it is essential to human survival and, as such, is a condition for the exercise of other human rights.*"[47]

It is extremely unlikely that the current regulatory frameworks that control waters in the United States will produce a sustainable supply for humans, whether locally or nationally, rurally or in urban situations; it has not yet occurred despite ever-greater complexity and intricacy of the regulatory system. Nor does privatization of water contribute to the solutions; on the contrary, it worsens the problems. *Laudato Si* also warns against privatizing water and "turning it into a commodity subject to the laws of the market."[48]

No current framework of water law comes even remotely close to ensuring that water systems can sustain themselves, nourish bays, estuaries and deltas, provide clean water, fish habitat, and a myriad of other benefits. Relatively recent attempts to focus on the biophysical requirements of a watercourse itself come through elaborate tweaks of the water-law systems to provide for "instream" integrity. It is possible, for example, to purchase an old water right in the Prior Appropriation system and retire it from use for a human purpose. It is thus returned to the stream to provide water essentially "belonging" to the stream. But the process of creating, purchasing or legally creating instream rights is slow, patchy, subject to prior laws, and not occurring in any over-arching framework that places the biophysical requirements of the watercourse *first*.

I (Cameron), working in environmental activism on the Oregon coast, frequently deal with water issues. Policymakers often think that the Oregon coast has no water-supply problems: the population is low, and the rainfall averages 65–90 inches per year. But, in fact, human water use peaks in the driest months of June through September, and summer water shortages in coastal communities are often acute. I have fought against three coastal destination resort proposals in Coos and Curry Counties on the south coast, all three of which had severe water constraints that contributed heavily to questions about their viability. In one case, the proposed water supply was from groundwater, though studies and bore holes indicated insufficient supply for a golf course. In a second instance, the main proposed potable water source was from small surface streams, the use of which would have caused them to run dry in summer months. The third resort proposed groundwater, and then use of a nearby town's sewage effluent, to find adequate water for golf course irrigation, though both options created potentially heavy costs for nearby rural well users and town residents.

Management Considerations for a Rights of Nature System

The primary, overarching consideration is that human use of water must take place inside a Rights of Nature system, in which the watercourse's ability to flourish has absolute primacy. That is, there must be enough water, both in quality and quantity, to sustain the watercourse itself and all the ecosystems that depend on them. This requires assessment of the actual, necessary flow of water for a stream or river—not the minimum flow that humans feel constrained to "allow" the stream or river so it will keep sacrificing itself for human needs, but rather the water inherent to the stream or river's own ability to flourish.

Once that framework is in place, there are many policies, changes, and repairs individuals and communities must make in order to restore the hydrological cycle, conserve water, restrict its use, and ensure a sustainable supply. As with land, the issues rapidly become local, and must be dealt with in part at local levels. Nevertheless, the Rights of Nature principles are similar, and must be applied across the social-environmental spectrum to repair and maintain freshwater for human use. There are no alternatives to water.

The major points for a Nature's Rights of water—for any beginning discussion, followed by detailed laws and systems of use, are:

1. Ensure that Rights of Nature laws enacted under federal and state Constitutional provisions focus on the entire hydrological cycle. This is critical, since rivers often flow through many states, and the snowpack of mountains in one state frequently provides the summer flow for rivers in states far away. The water cycle is highly vulnerable to overuse without considering the overall hydrological region.
2. Retool federal and state legal systems regulating the use of water to provide a hierarchy of responsibilities and obligations between humans and the hydrological cycle. This could be modeled in part on the feudal chain of obligation discussed earlier. Currently, the United States has a complex, hierarchical system of water regulation based only on human need, which proliferates down the chain and fails at every level to protect the inalienable rights of both surface waters and belowground waters to exist and replenish themselves.
3. Provide in statute the provision for baseline data on the river's needs, which in most instances will require historical research on the state of the watercourse before human overreaching degraded its ability to flourish. This is similar to the environmental assessment required under the National Environmental Policy Act for federal actions, and some similar statutes. However, unlike those requirements, this is water body-focused rather than human-focused.
4. To prevent abuse by self-interested polluters, levy a general water tax as part of a market pricing structure for use of water, either at local levels or state levels. This tax would be used for research on the needs of a given body of water, such as the requirements for restoration and the methods of undertaking them, as well as providing grants to aid in the work.
5. Create a statutory standard to be used in restoration of a harmed watercourse or aquifer. Restoration must aim to return the water body as close to its pristine state as can be managed, given settlement patterns and irreducible changes in the area—not just the cost of restoring the watercourse to its state before the injury.
6. Revise state and local water laws away from the Riparian and Prior Appropriation doctrines. The new standard must be that no upstream user may harm a downstream user, whether of a large interstate river, such as the Colorado, or a local stream and lake in Appalachia. Under a Rights of Nature system, a "downstream user" includes the river or lake itself, as well as human users. This will eliminate the unwieldy distinction between point and non-point pollution, and streamline the regulatory framework to focus on prohibiting damage and requiring the polluter to pay the full cost of restoration to the pristine state.

7. Protect, repair, and improve local water infrastructure. Provide generous federal grants and federal/state programs, financed by the water taxes, to aid communities and states in this gigantic task. The end of federal grants in the 1980s was devastating to the nation's water infrastructure, and such grant programs must be resurrected. Improvements must prioritize repairing and upgrading existing infrastructure—for which the need is exceedingly desperate.[49]
8. Encourage and fund via federal grant programs emerging and tested technologies such as water-cleansing membranes, equipment to remove phosphorus and nitrogen, irrigation sensors, green infrastructure, permeable paving to reduce impermeable surfaces, grassy roofs, bioswales, and water recycling to use water more effectively and purify it during its use cycle in human communities.
9. Secure protection of local water catchments through purchase, lease, rent, conservation easement, tax credit, or other agreement. With water in mind, it is wise to purchase as much of the local water catchments as possible and maintain them as open space expressly for the purpose of collecting and storing water in the ground, where it can purify itself, as it flows slowly toward the wells it recharges.
10. Research and implement market-rate pricing for domestic and agricultural water use, to encourage curtailment of wasteful use, and implementation of conservation strategies. This must go hand in hand with grants and technical expertise to find innovative techniques to conserve water, such as drip irrigation and reuse of municipal wastewater.
11. Provide statutory opportunities for both qualified nonprofit corporations and government agencies—whose mandate is protection of water resources—to sue in court for the right of the water body to flourish free of human disturbance. Also expand the right of qualified nonprofits to sue polluters, as currently exists under the Clean Water Act, to plead for full restoration of the damaged water body. The standards for mandating such restoration must be broad and clear, to show they are the preferred alternative in the statute, and to ensure courts interpret them generously.
12. Pass laws to encourage and provide grants for local experimentation with innovative pacts, agreements, and other instruments to protect watercourses and water catchments, along the lines used by New Zealand to define the Whanganui River as a living entity, and partner with indigenous people who historically considered the Whanganui part of their community.
13. Create a program similar to the "America's Most Endangered Rivers" report published annually by American Rivers.[50] Each year, they highlight 10 rivers around the country threatened by major development projects to be decided in the coming year, projects that—if implemented—will compromise the rivers' integrity. Publish these threats and initiate research to verify the danger to the river if the project goes ahead.
14. Pass laws setting up a national and state "River Ombudsman" or "Monitor" whose task is to identify major threats to waterways nationwide, research how deeply they compromise the rivers' right to life under the Rights of Nature system. Nationally, and perhaps also at state levels, give the River Ombudsman or Monitor power to commence the process to prohibit projects that damage the river's right to flourish under the Rights of Nature system.

Endnotes

1. [Ecuadorian] Constitution, full language. *op. cit.* Article 72.
2. *Ibid.* Article 72.
3. Language of the Bolivian law (compete text). *op. cit.* Chapter II, Article.
4. *Ibid.*
5. *Ibid.* Chapter 3, Article 7, I (5).
6. *Ibid.* Chapter 3, Article 7, I (6).
7. Universal Declaration of Rights of Mother Earth: Global Alliance for the Rights of Nature. *op. cit.* Article 2 (1) (a–b).
8. *Ibid.* Article 2 (1)(e).
9. Chapter 3. *op. cit.*
10. The foregoing two paragraphs are based on: David Hulse, Stan Gregory, and Joan Baker (eds.). *Willamette River Basin Planning Atlas: Trajectories of Environmental and Ecological Change.* Oregon State University Press, Corvallis, OR. (2002) 192 pp.
11. How much water is there on, in, and above the Earth? http://water.usgs.gov/edu/earthhowmuch.html (accessed February 18, 2016).
12. The foregoing discussion of stream-order is based on: Chris Maser and James R. Sedell. *From the Forest to the Sea: The Ecology of Wood in Streams, Rivers, Estuaries, and Oceans. op. cit.*
13. (1) Jianchu Xu, R. Edward Grumbine, Arun Shrestha, and others. The melting Himalayas: Cascading effects of climate change on water, biodiversity, and livelihoods. *Conservation Biology*, 23 (2009):520–530; and (2) Soutik Biswas. Is India facing its worst-ever water crisis? http://www.bbc.com/news/world-asia-india-35888535 (accessed March 5, 2015).
14. (1) David Hazony. How Israel Is Solving the Global Water Crisis. http://www.thetower.org/article/how-israel-is-solving-the-global-water-crisis/ (accessed August 10, 2016); (2) Ariel Rejwan and Yossi Yaacoby, Watech, Mekorot. http://www.oecdobserver.org/news/fullstory.php/aid/4819/Israel:_Innovations_overcoming_water_scarcity.html (accessed August 10, 2016); and (3) Karin Kloosterman. Israel's top ten water technology companies. http://www.israel21c.org/israels-top-ten-water-technology-companies/ (accessed August 10, 2016).
15. The preceding two paragraphs are based on: (1) Wells A. Hutchins and Harry A. Steele. Basic Water Rights Doctrines and Their Implications for River Basin Development. http://scholarship.law.duke.edu/cgi/viewcontent.cgi?article=2717&context=lcp (accessed February 13, 2016); and (2) Water Appropriation Systems. http://www.undeerc.org/Water/Decision-Support/Water-Law/pdf/Water-Appr-Systems.pdf (accessed February 13, 2016).
16. The preceding two paragraphs are based on: Water Appropriation Systems. *op. cit.*
17. *Ibid.*
18. Wells A. Hutchins and Harry A. Steele. Basic Water Rights Doctrines and Their Implications for River Basin Development. *op. cit.*
19. The foregoing two paragraphs are based on: Federal Reserved Water Rights. http://www.blm.gov/pgdata/etc/medialib/blm/co/field_offices/denca/DENCA_Advisory_Council/extra_materials_for.Par.33214.File.dat/FedResWaterRights.pdf (accessed February 13, 2016).
20. America's Neglected Water Systems Face a Reckoning. http://knowledge.wharton.upenn.edu/article/americas-neglected-water-systems-face-a-reckoning/ (accessed February 13, 2016).
21. The foregoing two paragraphs are based on: Brett Walton. Report: U.S. Water Systems, Deteriorated and Slow to Change, Need New Strategy—and Money. http://www.circleofblue.org/waternews/2014/world/u-s-water-systems-deteriorated-slow-change-need-new-strategy-money/ (accessed February 13, 2016).
22. America's Neglected Water Systems Face a Reckoning. *op. cit.*
23. Brett Walton. Report: U.S. Water Systems, Deteriorated and Slow to Change, Need New Strategy—and Money. *op. cit.*

24. The two paragraphs are based on: Dan Seligman. "Laws of the Rivers." http://crc.nv.gov/docs/Laws_of_the_Rivers.pdf (accessed February 15, 2016).
25. State of Kansas v. State of Colorado, (1902). http://caselaw.findlaw.com/us-supreme-court/185/125.html (accessed February 15, 2016).
26. The preceding discussion of interstate rivers and their management is based on: Dan Seligman. "Laws of the Rivers." *op. cit.*
27. Donald Worster. Good Farming and the Public Good in Meeting the Expectations of the Land. pp. 37–40. *In*: Meeting the Expectations of the Land: Essays in Sustainable Agriculture and Stewardship. W.B. Wes Jackson and Bruce Colman (eds.). North Point Press, San Francisco, CA (1984).
28. (1) Farming: Wasteful water use. http://wwf.panda.org/what_we_do/footprint/agriculture/impacts/water_use/ (accessed January 5, 2016); (2) Jason Clay. *World Agriculture and the Environment: A Commodity-by-Commodity Guide to Impacts and Practices.* Island Press, Washington, DC. (2004) 570 pp.; (3) David Pimentel, Bonnie Berger, David Filiberto, and others. Water Resources: Agricultural and Environmental Issues. *BioScience*, 54 (2004):909–918; (4) Jeff Guo. Agriculture is 80 percent of water use in California. Why aren't farmers being forced to cut back? *The Washington Post*, https://www.washingtonpost.com/blogs/govbeat/wp/2015/04/03/agriculture-is-80-percent-of-water-use-in-california-why-arent-farmers-being-forced-to-cut-back/ (accessed January 5, 2016); and (5) Farms Waste Much of World's Water. *Associated Press*, March 19, 2006 http://www.wired.com/2006/03/farms-waste-much-of-worlds-water/ (accessed January 3, 2016).
29. (1) E.P. Sauer, P.A. Bower, M.J. Bootsma, and S.L. McLellan. Detection of the human specific *Bacteroides* genetic marker provides evidence of widespread sewage contamination of stormwater in the urban environment. *Water Research*, 45 (2011):4081–4091; (2) R.M. Litton, J.H. Ahn, B. Sercu, and others. Evaluation of chemical, molecular, and traditional markers of fecal contamination in an effluent dominated urban stream. *Environmental Science & Technology*, 44 (2010):7369–7375; (3) Willard S. Moore, Jorge L. Sarmiento, and Robert M. Key. Submarine groundwater discharge revealed by 228Ra distribution in the Upper Atlantic Ocean. *Nature Geoscience*, 1 (2008):309–311; (4) Anonymous. Effects of Human Activities on the Interaction of Ground Water and Surface Water. http://pubs.usgs.gov/circ/circ1139/pdf/part2.pdfă (accessed June 13, 2013); and (5) Soutik Biswas. Is India facing its worst-ever water crisis? http://www.bbc.com/news/world-asia-india-35888535 (accessed March 29, 2016).
30. The discussion of ditches is based on: (1) Chris Maser. The humble ditch. *Resurgence*, 172 (1995):38–40; and (2) Chris Maser. Interactions of Land, Ocean and Humans: A Global Perspective. *op. cit.*
31. The discussion of pollution in the waters of Dixon Creek is based on: Scott MacWilliams. What's in the water? *Corvallis Gazette-Times*, Corvallis, OR (February 14, 1998).
32. The foregoing discussion of farm chemicals and pollution is based on: (1) Tad W. Patzek. Thermodynamics of the Corn-Ethanol Biofuel Cycle. *Critical Reviews in Plant Science*, 23 (2004):519–567; (2) Jason Hill, Erik Nelson, David Tilman, and others. Environmental, economic, and energetic costs and benefits of biodiesel and ethanol biofuels. *Proceedings of the National Academy of Sciences*, 103 (2006):11206–11210; (3) Alice Friedemann. Peak Soil. http://culturechange.org/cms/index.php?option=com_content&task=view&id=107&Itemid=1 (accessed October 28, 2012); (4) Lian Pin Koh. Potential habitat and biodiversity losses from intensified biodiesel feedstock production. *Conservation Biology*, 21 (2007):1373–1375; (5) Johan Rockström, Will Steffen, Kevin Noone, and others. A safe operating space for humanity. *Nature*, 461 (2009):472–475; and (6) David Tilman, Kenneth G. Cassman, Pamela A. Matson, and others. Agricultural sustainability and intensive production practices. *Nature*, 418 (2000):671–677.
33. The preceding discussion of dead zones is based on: (1) Dead zone. *Science Daily*. http://www.sciencedaily.com/articles/d/dead_zone_(ecology).htm (accessed January 10, 2015); (2) F. Chan, J.A. Barth, J. Lubchenco, and others. Emergence of anoxia in the California current large marine ecosystem. *Science*, 319 (2008):920; (3) Thomas O'Connor and David Whitall. Linking hypoxia to shrimp catch in the Northern Gulf of Mexico. *Marine Pollution Bulletin*, 54 (2007):460–463;

(4) Donald Scavia and Kristina A. Donnelly. Reassessing hypoxia forecasts for the Gulf of Mexico. *Environmental Science & Technology*, 41 (2007):8111–8117; and (5) Sarah C. Williams. Dead serious. *Science News*, 172 (2007):395–396.

34. (1) Janet Raloff. Sharks, dolphins store pollutants. *Science News*, 170 (2006):366; and (2) Christian Sonne, Pall S. Leifsson, Rune Dietz, and others. Xenoendocrine pollutants may reduce size of sexual organs in East Greenland Polar Bears (*Ursus maritimus*). *Environmental Science & Technology*, 40 (2006):5668–5674.
35. Eugene P. Odum. *Ecology and Our Endangered Life Support Systems*. Sinauer, Stanford, CT. (1989) 283 pp.
36. (1) Perrine Fleury, Michel Bakalowicz, and Ghislain de Marsily. Submarine springs and coastal karst aquifers: A review. *Journal of Hydrology*, 339 (2007):79–92; (2) D. Reide Corbett, William C. Burnett, and Jeffrey P. Chanton. Submarine Groundwater Discharge: An Unseen Yet Potentially Important Coastal Phenomenon. University of Florida IFAS Extension. http://edis.ifas.ufl.edu/sg060 (accessed February 19, 2016); (3) Sid Perkins. Invisible rivers. *Science News*, 168 (2005):248–249; (4) Takeshi Uemura, Makoto Taniguchi, and Kazuo Shibuya. Submarine groundwater discharge in Lützow-Holm Bay, Antarctica. *Geophysical Research Letters*, 38 (Issue 8), L08402 (2011): 6 pp. doi:10.1029/2010GL046394 (accessed February 19, 2016); and (5) Q. Liu, M. Dai, W. Chen, and others. How significant is submarine groundwater discharge and its associated dissolved inorganic carbon in a river-dominated shelf system? *Biogeosciences*, 9 (2012):1777–1795.
37. (1) Earl Bardsley. Conveying waste with water. *New Zealand Science Monthly*. http://nzsm.webcentre.co.nz/article449.htm; (2) Perrine Fleury, Michel Bakalowicz, and Ghislain de Marsily. Submarine springs and coastal karst aquifers: A review. *Journal of Hydrology*, 339 (2007):79–92; and (3) D. Reide Corbett, William C. Burnett, and Jeffrey P. Chanton. Submarine Groundwater Discharge: An Unseen Yet Potentially Important Coastal Phenomenon. University of Florida IFAS Extension. http://edis.ifas.ufl.edu/SG060 (accessed October 26, 2012).
38. The preceding two paragraphs are based on: (1) Matt McGrath. Mercury Exposure Linked to Dramatic Decline in Arctic Foxes. http://www.bbc.co.uk/news/science-environment-22425219 (accessed May 6, 2013); and (2) F. Wang, R.W. Macdonald, G.A. Stern, and P.M. Outridge. When noise becomes the signal: Chemical contamination of aquatic ecosystems under a changing climate. *Marine Pollution Bulletin*, 60 (2010):1633–1635.
39. The foregoing three paragraphs are based on: (1) Richard Lovett. Oceans release DDT from decades ago. *Nature News*, doi:10.1038/news.2010; (2) Irene Stemmler and Gerhard Lammel. Cycling of DDT in the global environment 1950–2002: World ocean returns the pollutant. *Geophysical Research Letters*, 36, L24602 (2009): 5 pp., doi:10.1029/2009GL041340; and (3) Asuncion Borrell and Alex Aguilar. Variations in DDE percentage correlated with total DDT burden in the blubber of fin and sei whales. *Marine Pollution Bulletin*, 18 (1987):70–74.
40. The preceding two paragraphs are based on: (1) A. Rabatel, B. Francou, A. Soruco, and others. Current state of glaciers in the tropical Andes: A multi-century perspective on glacier evolution and climate change. *The Cryosphere*, 7 (2013):81–102; (2) James Painter. Peru's Alarming Water Truth. March 12, 2007. http://news.bbc.co.uk/2/hi/americas/6412351.stm (accessed February 10, 2016); (3) Water Supply and Sanitation in Peru. 2007. http://en.wikipedia.org/wiki/Water_supply_and_sanitation_in_Peru (accessed January 11, 2016); and (4) Quelccaya Ice Cap. http://en.wikipedia.org/wiki/Quelccaya (accessed December 18, 2015).
41. Matt McGrath. Spanish water rights fight raises fears for Ebro delta. http://www.bbc.com/news/science-environment-35502084 (accessed February 17, 2016).
42. Federal Water Pollution Control Act (1948). http://www.encyclopedia.com/doc/1G2-3407400129.html (accessed February 19, 2016).
43. Codi Kozacek. United States Clean Water Rule Quandary Begins on Land. http://www.circleofblue.org/waternews/2015/world/united-states-clean-water-quandary-begins-on-land/ (accessed February 19, 2016).

44. (1) Michael Wines. Behind Toledo's Water Crisis, a Long-Troubled Lake Erie. http://www.nytimes.com/2014/08/05/us/lifting-ban-toledo-says-its-water-is-safe-to-drink-again.html?_r=0 (accessed February 19, 2016).
45. Codi Kozacek. United States Clean Water Rule Quandary Begins on Land. *op. cit.*
46. Understanding the Safe Drinking Water Act. http://www.epa.gov/sites/production/files/2015-04/documents/epa816f04030.pdf (accessed February 19, 2016).
47. Encyclical Letter *Laudato Si* of The Holy Father *op cit. Francis* on Care for Our Common Home [in English]. *op. cit.* (Section II, Chapters 28–30, emphasis in original).
48. *Ibid.* (Section 11, Chapter 30).
49. Brett Walton. Report: U.S. Water Systems, Deteriorated and Slow to Change, Need New Strategy—and Money. *op. cit.*
50. America's Most Endangered Rivers of 2015. http://act.americanrivers.org/page/content/mer-2015/#sthash.SwB4A71O.dpbs (accessed August 9, 2016).

12

Air and Climate

Air is perhaps the most difficult topic to consider in terms of environmental sustainability or (more specifically) in a Rights of Nature context because it is vast, ever shifting, generally invisible, and—even more than all the water on Earth—literally envelops the entire world. No part of the globe's surface is without air, and no system of human ownership can parcel it among users. Thus, to discuss the global health of the air with respect to life, or to a sustainable climate, is exceedingly challenging.

It is therefore necessary for nations and international organizations to have a vigorous, detailed, and multi-faceted discussion of what constitutes "rights" for air and climate. Such a discussion must explore how they would be recognized, measured, defended, and redressed, especially since air is, by definition, global. A first-order principal outcome must be that all nations are responsible for protecting the air and climate. But this will not be enough. Much of the discussion will be in technical, legal terms, as it should be, if Nature's Rights for air is to flourish through the world's legal systems. The discussion may result in new international pacts, international court structures, or international environmental crimes. Of all aspects of the environment needing attention, air is the most likely to bring about international structures with the capability to grapple with the problem.

But before and alongside those technical discussions must come the baseline understanding of environmental sustainability. We must think of the global ecosystem's ability to flourish as the overarching necessity that gives humans the basis from which to change legal systems and activities to align with the Rights of Nature.

Air and Climate: Rights of Nature

The 2008 Ecuadorian Constitution does not specifically mention air or climate in its Rights of Nature section, codified in Title II, Chapter Seven.[1] Clearly air and climate benefit from the statement: "Nature has the right to be restored"[2] and the State's mandatory obligation to "apply preventative and restrictive measures on activities that might lead to … the permanent alteration of natural cycles."[3]

Bolivia's law is much more explicit, stating that among Mother Earth's Rights is "The right to preserve the quality and composition of air for sustaining living systems and its protection from pollution, for the reproduction of the life of Mother Earth and all its components."[4] This is in addition to an overarching right to "maintain the integrity of living systems and natural processes that sustain them, and capacities and conditions for regeneration."[5]

Almost as explicit is the Universal Declaration of the Rights of Mother Earth. This is in part because by 2010, when it was promulgated, the problem of climate change had clearly taken the world stage as the greatest threat to ecosystemic integrity and human welfare ever known.

Thus, the Universal Declaration proclaimed that Mother Earth and all beings of which she is composed have the right to life and to exist, to be respected, the right to "continue their vital cycles and processes free from human disruptions," and to "maintain its identity and integrity as a distinct, self-regulating and interrelated being."[6] However, in addition to this, the Universal Declaration of the Rights of Mother Earth specifically mentions "the right to clean air" as a right of Mother Earth and all her beings.[7]

Humans have obligations under Section 3 of the Universal Declaration to "promote the full implementation and enforcement of the rights and obligations" of Mother Earth, including establishing effective norms for "defense, protection and conservation" of those rights. Restoration is called upon as necessary to "respect … the vital ecological cycles, processes and balances of Mother Earth."[8] The Declaration seeks to place human endeavor in the context of Earth's needs, to "ensure that the pursuit of human wellbeing contributes to the wellbeing of Mother Earth."[9] The foregoing means we humans must not only defend the rights of Mother Earth but also change our pursuits so they actually contribute to the Earth's well-being. Put differently, we must live in a way that creates and maintains true social-environmental sustainability for all generations.

Moreover, Pope Francis has emerged as a key defender of climate and the need to curb emissions. In his encyclical *Laudato Si,* he says boldly, "The climate is a common good, belonging to all and meant for all…. Humanity is called to recognize the need for changes of lifestyle, production and consumption, in order to combat this [climate system] warming or at least the human causes which produce or aggravate it."[10] He goes on to pinpoint the cause: "The problem is aggravated by a model of development based on the intensive use of fossil fuels…."[11] He then discusses the "grave implications" of climate change at every level—environmental, social, economic, and political.[12] *Laudato Si* does not flinch from blaming countries with economic and political power who have shirked responsibility.[13]

Air and How It Works

We take clean air and a stable climate for granted, as part of the global commons—a public trust that is everyone's birthright. To take something "for granted" is to be certain of the status quo and thus pay little or no attention to it.

In the Archaean Era (3.8 to 2.5 billion years ago), Earth was not the planet we recognize today. By that time, however, the Earth's crust had cooled enough for rocks and continental plates to begin forming. The atmosphere was likely composed of methane, ammonia, and other gases that would be toxic to most life today.

Yet, it was early in the Archaean Era that life first appeared on Earth, as attested by oldest fossils of cyanobacteria (often thought of as "blue-green algae") that date back to roughly 3.5 billion years ago, and are still among the oldest fossils known. Concentrations of atmospheric oxygen rose from negligible levels to about 21 percent of that present today and can be attributed to the cyanobacteria, which have also been tremendously important in shaping the course of evolution and ecological change throughout Earth's history.

This increase in oxygen is thought to have occurred in six steps, each measured in billions of years required for it to take place. The first step appeared to have been a decrease in the amount of dissolved nickel in the seawater, which could have stifled the methane-producing bacteria and set the stage for oxidation of the Earth's atmosphere, because the methane would have reacted with any oxygen and created carbon dioxide and water. The initial change in Earth's atmosphere took place 2.4 billion years ago, in what scientists today call the "Great Oxidation Event."

The timing of these steps coincides with the amalgamation of Earth's landmasses into supercontinents. The collisions of continents required to form supercontinents produced huge mountains, which eroded quickly and thereby released large amounts of nutrients into the oceans, such as iron and phosphorus. These nutrient pulses led to explosions of algae and cyanobacteria that, in turn, caused marked increases in photosynthesis and thus the production of oxygen. Enhanced sedimentation during these periods buried large amounts of organic carbon and pyrite, which not only prevented their reaction with free oxygen but also led to sustained increases in atmospheric oxygen.

In fact, much of the oxygen in the atmosphere we depend on was generated through the photosynthesis of cyanobacteria during the Archaean and Proterozoic Eras, the latter of which occurred 2.5 billion to 543 million years ago. Moreover, the beginning of the Middle Proterozoic (16 million years ago) saw substantial evidence of oxygen accumulating in the atmosphere.[14] Nevertheless, some of the first creatures to leave the ocean and venture onto land may have been sea-dwelling arthropods whose shells protected their delicate gills in a small reservoir of seawater, which prevented them from drying out, like the hermit crabs of today.[15]

The Gift of Oxygen

Everyone knows that land-dwelling creatures, from insects to amphibians, reptiles, birds, and mammals, require oxygen. What most people probably do not think about is that all terrestrial plants require air for life, whether they have chlorophyll and undertake photosynthesis or not.

In addition, plants need soil in which to grow and we need plants as the basis of our food chain. And both plants and animals require fertile soil. In turn, productive soil has spaces filled with air between the particles and chunks that comprise its matrix. These pockets of air are created by all the organisms living in the soil—from microbes to larger animals, as well as the roots of plants. Most of these organisms depend on the availability of air and water moving through the soil in order to perform their vital, ecological functions that, in concert, create and maintain the soil's fertility and so that of a forest, grassland, alpine meadow, or desert. In this sense, healthy soil acts more like than a sponge than a brick because air normally constitutes half or more of its total volume.

Compaction of soil also reduces its ability to absorb and store water, which simulates a drought for those organisms that do survive the initial compression of their habitat, particularly in fine-textured clays and silts. Over time, compacted soil is more prone to actual drought than is healthy, friable soil.[16]

Meanwhile, dust traveling in the rivers of air high above the ground initiates an incredible range of effects as it goes from place to place. Such airborne particulate matter as dust is often termed "aerosols" and refers to the tiny particles and droplets suspended in the atmosphere, which include fog, smoke, and sulfur dioxide, as well as an array of anthropogenic pollutants that scatter light.

Dust

Long before humans harnessed the wind, dust circumnavigated the globe in an ocean of air. The wind's variable strength—which can be thought of as circulating energy—determines both the amount and size of the airborne dust.[17] Today, the Sahara Desert is the largest source of mineral dust in the world, having experienced a sharp increase in the early 1970s, a change attributed primarily to drought in the Sahara/Sahel region owing to changes in global sea-surface temperatures. In addition, however, the onset of commercial agriculture at the beginning of the 19th century in the Sahel region has contributed to the atmospheric volume of dust for roughly 200 years.[18] Beyond that, Asian dust is a regular component of the troposphere over the eastern Pacific and western North America, and is common across North America, at least during spring.[19] (The troposphere is the lowest major layer of the atmosphere, extending to a height of 6–10 miles from the Earth's surface.) The same is generally true for such Pacific Islands as Midway, Mauna Loa, Guam, and Japan.[20]

On the other hand, most of the dust in Antarctic ice cores originates from the glacial outwash in Patagonia. (Glacial outwash refers to the sediments deposited by streams that are flowing away from glaciers.) Sedimentary evidence suggests that proglacial lakes provided an on/off switch for the flux of dust to Antarctica during the last glacial period. Peaks in the amount of dust coincide with periods when the rivers of glacial meltwater (water that comes directly from melting snow or ice) deposited sediment directly onto easily mobilized outwash plains, but no such peaks occurred when glacial meltwater went directly into proglacial lakes. (A proglacial lake is a lake formed either by the damming action of a moraine or ice during the retreat of a melting glacier. A moraine, in turn, is an accumulation of loose soil and rock carried by an advancing glacier and left at its front and side edges as it retreats.)[21]

Traveling dust initiates an incredible range of effects as it goes from place to place. For example, wind-blown, eolian dust from the Sahara Desert of North Africa landed in Florida around 4,600 years ago, where it enriched the nutrient-poor wetlands—that is until an abrupt shift in wind direction around 2,800 years ago stifled the supply of dust by shunting the tropical storms carrying it to the south of Florida and into the Gulf of Mexico.[22] (*Eolian dust* is composed of sand, silt, or clay-size, fragmented material transported and deposited primarily by wind.)

Today, however, the wind-scoured, nearly barren southern Sahara Desert of North Africa feeds the Amazonian jungle of South America with mineral-coated dust from the Bodélé Depression, which is the largest source of dust in the world. During the Northern hemisphere winter, winds routinely blow across this part of North Africa, where they pick up 700,000 tons of dust on an average day and sweep much of it across the Atlantic. Approximately 20 million tons (18.1 million metric tons) of this mineral-rich dust fall on the Amazon rainforest and enrich its otherwise nutrient-poor soils. The Bodélé Depression accounts for only 0.2 percent of the entire Saharan Desert and is only 0.05 percent of the size of the Amazon itself.[23]

On the other hand, dust storms in one region can change climate dramatically in another. Two springtime low-pressure systems in 1998 generated intense dust storms over the Gobi desert (straddling northern China and southern Mongolia), which crossed the Pacific Ocean in five days and reached the mountain ranges between British Columbia, Canada, and California in the United States. Once there, the dust had a severe impact on visibility in areas where it concentrated and simultaneously reduced the direct solar radiation, but doubled the diffuse radiation.

Yet in East Asia, the blowing dust increased the albedo effect over the ocean on a cloudless day because the dust was lighter than the ocean's surface and thus reflected the electromagnetic radiation back into space, thereby preventing it from being absorbed by the dark water.[24] ("Albedo effect" is the electromagnetic radiation reflected back into space by a light surface, such as snow; *albedo* is Late Latin for whiteness, from the Latin *albu* ["white"].)

Decreases in the amount of atmospheric dust over the past 30 years have contributed more to the warming of the Tropical North Atlantic Ocean than has climate change per se.[25]

The Effect of Climate

Weather is the day-to-day experience of the current atmospheric condition and its short-term variation in minutes to weeks (Figure 12.1). Weather is generally thought of as the combination of temperature, humidity, precipitation, cloudiness, visibility, and wind. People talk about weather in terms of current conditions or those in the near future: "Boy is it hot and humid today." "I wonder if it will be as stifling tomorrow?" or "Will we get the snowstorm that's been predicted for next week?"

Climate, on the other hand, is the cumulative weather of a location considered as an averaged over time. In essence, climate is how the atmosphere "behaves" in some part of the world throughout the years, decades, and centuries, such as cycles in the Asiatic monsoons.

People of civilizations that collapsed centuries ago were probably oblivious to the impact that could be wrought by long-term shifts in climate, such as the torrential monsoon rains and drought. However, new data from studying the stalactites of Soreq Cave in Israel suggest that a shift in climate may be partly responsible.

Stalactites are the most familiar, bumpy, relatively icicle-shaped structures found hanging from the ceilings of limestone caves. They are formed when water accumulates minerals as it percolates through soil before seeping into a cave. If the water's journey takes it through limestone, it typically leaches calcium carbonate and carbon dioxide in its descent. The instant the water seeps from the ceiling of a cave, some of the dissolved carbon dioxide in the fluid escapes into the cave's air. This gentle, soda pop–like fizzing process causes the droplet to become more acidic and so results in some of the calcium carbonate crystallizing

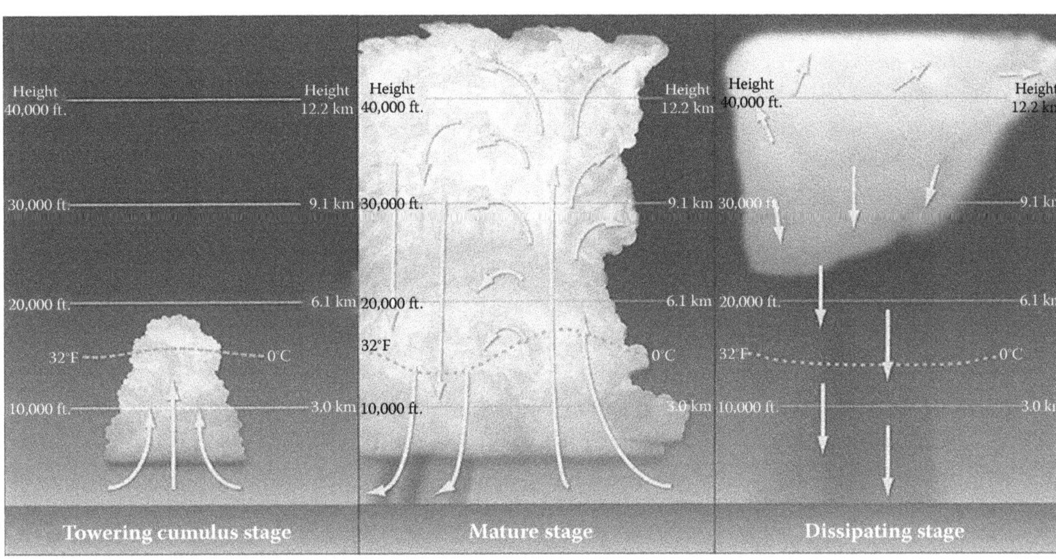

FIGURE 12.1
Formation and dissipation of a thunderstorm. Illustrations by Jonathan Lamb. Diagram from the U.S. National Oceanic and Atmospheric Administration, National Weather Service training materials (https://commons.wikimedia.org/wiki/File:Thunderstorm_formation.jpg).

on the cave's ceiling, thereby initiating a stalactite. As this process is performed over and over, the separation of calcium carbonate from within the thin film of fluid flowing down its surface allows the stalactite to grow. The procedure is so slow that it typically takes a century to add four-tenths of an inch to a stalactite's growth.[26]

By using an ion microprobe, it has become possible to read the chemical-deposition rings of the stalactites in the Soreq Cave with such precision that even seasonal increments of growth can be teased out of a given annual ring. The results indicate that a prolonged drought, beginning in the Levant region as far back as 200 years BCE and continuing to 1100 CE, coincides with the fall of both the Roman and Byzantine empires.[27] (Levant is the former name of that region of the eastern Mediterranean that encompasses modern-day Lebanon, Israel, and parts of Syria and Turkey.)

Air Pollution and Its Sources

Although air currents carry life-giving oxygen, water, and life-sustaining dust, they also transport the "key of death"—human legacy made visible. Persistent, organic pollutants are organic compounds of artificial origin that not only resist degradation but also accumulate in the food chain worldwide, in part by riding the airways to lands far distant from their origin. The best known of these is DDT, but there are many others equally dangerous.

Moreover, these pollutants are not only airborne but also prone to long-distance, atmospheric transport resulting in widespread distribution across the Earth—including regions where they have never been used, such as oceans. Because of their toxicity, they pose a long-term threat to humans and the environment.

Despite several decades of working to control air pollution, millions of tons (metric tons) of such toxins as nitrogen oxide, sulfur dioxide, and mercury continue to rain down on the nation's water catchments, rivers, and lakes. In addition, nitrogen oxide and sulfur dioxide react with the atmosphere to form nitric acid and sulfuric acids—acid rain. Mercury, on the other hand, is absorbed by fish and thereby causes a health hazard when humans, especially children, eat those fish. According to the U.S. Government Accountability Office:

- 53,000 square miles of the Great Lakes, or 88 percent of those lakes, are impaired by acid rain.
- 550 lakes in the Adirondack Mountains are affected by acid rain.
- 21,000 miles of streams in the central Appalachian Mountains are tainted.
- An excess of nitrogen from acid rain and other sources impairs most of the Chesapeake Bay and its tidal waters.
- In the Northeast, the fish in more than 10,000 lakes, ponds, reservoirs, and 46,000 miles of rivers, are designated as unfit for human consumption because of high levels of mercury.[28]

In addition, toxins from such areas as the notoriously polluted air of Mexico City hitchhike on the wind across the Gulf of Mexico to the United States, which says nothing about the 5.5 million people worldwide who are dying prematurely every year as a result of air pollution.[29]

And this is just Mexico. "Faster than mail traveling from Beijing to Seattle, air pollution and dust from China can speed across the Pacific Ocean and blanket broad swaths of North America."[30] Although homegrown pollution is clearly the most potent, everyone's aerial garbage goes somewhere. For example, Asian dust crosses the Pacific to North American shores in 4 to 10 days, and carries with it such pollutants as arsenic, copper, lead, and zinc. In one case, at least, the heavy metals were traced to smelters in Manchuria because the dust passed over the smelters on its way to North America.

Moreover, climate change may significantly accelerate the release of "old" Lindane—an agricultural insecticide used in Asia—from continental storage in soil, vegetation, and high mountains and initiate long-range transport from its terrestrial sources to deposition in the open oceans.[31]

Thus, no matter how far people travel from the centers of civilization, they are still breathing pollution. It is everywhere, and will continue to compound as long as decisions are made to placate corporate industry, thereby sabotaging the global pursuit of baseline standards for good quality air.

We dare not kid ourselves about the importance of air quality. Our Earthly survival—and progressively that each generation into all of the future—ultimately depends on clean air, as do all living things.

Internal Combustion Engines

There is no doubt that the internal combustion engines of automobiles pollute the air and, through the air, the oceans of the world.[32] It works like this: Carbon monoxide, nitrogen oxides, and hydrocarbons are released when fuel is burned, as well as when air and fuel residuals are emitted through the vehicle's tailpipe. In addition, gasoline vapors escape into the atmosphere during refueling and when fuel vaporizes from a vehicle's engine and fuel system, both when it is driven and when it is outside in hot weather.

The emissions are known to damage lung tissue, and can lead to and aggravate respiratory diseases, such as asthma. They also contribute to the formation of acid rain and add to the greenhouse gases involved in causing climate change. However, proper maintenance of emission-control systems both limits harmful pollution and can improve a vehicle's fuel efficiency, performance, and extend it useful life. Beyond that, care in storing and handling gasoline and other solvents also reduces evaporative losses to the atmosphere.[33]

Concern about automobiles as a source of air pollution has been expressed periodically, but concrete concern was first demonstrated in the 1960s, when California established the initial emission standards for new cars. The scientific basis of this effort was the pioneering, atmospheric-chemistry research of A.J. Haägen-Smit, who demonstrated that photochemical reactions among hydrocarbons and nitrogen oxides formed the myriad secondary pollutants responsible for the reduced visibility in the Los Angeles area, which was accompanied by the irritation of the eyes and nose.[34]

Although electric cars are an intermediate solution, they still require hydroelectric power or power from the combustion of fossil fuels. The ultimate solution to vehicular emissions is solar-powered cars and trucks. Much like solar-powered homes, solar cars harness energy from the sun and convert it into electricity, which then fuels the battery that runs the car's motor. Some solar cars bypass the battery, however, and direct the power straight to the electric motor.[35]

Stella is the first-ever, family-sized, solar-powered, road vehicle. A large solar panel sits atop the roof to power the car up to 500 miles on a single charge. The car won the World

Solar Challenge and the Michelin Cruiser Class for completing a 1,864-mile journey from Darwin to Adelaide, Australia, in 2015.[36]

Coal-Fired Power Plants

Coal-fired power plants are an enormous cause of air pollution (Figure 12.2). For example, the Union of Concerned Scientists found in a case study that a single 500-megawatt coal-fired power plant produces 3.5 billion kWh per year, enough to power a city of about 140,000 people. To produce this amount of electricity, it burns 1,430,000 tons of coal, uses 2.2 billion gallons of water, and 146,000 tons of limestone. It also puts out annually:

1. 10,000 tons of sulfur dioxide. Sulfur dioxide (SO_x) is the main cause of acid rain, which damages forests, lakes, and buildings.
2. 10,200 tons of nitrogen oxide. Nitrogen oxide (NO_x) is a major cause of smog, and also a cause of acid rain.
3. 3.7 million tons of carbon dioxide. Carbon dioxide (CO_2) is the main greenhouse gas and is the leading cause of global warming. There are no serious U.S. or world-wide regulations limiting carbon dioxide emissions.
4. 500 tons of small particles. Small particulates are a health hazard, causing lung damage. Particulates smaller than 10 μm have not traditionally been regulated.

FIGURE 12.2
Alma coal-fired electrical power station (right) and John P. Madgett coal-fired electrical power station (left) on the Mississippi River at Alma, Wisconsin. Photograph by the U.S. Geological Survey (https://commons.wikimedia.org/wiki/File:Alma_Wisconsin_Coal_power_plants.jpg).

5. 220 tons of hydrocarbons. Fossil fuels are made of hydrocarbons; when they do not burn completely, they are released into the air, where they are a cause of smog.
6. 720 tons of carbon monoxide. Carbon monoxide (CO) is a poisonous gas and a contributor to global warming.
7. 125,000 tons of ash and 193,000 tons of sludge from the smokestack scrubber. A scrubber uses powdered limestone and water to remove pollution from a plant's exhaust. Instead of going into the air, the pollution goes into a landfill or into products like concrete and drywall. This ash and sludge consists of coal ash, limestone, and many pollutants like such toxic metals as lead and mercury.
8. 225 pounds of arsenic, 114 pounds of lead, 4 pounds of cadmium, and many other toxic heavy metals. Mercury emissions from coal-fired power plants are suspected of contaminating lakes and rivers in northern and northeast United States and Canada. In Wisconsin alone, more than 200 lakes and rivers are contaminated with mercury. Health officials warn against eating fish caught in these waters, since mercury can cause birth defects, brain damage, and other ailments. Acid rain also causes mercury poisoning by leaching mercury from rocks and making it available in a form that can be taken up by organisms.[37]

Hydraulic Fracturing

The process of hydraulic fracturing, or "fracking," the technique for releasing natural gas from belowground rock formations, is a more dangerous version of natural gas and oil extraction than the traditional wells. Designed to enable the extraction of previously untapped gas and oil reserves, fracking pumps a high-pressure mixture of toxic chemicals (known as PAHs or polycyclic aromatic hydrocarbons) and water underground to fracture deep shale rock formations.

The shale-gas boom of the past 15 years has brought heavy industry into proximity with more than 15 million Americans. The process uses extremely toxic chemicals in the hydraulic fracturing fluids and releases volatile organic compounds during gas drilling. Part of the risk to human and animal health from fracking-related activities is based on the lack of anything resembling *definitive* data or proof of harm owing to the newness of the technology, which has resulted in so-called medical "gag rules." These "gag rules" are simply nondisclosure agreements in private settlements between victims and industry, in addition to the refusal of oil and gas companies—to some extent protected in federal law—to disclose the identity of chemicals they use in hydraulic fracturing.

Nevertheless, some of the air contaminants associated with hydraulic fracturing are known:

1. Benzene: A known carcinogen, benzene can cause anemia, can lessen count of white blood cells, and can weaken the immune system. Prolonged exposure may result in such blood disorders as leukemia, as well as reproductive and developmental disorders, and other cancers.
2. Toluene: Long-term exposure to toluene can affect the nervous system; cause irritation of the skin, eyes, and respiratory tract; and cause birth defects.
3. Ethyl-benzene: Long-term exposure to ethyl-benzene can result in blood disorders.
4. Xylenes: Short-term exposure to high levels of xylenes can cause irritation of the nose and throat, nausea, vomiting, gastric irritation, and neurological effects. Long-term exposure at high levels can also damage the nervous system.

5. Nitrogen oxides: Short-term exposure to nitrogen oxides causes inflammation of the air passage and aggravates asthma. It also combines with volatile organic compounds to form ozone.
6. Methane, ethane, propane: Exposure to methane, ethane, or propane can cause rapid breathing, rapid heart rate, clumsiness, emotional upset, and fatigue. At greater exposure, any one of them can cause vomiting, collapse, convulsions, coma, and death.[38]

Laws Regulating Air Quality and Air Pollution

There are just beginning to be American laws regulating the climate, since realization of climate change is very new, and in any event is global in scope, thus making climate change the perfect arena for strong international action. However, the United States, like other nations, has a framework of laws to limit—though only rarely prohibit—air pollution of various kinds.

As with nearly all environmental laws, the system is regulatory, providing for permits to commit various polluting activities, but recognizing limits on allowable discharges. Federal laws must be upheld by all the states; indeed, the Environmental Protection Agency (EPA), the federal oversight agency, delegates many aspects of air quality regulation to the states.

The fundamental law regulating air pollution is the Clean Air Act of 1970, which authorized the development of comprehensive federal and state regulations to limit emissions from both stationary (industrial) sources and mobile sources. The law was substantially amended in 1990. The main impetus for the original legislation was the heavy pall of smog hanging over major U.S. cities, and visibly worsening in the 1970s. The first federal air pollution law was passed in 1955, which funded research on the scope and sources of air pollution. A key 1963 statute first authorized the development of a national program to address air pollution, and initiated research into methods of monitoring, and then controlling, air pollution. Under the 1967 Air Quality Act, the federal government began to initiate enforcement proceedings in situations involving the transport of interstate pollution.

But the Clean Air Act of 1970 created the watershed change that made the federal government the major actor in air pollution control. The Clean Air Act directed the federal government to develop comprehensive regulations to limit emissions from both industrial and mobile sources. The Environmental Protection Agency was created at about the same time to implement the Clean Air Act, among other laws. The initial Clean Air Act established the National Ambient Air Quality Standards, emission standards for airborne hazardous pollutants, and various performance standards that states would have to meet under federal direction.

The Clean Air Act was amended in 1977 to strengthen national standards for federal air quality, and establish permit review requirements to help ensure that local governments met the federal standards. The most significant and wide-ranging amendments came in 1990, which greatly increased federal authority and responsibility for air quality. The new legislation directed the Environmental Protection Agency to implement new regulations on acid rain, and an expanded program controlling toxic air pollutants. The

1990 amendments also initiated a program of ozone protection, by phasing out chemicals that deplete the ozone layer, and initiating proper recycling programs. They also commit the United States to implement the Montreal Protocol, which aims to eliminate all ozone-depleting chemicals.[39]

Under the Clean Air Act, states are responsible for developing plans to meet and implement air quality standards. The Environmental Protection Agency provides guidance and technical assistance, and reviews state plans for compliance. This includes problem areas, where so-called "regional haze" has developed, damaging air quality in National Parks and other protected areas. States are required to adopt enforceable plans to reduce pollutants that cause regional haze, for which the Environmental Protection Agency provides guidance on state planning and required controls.

Has it worked? In many ways, yes—the nation's air quality is decidedly better than it was in 1970. But many hurdles remain. For example, while visible air pollution is much less widespread, particle pollution remains widely dispersed in our air. The United States also has problems with sulfur dioxide, nitrogen dioxide, and airborne lead. This last-named pollutant is particularly acute in neighborhoods near airports hosting small aircraft, whose fuel still contains lead.[40]

Climate Change

Climate change, or global warming, is the result of a buildup of carbon dioxide and other long-lived "greenhouse gases" such as methane that trap heat in the atmosphere and make the climate warmer. Among other ills, climate change causes more intense hurricanes and storms, heavier and more frequent flooding, increased drought, increased heat waves, and severe wildfires. In addition, it is a chief cause of melting glaciers, which in turn triggers sea level rise worldwide, endangering coastal communities and ecosystems.[41]

Pope Francis, in his encyclical *Laudato Si*, made the challenge clear, "There is an urgent need to develop policies so that, in the next few years, the emission of carbon dioxide and other highly polluting gases can be drastically reduced, for example, substituting for fossil fuels and developing sources of renewable energy. Worldwide there is minimal access to clean and renewable energy. There is still a need to develop adequate storage technologies."[42]

Progress in dealing with the issues of climate change is slow internationally, and in the United States as well, because so many multipronged initiatives must be put in place to have any real effect. Between 2010 and 2012, the Environmental Protection Agency began issuing the initial national greenhouse-gas emission standards and fuel-economy standards for cars and trucks. In 2011, the Agency first limited greenhouse gases from large, new industrial sources, such as power plants and refineries being built or undergoing major modification.

In 2015, President Obama announced the Clean Power Plan, which contains state goals to slash the carbon pollution that causes climate change. It also contains strong, achievable standards for power plants' pollution. Methane is a very strong global warming gas, and the Environmental Protection Agency is slowly beginning to tackle this challenge, unveiling a new goal to cut methane emissions from the oil and gas sector by 40 percent to 45 percent of 2012 levels by 2025, through an unfolding rulemaking process.[43]

The Kyoto Protocol

The "Kyoto Protocol" is a shorthand term for the United Nations Framework Convention on Climate Change, of which the Kyoto Protocol is a part. Negotiators from many nations agreed to the United Nations Framework Convention on Climate Change in 1994, but countries were aware that its provisions were too weak to actually address the issue. A new round of talks began the following year, which led to the Kyoto Protocol in 1997. This treaty, for the first time, established legally binding emission targets of greenhouse gases for industrialized nations. The Kyoto Protocol went into effect in 2004, after 55 nations ratified it.

Unfortunately, the United States, though a party to the negotiations at Kyoto, refused to ratify the treaty under the Bush Administration, stating that developing countries should also have emissions targets, and that cutting emissions would "seriously harm the economy of the United States." Since the United States accounted for 36 percent of the emissions in 1990, its refusal to ratify the Protocol was a major blow. Canada withdrew from the Kyoto Protocol in 2012, because its government wanted to prioritize development of tar sands in Alberta over environmental protection. Its emissions in 2009 were 17 percent higher than in 1990, and it was clear Canada would not meet the Kyoto emissions targets.

The goal of the United Nations Framework Convention on Climate Change is to stabilize atmospheric concentrations of greenhouse gases to a level that would prevent harm to the worldwide climate system. All parties are required to implement national programs to control greenhouse-gas emissions and promote the use of climate-neutral technologies. Industrialized countries have additional commitments of limiting greenhouse gases; the wealthier developed (industrialized) countries are required to transfer "climate friendly technologies" to developing countries.

The 1997, the Kyoto Protocol focused on strengthening the commitment to reducing greenhouse gases, creating three new mechanisms for the purpose, the best known of which is "emissions trading."[44]

The European Union led the way. Its emissions-trading system is both a cornerstone of its policy to combat climate change and its primary tool for cost-effectively reducing industrial greenhouse-gas emissions. The first—and still by far the most extensive—international system for trading allowances for greenhouse-gas emissions, it covers more than 11,000 power stations and industrial plants in 31 countries, as well as airlines.

The Union's emissions-trading system works on the "cap and trade" principle, where a "cap" or limit is set on the emission "allowances"—the allowable volume of greenhouse gases that can be annually emitted by the factories, power plants, and other installations. The initial cap established the baseline, after which the total, allowable, annual emissions will be progressively reduced over time, which means that emissions from fixed installations in 2020 will be 21 percent lower than they were in 2005.

Within the European cap system, companies receive or buy emission "allowances," which they can trade as needed. Each "allowance" gives the holder the right to emit one metric tonne (1.10 short tons) of carbon dioxide (CO_2), the main greenhouse gas, or the equivalent amount of two more powerful greenhouse gases: nitrous oxide (NO_2) and perfluorocarbons (PFCs).

Power stations and other fixed installations have a separate emission cap from aviation. Unlike the caps for fixed installations, however, caps for the aviation sector will remain the same in each year of the 2013–2020 trading period.

After each year, a company must surrender enough allowances to cover all its overall emissions, or face heavy fines. If a company reduces its emissions, it can either keep the spare allowances to cover future needs or sell them to another company that is short of allowances. The flexibility that trading brings ensures that emissions are cut in the most cost-effective manner.[45]

The Cancun Agreements

The next step in international regulation of climate change came with the Cancun Agreements, which ended in 2010. Under the agreements, industrialized countries had to devise low-carbon development plans and strategies, as well as ways to meet them. In addition, the actions taken by non-industrialized countries to reduce emissions were recognized, as were the technological means to support them. Consequently, the negotiations centered, in part, on technology transfer to increase cooperation in support of climate mitigation.

The agreements also provided $30 billion in fast-start financing from industrialized countries to augment climate action in non-industrialized countries up to 2012—with the intention of raising $100 billion in long-term funds by 2020. The Cancun Adaptation Framework was established to improve climate-adaptable projects with increased financial and technical support.

Among the successes of the Cancun Agreements are the voluntary pledges by 76 nations to control their greenhouse gases—nations that, collectively, were responsible for 85 percent of the worldwide emissions.[46]

Later Efforts and the Paris Talks

In view of the fact that air is a global commons, the goals to improve air quality are critically important and international cooperation is essential; so the talks on climate change have continued.

For example, parties present at the Bali Action Plan, adopted in 2007 by parties to the United Nations Framework Convention on Climate Change, set up a comprehensive process to enable full implementation of the convention. Nevertheless, in 2012, the United States, Japan, Russia, and Canada refused to continue participation in Kyoto. Even so, at the end of the 2012 United Nations Climate Change Conference, the negotiators agreed to extend the Protocol to 2020, an agreement that has been heavily criticized because it includes so little of the total greenhouse gases.[47]

The Paris talks, which took place in December 2015, led to the Paris Agreement, a global agreement on the reduction of climate change. It reached a consensus of the 196 nations that sent representatives. The goal was to limit global warming to less than two degrees Celsius compared to preindustrial levels, with zero net greenhouse-gas emissions by the second half of the 21st century. The underlying goal was to achieve a binding, universal agreement on climate by all nations worldwide. The Agreement becomes legally binding once joined by at least 55 of the countries representing 55 percent of global greenhouse emissions.

Because the two largest greenhouse-gas emitters are China and the United States, their participation was crucial to the successful outcome. A November 2014 agreement between the two countries to limit greenhouse gases set the stage. However, while the Paris goals were successfully adopted on paper, it remains to be seen if the Paris Agreement in practice will provide substantive reductions in greenhouse gases over time.

The best exemplar for other nations is France, which, as of 2012, generated 90 percent of its electricity from non-fossil fuel sources, such as nuclear, hydroelectric, and wind. Even if these sources have serious environmental issues, ranging from the contamination of rivers to nuclear waste, France's model aptly illustrates that dependence on fossil fuels is not a necessary corollary of a good quality of life and desirable economic development.[48]

Has It Worked?

There seems to be general agreement that the United Nations Framework Convention on Climate Change, Kyoto Protocol, and the Cancun Agreement have had a negligible effect on greenhouse gas emissions. The World Bank, in a 2010 study, noted that, although Kyoto was finalized in 1997, carbon dioxide emissions had increased 24 percent by 2006. Kyoto also had only modest success in providing financial support to non-industrialized countries.

The problems are many, because air is the most evenly distributed of the commons resources; unlike the ocean, it has no borders at all. One of the pivotal challenges continues to be whether the warming climate should be approached as a problem of environmental change or of economic equity. Although all agree that there is a serious ecological problem, the equity perspective argues that industrialized countries use most of the resources affecting climate. As such, they should bear the greatest burden of reducing emissions while aiding nonindustrialized countries with financial and technology assistance. On the other hand, large industrializing nations, with increasing emissions, are uncomfortable with negotiating absolute emissions targets that will hamper their continual economic development.

These two perspectives have, to a large extent, paralyzed international climate talks. Consequently, negotiators are seeking ways around the polarizing debates that have characterized many discussions. Possible ideas include focusing on development rather than emissions, so that initiatives for domestic development could be undertaken with low-carbon footprints, and climate goals would be integrated with development. Other possibilities include greenhouse-development rights, carbon markets, and other similar ideas.[49]

Other problems slowing the progress of climate talks include:

1. Disagreement over what strategy to pursue, since each has significant trade-offs in terms of costs: Is it better to pursue emissions reduction, or adaptation to a climate-changing world?
2. The cost of reducing greenhouse gases, which continue to increase, is always a problem: How much should a nation pay, and how much is it worth in the balance of other needs?
3. If reduction is a major goal, how best can it be achieved? A nation could raise the price of emissions through a tax, or place a lid on total allowable emissions and set up some kind of cap-and-trade market. That leads to questions of how many permits to be issued, to whom, in which nations, or for which industries.
4. The perennial problem of the commons is also a stumbling block: Each nation has incentive to ride on the efforts of others, and take no responsibility for reducing greenhouse gases itself. That nation gets a free ride that makes its industries more competitive in the world market.[50]

As an example of national difficulties in meeting even local or regional air quality standards, and how air pollution is an inextricable part of the even larger problem of climate change, take the case of London and the United Kingdom. The UK has been struggling to meet European Union limits for nitrogen dioxide, and is the target of an ongoing lawsuit by

ClientEarth over these violations and lagging government actions to solve the problems. Bad air quality contributes to up to 40,000 premature deaths each year in the United Kingdom.

Despite creating anti-pollution zones in several English cities, and committing funds to improving standards for buses and other heavy vehicles, the United Kingdom does not see itself meeting European Union standards until 2025 in London, and 2020 in the nation as a whole. This is partly because of the problem of getting older vehicles off the roads, and partly because the government encouraged diesel cars. The government provided incentives for UK residents to purchase diesel cars because they spew less carbon dioxide, a prime global warming gas—but, unfortunately, more nitrogen dioxide and other pollutants. Furthermore, ClientEarth claims the government exacerbated the problems by lobbying the European Union to soften the emission limits for them. Air pollution and climate change cannot be solved separately.[51]

Management Considerations for a Rights of Nature System

As mentioned before, there is no resource more of a global commons than air. Thus, both strong national and international efforts are essential to reduce both air pollution and, more specifically, greenhouse gases responsible for climate change.

What could the Rights of Nature do to solve the dilemmas that bedevil international climate talks, and the surge of fossil fuel development via hydraulic fracturing that is taking place in the United States?

Placing a Rights of Nature provision in the federal and state Constitutions would immediately change the nature of the debate from an anthropocentric one to questions of what the ecosystems of the world need in order to flourish—not just to survive. It is certain that many and vast changes would be required.

Here are a few initial possibilities. Some have already been launched, at least in part, and most have been discussed. However, climate litigation is in its infancy and just beginning to forge workable legal doctrines. For these suggestions, we assume that Rights of Nature is already embedded in federal and state Constitutions as the primary rights-holder. Clearly for some of these suggestions to be viable, Rights of Nature would need to be at the heart of the international legal system as well.

1. Create a National Climate Ombudsman. This office would be responsible for synthesizing and initiating needed air-quality research to determine national baselines for air quality—not a minimal standard based on maximizing human health and climate stability, but rather the optimum needed for the air and all ecosystems that depend on it to flourish. Insulate this office from partisan and corporate politicking as well as from corrupting financial considerations. It would be useless, for example, if funded by fees paid by polluters, because that provides perfect incentive to protect polluters. Such an office could be funded by a national or regional "environmental tax" paid by all citizens.
2. Create a Climate Guardian Office, and give it wide-ranging powers. Set up a national and international mechanism to allow an appointed Guardian to both advocate for air and the climate in national and international forums and sue, in American and international forums, for the climate's own right to exist and

flourish. Existing law already provides many examples of such guardianship or trusteeship, most specifically in the arena of endangered species. For example, the National Oceanic and Atmospheric Administration is the trustee for migrating whales; as such, it can institute civil penalties, and ask the Department of Justice to sue against pollutants that harm whales.[52]

3. Give environmental organizations advocacy power over climate guardianship. If a Guardian is dragging its feet on legislating or suing on behalf of the climate, expand the German system (*altruistische Verband* = "Altruistic Association" or *altruistische Verbandsklage* = "altruistic class action") whereby qualified, environmental nonprofit organizations are designated by the government to participate in environment-affecting activities, ranging from initial planning to litigation.[53]
4. Allow and encourage lawsuits by those most affected by climate change. Set up a national, and international, mechanism to allow lawsuits based on the rights of those most likely to be affected by climate change. For example, in 2005, the Inuit Circumpolar Conference and Earthjustice filed a petition before the Inter-American Commission on Human Rights, requesting relief from the impacts of climate change to the Inuit, including right to residence and movement, right to inviolability of the home, right to health and well-being, and the rights to a culture.[54]
5. Give citizens wide-ranging climate litigation opportunities. If the Trustee or Guardian fails to take action to protect the climate, then laws under a Rights of Nature system would empower citizens to force the Guardian's hand and institute rulemaking or penalties to stop the harmful activity. Again, this already exists in the law, through the provisions of the Endangered Species Act, which empower "any person" to sue.[55] Climate litigation, however, has not fared well thus far.
6. Expand the Public Trust Doctrine to cover climate. Expansion of the Public Trust Doctrine would allow lawsuits by citizens on behalf of natural objects or beings, where there was no specific law, such as the Endangered Species Act, that expressly permits such suits. The American legal system already provides many existing, nonhuman plaintiffs with powers to sue to protect their interests—such as corporations and municipalities, created groups composed entirely of human beings.[56]

But climate litigation, specifically, has faced significant roadblocks. Federal courts have thus far turned down cases arguing that there is a federal Public Trust Doctrine, holding that it exists only in state common law. Even if there were a federal Public Trust Doctrine, say the courts, it has been displaced by the Clean Air Act and the Environmental Protection Agency's climate regulations promulgated under it.[57]

Despite these setbacks, Our Children's Trust filed a climate change lawsuit against the federal government in 2015 in the U.S. District Court of Oregon. The lawsuit argues that the government's promotion of fossil fuels has violated the younger generation's constitutional rights to life and liberty and has failed to protect the climate—a public trust resource. The plaintiffs directly challenged the federal government's encouragement of fossil fuels and national programs that continue to advocate for and stimulate fossil fuel production and use. The judge, however, granted the motion to intervene by three trade associations, representing most of the largest fossil fuel companies. Promptly on the heels of the interventions, the Global Catholic Climate Movement and the Leadership Council of Women Religious filed briefs with the court in support of the lawsuit, arguing from Pope Francis's encyclical *Laudato Si* in favor of the need to protect the climate. As expected, the case faced a motion to dismiss by the defendants. In November 2016, U.S. District Judge Ann Aiken, in a historic holding, denied the U.S. government's and fossil fuel industry's motions to dismiss the case, and ruled the young plaintiffs have standing to sue. The case has proceeded to trial. In her holding, Aiken stated, "Exercising my 'reasoned judgment,' I have no doubt that the right to a climate system capable of sustaining human life is fundamental to a free and ordered society."[58]

This is a highly significant victory in the court system for a novel challenge: the rights of the upcoming generations to have a climate they can depend on. Regardless of ultimate outcome, this challenge—arguing the government had failed to protect the climate *as a public trust resource*—will have positive legal repercussions for efforts to expand the Public Trust Doctrine to cover issues of climate and climate change.

7. Pass legislation to mandate, incentivize, and fund the change to solar energy for all purposes. Solar radiation is the only true renewable energy, for cars, manufacturing, home and office heating and cooling, and every other use. Doing this would immediately aid both air quality and climate, if done sustainably—that is, not simply aggravating a cycle of increased growth because of greater low-cost energy availability.

Endnotes

1. [Ecuadorian] Constitution, full language. *op. cit.*
2. *Ibid.* (Section 72).
3. *Ibid.* (Section 73).
4. Language of the Bolivian law (compete text). *op. cit.* (Chapter III, Article 7 (4)).
5. *Ibid.*
6. Universal Declaration of Rights of Mother Earth: Global Alliance for the Rights of Nature. *op. cit.* (Article 2 (1) (a)–(d)).
7. *Ibid.* (Article 2 (1)(e)).
8. *Ibid.*
9. *Ibid.*
10. Encyclical Letter **Laudato Si** of The Holy Father Francis on Care for Our Common Home [in English]. *op. cit.* (Chapter 1, Section 23).
11. *Ibid.*
12. *Ibid.* (Section 25).
13. *Ibid.* (Section 26).
14. The preceding discussion of the origin of oxygen on Earth is based on: (1) Ian H. Campbell and Charlotte M. Allen. Formation of supercontinents linked to increases in atmospheric oxygen. *Nature Geoscience*, 1 (2008):554–558; (2) Introduction to the Archaean. http://www.ucmp.berkeley.edu/archaea/archaea.html (accessed January 6, 2016); (3) Bacteria: Fossil Record. http://www.ucmp.berkeley.edu/bacteria/bacteriafr.html (accessed January 6, 2016); (4) The Proterozoic Era. http://www.ucmp.berkeley.edu/precambrian/proterozoic.html (accessed January 6, 2016); (5) Deep Sea Rocks Point To Early Oxygen On Earth. http://live.psu.edu/story/38514 (accessed January 6, 2016); and (6) Ernesto Pecoits, Stefan V. Lalonde, Dominic Papineau, and others. Oceanic nickel depletion and a methanogen famine before the great oxidation event. *Nature*, 458 (2009):750–753.

15. James W. Hagadorn and Adolf Seilacher. Hermit arthropods 500 million years ago? *Geology*, 37 (2009):295–298.
16. The foregoing discussion about soil is based on: (1) Elaine R. Ingham. Organisms in the soil: The functions of bacteria, fungi, protozoa, nematodes, and arthropods. *Natural Resource News*, 5 (1995):10–12, 16–17; and (2) Michael Snyder. Why is soil compaction a problem in forests? *North Woodlands*, 11 (2004):19.
17. P.E. Biscaye, F.E. Grousset, M. Revel, and others. Asian provenance of glacial dust (stage 2) in the Greenland Ice Sheet Project 2 Ice Core, Summit, Greenland. *Journal of Geophysical Research*, 102 (1997):26765–26781.
18. Stefan Mulitza, David Heslop, Daniela Pittauerova, and others. Increase in African dust flux at the onset of commercial agriculture in the Sahel region. *Nature*, 466 (2010):226–228.
19. Richard A. VanCuren and Thomas A. Cahill. Asian aerosols in North America: Frequency and concentration of fine dust. *Journal of Geophysical Research*, 107, 4804 (2002): 16 pp. doi:u10.1029/2002JD002204.
20. Daniel A. Jaffe, Alexander Mahura, Jennifer Kelley, and others. Impact of Asian emissions on the remote North Pacific atmosphere: Interpretation of CO data from Shemya, Guam, Mid-Way, and Mauna Loa. *Journal of Geophysical Research*, 102 (1997):28627–28636.
21. (1) Robert P. Ackert Jr. Patagonian dust machine. *Nature Geoscience*, 2 (2009):244–245; and (2) David E. Sugden, Robert D. McCulloch, Aloys J.-M. Bory, and Andrew S. Hein. Influence of Patagonian glaciers on Antarctic dust deposition during the last glacial period. *Nature Geoscience*, 2 (2009):281–285.
22. Paul H. Glaser, Barbara C.S. Hansen, Joe J. Donovan, and others. Holocene dynamics of the Florida Everglades with respect to climate, dustfall, and tropical storms. *Proceedings of the National Academy of Sciences*, 110 (2013):17211–17216.
23. Koren, Y. Kaufman, R. Washington, and others. The Bodélé depression: A single spot in the Sahara that provides most of the mineral dust to the Amazon Forest. *Environmental Research Letters*, 1 (2006):1–5.
24. R.B. Husar, D.M. Tratt, B.A. Schichtel, and others. Asian dust events of April 1998. *Journal of Geophysical Research*, 106 (2001):18317–18330.
25. Amato T. Evan, Daniel J. Vimont, Andrew K. Heidinger, and others. The role of aerosols in the evolution of tropical North Atlantic Ocean temperature anomalies. *Science*, 324 (2009): 778–781.
26. (1) Sid Perkins. Buried treasures. *Science News*, 169 (2006):266–268; (2) Martin B. Short, James C. Baygents, and Raymond E. Goldstein. Stalactite growth as a free-boundary problem. *Physics of Fluids*, 17 (2005): 083101. 12 pp. (accessed December 17, 2008); and (3) M.B. Short, J.C. Baygents, J.W. Beck, and others. Stalactite growth as a free-boundary problem: A geometric law and its platonic ideal. *Physical Review Letters*, 94 (2005): 018510. 4 pp. (accessed December 17, 2008).
27. (1) Ian J. Orland, Miryam Bar-Matthews, Noriko T. Kita, and others. Climate deterioration in the Eastern Mediterranean as revealed by ion microprobe analysis of a speleothem that grew from 2.2 to 0.9 Ka in Soreq Cave, Israel. *Quaternary Research*, 71 (2009):27–35; and (2) A. Kaufman, G.J. Wasserburg, D. Porcelli, and others. U-Th isotope systematics from the Soreq Cave, Israel and climatic correlations. *Earth and Planetary Science Letters*, 156 (1998):141–155.
28. The foregoing three paragraphs are based on: (1) R.W. Macdonald, D. Mackay, Y.-F. Li, and B. Hickie. How will global climate change affect risks from long-range transport of persistent organic pollutants? *Human and Ecological Risk Assessment*, 9 (2033):643–660; (2) Paul Recer. 1995. Old pesticides spread across globe. *Corvallis Gazette-Times*, Corvallis, OR. (October 13, 1995); and (3) Mark Mooney. EPA Can't Stop the (Acid) Rain. http://abcnews.go.com/US/epa-stop-acid-rain/story?id=18643424 (accessed March 4, 2013).
29. (1) Sid Perkins. What goes up. *Science News*, 172 (2007):152–153, 156; and (2) Jonathan Amos. Polluted air causes 5.5 million deaths a year new research says. http://www.bbc.com/news/science-environment-35568249 (accessed February 14, 2016).
30. R. Monastersky. Asian pollution drifts over North America. *Science News*, 154 (1998):374.

31. (1) Dan Jaffe, Theodore Anderson, Dave Covert, others. Transport of Asian air pollution to North America. *American Geophysical Research Letters*, 26 (1999):711-714; (2) Z. Xie, B.P. Koch, A. Möller, and others. Transport and fate of hexachlorocyclohexanes in the oceanic air and surface seawater. *Biogeosciences*, 8 (2011):2621–2633; and (3) R. Monastersky. Asian pollution drifts over North America. *op. cit.*
32. Chris Maser. *Interactions of Land, Ocean and Humans: A Global Perspective.* CRC Press, Boca Raton, FL. (2014) 308 pp.
33. The foregoing discussion of auto emissions is based on: (1) Controlling Air Pollution from Motor Vehicles. http://www.dec.ny.gov/chemical/8394.html (accessed 17 January, 2016); (2) Ann Y. Watson, Richard R. Bates, and Donald Kennedy. *Air Pollution, the Automobile, and Public Health.* National Academies Press, Washington, DC. (1988) 704 pp.; and (3) Automobile Emissions: An Overview. http://www3.epa.gov/otaq/consumer/05-autos.pdf (accessed January 17, 2016).
34. (1) John H. Johnson. Pollution from Automobiles—Problems and Solutions. http://www.ncbi.nlm.nih.gov/books/NBK218144/ (accessed January 17, 2016); and (2) Douglas Smith. Fifty Years of Clearing the Skies. http://www.caltech.edu/news/fifty-years-clearing-skies-39248 (accessed January 17, 2016).
35. Cristen Conger. How Can Solar Panels Power a Car? http://auto.howstuffworks.com/fuel-efficiency/vehicles/solar-cars.htm (accessed January 17, 2016).
36. Sarah Buht. The First Four-Seater, Solar-Powered Vehicle Hits The U.S. Road. http://techcrunch.com/2014/09/24/the-first-four-seater-solar-powered-vehicle-hits-the-u-s-road/ (accessed January 17, 2016).
37. (1) Union of Concerned Scientists. How Coal Works. http://www.ucsusa.org/clean_energy/coalvswind/brief_coal.html#.VqOpZhxm3TM (accessed January 23, 2016); and (2) Julie Kerr Casper. *Fossil Fuels and Pollution: The Future of Air Quality.* Infobase Publishing, New York. (2010) 268 pp.
38. The discussion of hydraulic fracturing and air pollution is based on: (1) L. Blair Paulik, Carey E. Donald, Brian W. Smith, Lane G. Tidwell, Kevin A. Hobbie, Laurel Kincl, Erin N. Haynes, Kim A. Anderson. Impact of natural gas extraction on PAH levels in ambient air. *Environmental Science & Technology*, 49 (2015):5203, doi: 10.1021/es506095e; (2) Fracking may affect air quality, human health. *ScienceDaily*, https://www.sciencedaily.com/releases/2015/05/150513093611.htm (accessed March 13, 2016); (3) Physicians for Social Responsibility. http://www.psr.org/assets/pdfs/fracking-and-air-pollution.pdf (accessed March 13, 2016); (4) Physicians for Social Responsibility. Health Risks of Hydraulic Fracturing: Food, Water, and Animals. http://static1.squarespace.com/static/50804b1484ae863ca6c1e36a/t/53beb78ce4b0ea34ad0a772d/1405007756823/PSR+Fracking%2C+Health+Impacts+on+Farm+Animals.pdf (accessed March 13, 2016); (5) Shale Gas Issues from Various Jurisdictions. https://6d7ad352d6bc296a468e63f5b74324c51bfc9d11.googledrive.com/host/0B3QWwx_US206WG1ocHAxdUtNRjA/Shale%20Gas%20Links%20Part%2017.pdf (accessed March 13, 2016); (6) Tanja Srebotnjak and Miriam Rotkin-Ellman. Fracking Fumes: Air Pollution from Hydraulic Fracturing Threatens Public Health and Communities. http://www.nrdc.org/health/files/fracking-air-pollution-IB.pdf (accessed March 13, 2016); and (7) Agency for Toxic Substances and Disease Registry. http://www.atsdr.cdc.gov/ (accessed March 15, 2016).
39. The foregoing discussion of the Clean Air Act is based on: Evolution of the Clean Air Act. http://www.epa.gov/clean-air-act-overview/evolution-clean-air-act (accessed March 10, 2016).
40. (1) Wolfram Schlenker and W. Reed Walker. Airports, Air Pollution, and Contemporaneous Health. http://faculty.haas.berkeley.edu/rwalker/research/SchlenkerWalker_Airports_2012.pdf (accessed March 10, 2016); (2) Susan Perry. Airport pollution may have been 'seriously underestimated,' study suggests. https://www.minnpost.com/second-opinion/2014/06/airport-pollution-may-have-been-seriously-underestimated-study-suggests (accessed March 10, 2016); and (3) Issue Briefing: Impacts of Airplane Pollution on Climate Change and Health. http://www.flyingclean.com/impacts_airplane_pollution_climate_change_and_health (accessed March 10, 2016).

41. Air Pollution: Current and Future Challenges. http://www.epa.gov/clean-air-act-overview/air-pollution-current-and-future-challenges (accessed March 10, 2016).
42. Encyclical Letter **Laudato Si** of The Holy Father Francis on Care for Our Common Home [in English]. *op. cit.* (Chapter 1, Section 26).
43. Regulatory Initiatives. http://www3.epa.gov/climatechange/EPAactivities/regulatory-initiatives.html (accessed March 10, 2016).
44. The preceding four paragraphs are based on: (1) Kyoto Protocol. https://en.wikipedia.org/wiki/Kyoto_Protocol (accessed March 10, 2016); and (2) UNFCCC [United Nations Framework Convention on Climate Change] and the Kyoto Protocol. http://www.un.org/wcm/content/site/climatechange/pages/gateway/the-negotiations/the-un-climate-change-convention-and-the-kyoto-protocol (accessed March 10, 2016).
45. The discussion of "emissions trading" is based on: (1) The EU Emissions Trading System (EU ETS). http://ec.europa.eu/clima/policies/ets/index_en.htm (accessed March 10, 2016); (2) EU Emissions Trading Scheme (EU ETS): Guide to the EU Emissions Trading Scheme (EU ETS) and its impact on business. https://www.carbontrust.com/resources/reports/advice/eu-ets-the-european-emissions-trading-scheme/ (accessed March 10, 2016); and (3) Allowances and caps. http://ec.europa.eu/clima/policies/ets/cap/index_en.htm (accessed March 11, 2016).
46. The Cancún Agreements. http://www.un.org/wcm/content/site/climatechange/pages/gateway/the-negotiations/cancunagreement (accessed March 11, 2016).
47. Kyoto Protocol. *op. cit.*
48. (1) The preceding three paragraphs are based on: 2015 United Nations Climate Change Conference. https://en.wikipedia.org/wiki/2015_United_Nations_Climate_Change_Conference (accessed March 11, 2016); and (2) Matt McGrath. CO2 data is 'wake-up call' for Paris climate deal. http://www.bbc.com/news/science-environment-35778464 (accessed March 10, 2016).
49. Integrating Development into the Global Climate Regime. http://siteresources.worldbank.org/INTWDRS/Resources/477365-1327504426766/8389626-1327510418796/Chapter-5.pdf (accessed March 11, 2016).
50. Christopher D. Stone. *Should Trees Have Standing, Third Edition*. Oxford University Press, New York. (2010) 248 pp.
51. Roger Harrabin. Law firm in new legal threat over UK air pollution. http://www.bbc.com/news/uk-35689427 (accessed March 11, 2016).
52. Christopher D. Stone. *Should Trees Have Standing, Third Edition. op. cit.*
53. *Ibid.*
54. *Ibid.*
55. Full Text of the Endangered Species Act (ESA). http://www.nmfs.noaa.gov/pr/laws/esa/text.htm (accessed March 11, 2016).
56. Christopher D. Stone. *Should Trees Have Standing, Third Edition. op. cit.*
57. Plaintiffs Ask Supreme Court to Review Novel Climate 'Trust' Litigation. http://ourchildrenstrust.org/sites/default/files/2014.10.03insideepa.pdf (accessed March 11, 2016).
58. Our Children's Trust: Landmark U.S. Federal Climate Lawsuit. http://ourchildrenstrust.org/us/federal-lawsuit (accessed November 28, 2016).

Section V

Rights of Nature in Practice

Redesigning the American food supply, energy sources, mineral sources, recycling markets, and trade practices to align with a Rights of Nature system is a large undertaking that will change the underpinning of current society. In order to live sustainably, however, this is a task that must be accepted, willingly pursued, and ultimately brought to successful completion. As with other subjects we have surveyed, the philosophies, technologies, and the necessary research to make the transition a smooth one are already in place, or very nearly so. All that is lacking is the political will to succeed. The biophysical constraints of the world's ecosystems, which are now increasingly obvious, make it clear that the need to completely revamp society's use of natural resources is urgent if humans are to live, as they must, inside Nature's Laws of Reciprocity. To explore how we can retool society to ensure Nature's right to life has primacy, we assume, as stated before, that Rights of Nature is in the federal and state Constitutions, with Nature itself as the primary rights-holder.

13

Food Supply

Food supply is essential for all living beings, including humans. This chapter is entitled "Food Supply" rather than "Agriculture" because procuring food is not always via agriculture. It may, and often does, include herding, gathering, or harvesting wild foods, and diverse forms of cropping or semi-cropping. However, agriculture is a central part of many food-supply systems, including in the United States, and will form the bulk of the discussion.

It is widely recognized that current forms of agriculture are poorly suited to meet the requirements of a world with an expanding human population, higher energy costs, and an increasingly unpredictable climate characterized by declining supplies of water, lower biodiversity, and a lessening hospitality to life—all emerging global trends over which humans have little or no direct control.

The most damaging aspects of many agricultural systems, in which the United States is a leader, include high inputs of fossil fuel, high levels of mechanization, high levels of wasteful irrigation, and progressive dependence on the use of pesticides and herbicides. In addition, there are numerous, long supply lines of imported foods from far afield, even when the same foods grow locally, as well as numerous corporate food systems with an organizational bias geared toward maximum profitability. None of these activities are compatible with Nature's Laws of Reciprocity, which distill the requirements necessary for ecosystems to flourish. Moreover, none are compatible with a Rights of Nature framework, which focuses on placing nature first to provide a resilient world that supports all life forms.

The Rights of Nature

All the Rights of Nature provisions currently implemented or being used as policy templates contain sections focusing specifically on the food supply—indicative of both of its importance and the many challenges in creating and maintaining an equitable, sustainable supply of food. Let us look at these for guidance in the discussion to follow.

In addition to its now-famous section on Nature's Rights, the Ecuadorian Constitution also contains a chapter entitled, "Rights of the good way of living." Article 13 proclaims, "Persons and community groups have the right to safe and permanent access to healthy, sufficient and nutritional food, preferably produced locally and in keeping with their various identities and cultural traditions. The Ecuadorian State shall promote food sovereignty."[1]

A section immediately follows these statements on "Healthy environment," which guarantees "The right of the population to live in a healthy and ecologically balanced environment that guarantees sustainability and the good way of living is recognized."[2] This "right" includes the protection of Ecuador's biodiversity and genetic assets, as well as preventing environmental damage, as "matters of public interest."

The new Ecuadorian Constitution also has a lengthy chapter guaranteeing the rights of communities, peoples, and nations, which is critical for a discussion of changing agriculture. These range from the right of indigenous peoples to maintain and develop their ancestral traditions and forms of social organization to the right of keeping ownership of their community and ancestral lands. They also have the right to "keep and promote their practices of managing biodiversity and their natural environment."[3] Most importantly, the Constitution upholds the peoples' right to protect and develop collective knowledge, sciences, ancestral wisdom, "genetic resources that contain biological diversity and agricultural biodiversity," and the right to restore plants, animals, and ecosystems.[4]

The Constitution does not stop there, however. In laying out the "Development of Systems," the Constitution's language makes food sovereignty "a strategic objective and an obligation of the State," so that peoples "achieve self-sufficiency with respect to healthy and culturally appropriate good on a permanent basis."[5] This "self-sufficiency" includes bolstering diversification and introducing organic farming techniques. The overall thrust of this section is to support small and medium-size producers, and set up equitable, rural-urban supply networks so that urban residents have access to locally grown foods. In addition, "self-sufficiency" includes creating and promoting policies of redistribution that will enable small farmers access to land, water, and materials needed for farming.[6]

The Bolivian Rights of Nature law contains less direct language to guide the change in food-supply technologies, but clearly the vision of the law is to provide a healthy environment in which food production can flourish. This includes Nature's right to the diversity of life and the "functionality of the water cycle"—including its protection from pollution. Moreover, Nature's rights include not only the "right" to equilibrium but also the "right to maintenance or restoration of the interrelationship, interdependence and functionality of the components of Mother Earth in a balanced way for the continuation of their cycles and reproduction of their vital processes."[7]

Pope Francis also discusses the role of agriculture in his *Laudato Si*, pointing out that Christian scripture tells humans to "till and keep the garden of the world." He explains this as follows: "Tilling refers to cultivating, ploughing or working, while 'keeping' means caring, protecting, overseeing and preserving. This implies a relationship of mutual responsibility between human beings and nature. Each community can take from the bounty of the earth whatever it needs for subsistence, but it also has the duty to protect the earth and to ensure its fruitfulness for coming generations."[8]

Mirroring the Ecuadorian concern for small-scale producers and diversification of crops and farmers, Pope Francis points out that small-scale systems of food production feed a majority of the world's population less wastefully than large-scale agriculture; but economies of scale force small land holders to abandon traditional lands and crops, to the detriment of community and self sufficiency. (Economies of scale means that a proportionate saving in costs will be gained by an increased level of production.) He supports the use of laws to protect small farmers from larger-scale agricultural production.[9]

From Family Farms to Corporate Farms

Colonial America was primarily rural and agricultural; some 90 percent of residents made their living from farming. The produce went mainly to feed the family and create small, local trading networks. By 2008, only approximately 2 percent of the American population, if that, was directly involved in farming. What changed?

The History of a Change in American Agriculture

Although this is not the place for a detailed history of American agriculture, the major change began in the 1920s, with such new technologies as the combine harvester, tractor, and other heavy equipment. Larger farms became more economical. Though American agriculture produced vast amounts of crops and farmers received relatively high prices during WWI, an agricultural depression settled in after the War.

In countering the Depression of the 1930s, the New Deal focused on reducing commodity supply to raise prices of food slightly to the consumer but greatly to the farmer. The 1933 Agricultural Adjustment Act implemented farm subsidies, under which farmers were paid to leave some lands idle and to kill excess livestock.[10] The idea was that the less produced, the higher the wholesale prices, and this did happen.

By 1936, farm incomes rose significantly. The law's subsequent iteration provided subsidies for farmers to plant soil-enriching crops, such as alfalfa, that would not be sold on the market. The law of price supports and crop subsidies is now staggeringly complex, but it is still the fundamental, national law structuring federal intervention in agricultural policy. Over the decades since the inauguration of the 1933 Agricultural Adjustment Act, agricultural policy has grown more comprehensive, ranging into soil conservation, use of surplus crops as food aid, organic food labeling, and incentives to conserve wetlands and habitats through such programs as the Wetlands Reserve Program and the Environmental Quality Incentive Program.

After World War II, farm productivity continued to increase, as did the size of farms and the trend toward mechanization. Rural electricity was especially helpful in this regard. But it was under Agriculture Secretary Earl Butz in the 1970s that U.S. policy stressed large-scale, mechanized, capital-intensive farms at the expense of local family farms, as lucidly described by Kirkpatrick Sale:

> The Department of Agriculture adopted a specific policy of "get big or get out" … and massive subsidies … ensured that the policy would be followed. From 1950 to 1990 the American farm population declined from 23 million to less than 5 million, the number of one-family farms from 3 million to 1.4 million…. The driving force by which this was accomplished was largely technological, a mechanization of the farm on a scale so vast, if invisible to the urban dweller, that by the 1980s American agriculture was actually much more mechanized than the manufacturing sector. Factory husbandry (particularly of chickens), center-pivot irrigation, mechanical tomato pickers (three workers do the work of sixty), cotton strippers (one worker does the work of fifty-six), grain and beet combines (one worker does the work of eighty), automated milking machines, cattle feedlots, computer bookkeeping, helicopter spraying—American farms have become almost entirely industrialized, corporations of great size and complexity. But, as always, mechanization serves the interests primarily of the largest firms, and those who cannot afford the latest expensive technology … inevitably fall behind, and the number of farms inexorably dwindles.[11]

Subsequent developments include increasing mechanization and concentration, leading to large-scale, confined, animal-feeding operations; genetically modified organisms (= seeds, among other things) commonly known as "GMOs"; vast monocultures, such as wheat, covering hundreds or thousands of acres; and the increasing use of both pesticides and herbicides in crops and growth hormones in the production of animal products, such as beef, milk, and poultry (Figure 13.1).

FIGURE 13.1
Rice harvest in Northern California. Photograph by Gary Kramer, U.S. Department of Agriculture, Natural Resources Conservation Service (https://photogallery.sc.egov.usda.gov/netpub/server.np?find&catalog=catalog&template=detail.np&field=itemid&op=matches&value=899&site=PhotoGallery).

Today's Worldwide Trends in Agriculture

The "green revolution" and, more recently, genetic engineering have increased food production by large corporations while simultaneously relegating the small- to mid-sized, independent farmer ever closer to the halls of antiquity.

Agriculture is increasingly *big business* controlled by corporations that, in many cases, strangle the small, family farmer by controlling the market and by making and keeping the small farmer dependent on expensive products. This type of corporate domination has led to a drive for "economic efficiency" in agriculture, which translates into highly capitalized technology that has largely killed the soul once found in the agrarian way of life, both at home and abroad.[12]

The signs that industrial agriculture is incompatible with Nature's Laws of Reciprocity, and thus any measure of ecosystem resilience and sustainable productivity, are many. For example, colossal amounts of methane—a potent greenhouse gas—are produced in confined animal-feeding operations, such as gigantic cattle feedlots. Confined animal-feeding operations routinely administer antibiotics that favor antibiotic-resistant, methane-producing organisms in the gut that, in turn, change the microbes in the digestive system of dung beetles, which are vital to the cycling of carbon and improvement of the soil.[13]

In addition to methane, livestock fecal material creates enormous problems of water pollution, including vast streams of nutrient-rich runoff from fertilizers and other agricultural inputs that endanger waterways. Once the nutrient runoff reaches the coastal waters and estuaries, it produces algal blooms that suck the oxygen from the waters, killing all oxygen-dependent life, thereby creating "dead zones."[14]

Wasteful overuse of water is another staggering problem. Agriculture in the United States is responsible for between 80 percent and 90 percent (in the Western states) of both surface and groundwater consumptive use of freshwater nationwide.[15] Nearly three-quarters of irrigated cropland is in the 17 Western states, where water is most frequently at a premium. Nationwide, there were around 55.8 million acres of irrigated cropland in the United States in 2012, a slight drop from the peak in 2007 of 56.6 million acres. Irrigated cropland is concentrated in the Columbia and Snake basins of the Northwest, California's Central Valley, the Ogallala region of the Great Plains, and the Mississippi Delta.[16] Irrigation for agriculture creates by far the largest demand for withdrawal of freshwater nationwide and has contributed heavily to surface water pollution and loss of groundwater in many regions of the country.

Moreover, industrialized agricultural areas not only are rife with pesticides and herbicides but also devoid of feeding corridors and thus cause the large-scale collapse of pollinators, especially bees.[17] Added to this is the decline in seed diversity around the world, replaced by a few, small, corporate varieties of genetically modified seeds. In addition, the economic drive for monocultures relies on a growing use of pesticides and herbicides, which saturate the soil. This in turn leads to large-scale poisoning of agricultural lands, streams, groundwater, and ultimately the oceans of the world.[18]

These problems are not confined to the United States. China, for example, recently admitted that at least one-fifth of its agricultural land was contaminated—the culprits being both rapid industrialization and indiscriminate use of industrial agricultural processes. Most of the contaminated land lies on the east coast, where the country's heavy industry is situated, but there was also extensive heavy-metal pollution in southwest China. At least 3.3 million hectares (81,544,776 acres) of land is polluted, especially in grain-producing regions. Chinese officials are considering withdrawing some of the worst contaminated lands from farm production.[19]

Ecosystems and Agricultural Systems

Both widespread practice and research show that small-scale agriculture—with a variety of crops, minimally mechanized and industrialized—provides what agriculture most needs in order to flourish as part of a functional ecosystem. A Rights of Nature system makes it even clearer that agriculture must be part of an ecosystem, not supplant it. A flourishing ecosystem can provide food from both gathered and farmed sources, and will, in its nature, protect the time-honored varieties of domesticated plants.

Plants

The integration of peasant farms and natural ecosystems into agro-ecosystems forms a continuum in which plant gathering and crop production are actively practiced. Many of these traditional, agro-ecosystems are still found throughout the non-industrialized countries, where they constitute major repositories of germplasm for both crop plants and wild plants. (*Germplasm* is a collection of genetic material for an organism. For non-arboreal plants, germplasm may be stored as a collection of seeds; for trees, the germplasm may be maintained by growing them in a nursery.)

Domesticated plants have evolved, in part, under the influence of farming practices shaped by various cultures and thus are directly dependent on the care given them by humans. Because genetic material is archived more effectively as living (rather than non-living) systems, maintaining traditional agro-ecosystems is a realistic strategy for protection of the genetic properties of both crop plants and wild plants. In addition, it would be wise to link the knowledge of a peoples' ethnobotany with their self-sufficiency.[20] (*Ethnobotany* is an inquiry into the complex relationship between plants and people and how a given culture uses plants.) Protecting traditional agro-ecosystems can be achieved only in conjunction with the maintenance of a local people within their culture, which, in turn, protects cultural diversity. Over time, however, the ecological effectiveness of small, diversified, family farms has been replaced by the economic efficiency of larger and larger monocultural farms. But bigger is not necessarily better, as exemplified in the Ghibe Valley of Ethiopia.

Successful control of tsetse fly–transmitted sleeping sickness in the Ghibe Valley appears to have accelerated the conversion of wooded grassland into farmland, which in turn has affected wildlife habitat. To assess the influence of this expanded agricultural land use, researchers looked at the species richness of and composition of bird communities as indicators of environmental impacts.

At the height of the growing season, the number of bird species and the associated vegetative complexity were greater in the oxen-plowed fields of small farms and riparian woodlands than in wooded grasslands or in the tractor-plowed fields of large farms. Species composition differed greatly among the types of land use—with many species only having a single habitat type. Although this result implies that converting land from wooded grasslands to small farms may not adversely affect the numbers of bird species in this region, it will alter the composition of their communities within different habitats.

Moderate land use, which creates a mosaic of small-farm fields in the Ghibe Valley, increases habitat heterogeneity and bird-species richness, as opposed to land use in the large, tractor-plowed fields of intensively used farms. But if the small farms significantly exceed their current number, the bird-species richness and community composition may be negatively affected.[21] Thus, once again, it is relationships and patterns—not numbers—that convey relative stability to ecosystems.

The farms of north-central Florida in the United States form a counterpart story to those of the Ghibe Valley with respect to their attractiveness for birds. A two-year analysis was conducted on paired organic and conventional farms, during which the diversity, distribution, and insect-foraging activity of native birds on farms were assessed. The results indicate not only that farms supported 82 percent to 96 percent of the land birds known to breed in the region but also that species richness and abundance varied significantly with the habitat matrix in which the fields are situated, as well as with the kind of border habitat surrounding the fields (but not with the year or type of farm management). The highest numbers of birds were associated with mixed crops, field borders, and the adjacent habitat, which was composed of forest and hedge, and the 10 species identified as functional insectivores (because they are most likely to contribute to the control of unwanted, crop-damaging insects) were attracted significantly more to mixed crops than to monocultures.[22]

Livestock

Causes for the decline in farm breeds, according to the United Nations Food and Agriculture Organization, include loss of habitat attributed to growth in the human population and wars. But the greatest threat comes when farmers in poor nations discard native breeds and switch to Western commercial livestock, which is highly productive.

FIGURE 13.2
A commercial meat chicken production house in Florida. Photograph by Larry Rana, U.S. Department of Agriculture (https://commons.wikimedia.org/wiki/File:Florida_chicken_house.jpg).

Western breeds like Holstein cows, Rhode Island Red chickens, and Yorkshire pigs, are alluring to Asian farmers because of the great quantity of milk, eggs, and meat they produce. For a poor farmer barely able to feed his family, they seem heaven-sent. The problem is that Asian farmers invariably cannot afford the high cost of maintaining and feeding such specialized Western breeds.

Livestock that are bred for the technology- and money-intensive agriculture of Western industrialized countries may not be suited to other environments, cultures, or methods of farming. Local breeds may well prove better and more profitable in the long run than those marketed by the West, but in the meantime, local breeds are increasingly dying out because those from the West are displacing them (Figure 13.2).

Small numbers of livestock is a frequent feature of small and medium-sized farms worldwide: a few chickens, a small herd of milk cows, a handful of pigs. But indigenous livestock, having also been caught in the "Green Revolution," move toward more Western-style livestock efficiency at the expense of local breeds adapted to place and local use. Every week, the world loses two breeds of its valuable varieties of domestic animals, such as the "Taihu pig," "gembrong goat," and "choi chicken."[23] These domestic Asian animals, along with as many as 1,500 other farm breeds worldwide, are as endangered as their wild relatives. For example:

1. In Madagascar, Renitelo cattle are nearly extinct. They are particularly well adapted to the different climate zones in Madagascar and provide meat and pulling power.

2. In Mexico, Chiapas sheep have been reared for almost 500 years in the highlands of the state of Chiapas. Indigenous women use the wool to produce clothing, both their own and for sale. The sheep are considered sacred, and people do not consume lambs or mutton.
3. In Vietnam, the importance of H'Mong cattle was only discovered in 1997 because these animals have been kept isolated for many years. The breed is well adapted to mountain regions up to 3,000 meters (9,843 feet). The current population is estimated at 14,000 individuals.
4. In Germany, Hinterwälder Rind cattle, found primarily in the Black Forest, are not only robust and highly fertile but also endangered.
5. In the Russian Federation, Yakut cattle are adapted to the freezing climate in Siberia. Their numbers, however, are estimated to be less than 1,000.[24]

Local breeds are the result of successful adaptations to particular environments that began when people started domesticating animals more than 10,000 years ago for food, fiber, the power to work, and for their droppings as fertilizer. For example, China's min pig tolerates extreme temperatures, the pygmy hog of northern India is ideal for small villages, and the zebu cattle of Java are disease resistant and prolific. But now, 105 such domestic animals are endangered in Asia—and Asia is not alone.

The United Nations Food and Agriculture Organization estimates that 30 percent of the 4,000 to 5,000 breeds of domestic animals thought to exist in the world are threatened with extinction. Half of all the domestic breeds that existed in Europe at the beginning of the 1900s have vanished, and more than one-third of all breeds of poultry and livestock in North America are rare, which may have a tremendous impact on rural economies in the future.[25]

In fact, 190 breeds of farm animals have gone extinct worldwide within the past 15 years, and there are currently 1,500 others at risk of becoming extinct. In the past 5 years alone, 60 breeds of cattle, goats, pigs, horses, and poultry have become extinct.[26]

In the United States, for example, a few main breeds dominate the livestock industry:

1. 83 percent of dairy cows are Holsteins, and five main breeds comprise almost all of the dairy herds.
2. 60 percent of beef cattle are of the Angus, Hereford, or Simmental breeds.
3. 75 percent of pigs come from three main breeds.
4. Over 60 percent of sheep come from four breeds, and 40 percent are Suffolk-breed sheep.[27]

Reinvigorating Family Farms and Traditional Methods of Farming

There are many converging efforts worldwide to change the industrial model of agriculture. Though the trend toward large corporate monocultures continues to increase in the United States, as does increasing mechanization and "hedgerow to hedgerow" farming, the picture is quite surprising in other ways. According to the U.S. Department of Agriculture, 97 percent of farms were family-owned as of 2012, and 88 percent of all farms were small, family farms.

These small, family farms own nearly half of all farmland (48 percent), but only accounted for 20 percent of agricultural sales, and 5 percent of the country's net farm income.

Only 3 percent of U.S. farms are non-family-owned—but accounted for 16 percent of all U.S. agricultural products in 2012. West Virginia, Oklahoma, Tennessee, and Alabama have the largest percentage of family farms, coming in at 98 percent. This was closely followed by a raft of other states through the Midwest and East—and Oregon, the only state in the West—with 96 percent to 97 percent of their farms being family operations. The American South and Southeast had the highest percentage of *small* family farms—once again along with Oregon—between 90 percent and 95 percent of all farms.[28]

As climatic changes become more intense and sustained, systemic changes in the allocation of agricultural resources must be considered, such as diversification of crops and production systems at all levels, as well as of the type of livelihood they create. Fortunately, though battered by more than half a century of relentless mechanization and industrialization, the family farm is still the majority in the United States, and rural infrastructure, towns, and communities are still to some degree intact. It is here that the re-tooling of agriculture can begin.

Dealing with the many barriers to effective adaptation will require a comprehensive rethinking of current approaches to agriculture, including providing educational opportunities for farmers to help them deal with new, weather-related risks to their chosen crops. In addition, new marketing strategies must be established to facilitate effective responses to the ongoing adaptations in the fields.

Science also has to adapt. Multidisciplinary problems require multidisciplinary solutions, and reorganizing agriculture will require much research, ranging across diverse subject areas, such as:

1. Organic small-scale farming and riparian management
2. Adaptation of traditional indigenous techniques to various crops
3. How best to rotate organic crops in different locales to increase soil fertility
4. The most effective agricultural uses of microclimates
5. Plant and animal management to benefit bees and their pollination
6. Conservation techniques for irrigation
7. Restructuring agricultural markets to create and sustain local supply networks

A crucial component of this approach is the implementation of interdisciplinary frameworks through which potential adaptations can be assessed for their relevance, robustness, and ease of use by all necessary and interested parties.[29] Part of this robustness and ease of use must include an interface with urban areas that protect farmlands, forestlands, and open spaces for the infiltration and storage of water—protection that is now missing.

Another essential tool of changing the face of farming is re-localizing the food supply economy. This allows communities to reduce dependence on the globalized international trade system, nurture local food supplies and support agriculture with urban consumption. The best known such effort is Community-Supported Agriculture, which has grown immensely in the United States since the 1990s. Community-supported agriculture operations vary, but typically a community or group of individuals become "shareholders" of a farm or garden, and pay in advance for a share of the harvest. This essentially creates a pledge that covers, in advance, costs of farm operation and farmer salaries, including the risks of inclement weather or pest outbreaks. In return, the members share in the farm's bounty during the harvest. For farmers, community-supported agriculture provides better prices for the crops, financial security, and relief from the burdens of marketing.

As of 2007, there were at least 12,500 community-supported agriculture farms in the United States. They generally share an emphasis on community and local produce, close connection between producers and consumers, and use of organic farming techniques. Community-supported agriculture continues to play an important role in reinvigorating family farms in diverse regions of the country.[30]

Rights of Nature and Emerging Agriculture

A Rights of Nature system must integrate agriculture into the surrounding ecosystem. It must use and expand some version of the feudal chain of obligation discussed earlier in this book, with rights and obligations extended, in a web rather than a line, to the streams, woods, hedgerows (= naturally vegetated fencerow of woody and herbaceous plants), savannahs, and other aspects of the ecosystem integrated with a mosaic of small fields. Any functional agricultural system will, as indigenous Northwestern cultures have done for millennia, cultivate indicator species (such as huckleberries, salmon, herring, or clams) to both increase the food supply for humans and make the ecosystem more resilient, which is a benefit to all its inhabitants. An "indicator species" is an organism whose presence, absence, or abundance reflects a specific environmental condition. As such, the species' abundance can indicate a change in the biophysical condition of a particular ecosystem, and thus may be used as a proxy to diagnose the system's sustainability and its resulting productivity.

To understand the importance of changing to a Rights of Nature agricultural system, we must remember that ecological sustainability flows from the biophysical *relationships* that have evolved among the various species, which in turn is the consequence of a landscape's variety of connected habitats. A seemingly stable cultural system—even a very diverse one—that fails to support these ecologically coevolved (or personally developed) relationships has little chance of being sustainable, especially when a long-term time scale is taken into account. As a human community in its living changes the landscape, the landscape in reaction alters the community.[31]

To create and maintain adaptable landscapes with desirable cultural capacities to pass to our heirs, we must focus on three primary things: (1) being conscious of and acting within the limitations of Nature's Laws of Reciprocity; (2) caring for and "managing" for a sustainable connectivity and biological richness between and among the landscape's natural and human-altered habitats, such as forests, agricultural fields, floodplains, riparian areas, urban areas, wildernesses and natural areas, pastures for livestock grazing, and many others within the context of the landscape as a whole; and (3) protecting existing biodiversity—including a diverse mix of connected habitats—and increasing it at every opportunity by restoration and a Rights of Nature-based pattern of land use. This not only protects ecosystem processes but also strengthens ecosystem resilience.

Organic and relocalized farming is a good beginning to changing the face of agriculture, but it is not enough, because organic crops can be grown in large-scale monocultures that disregard ecological diversity. Monocultures, even organic ones, have far-ranging, but poorly understood, effects. For example, monocultures reduce the number of foraging flowers for honeybees, which weakens bee colonies and lowers bee health, perhaps contributing to Colony Collapse Disorder, as well as overall higher bee mortality.[32]

Permaculture

Agricultural changes that go deeper and align more closely to a Rights of Nature system include, especially, Permaculture. Unlike some other solutions, Permaculture is based on whole-systems thinking and has developed an entire set of principles that can be implemented across diverse landscapes. It fits well with parallel efforts around the world to move away from industrial agriculture and revive and strengthen traditional and indigenous agricultural methods, adapting them to modern conditions as needs be.

The Permaculture system revolves around three ethics: Earth care, people care, and fair share,[33] all of which are designed to regulate self-interest for the good of the larger community. Permaculture thinkers have distilled ideas and ideals from cultures that have managed to co-exist with their environments for centuries. Earth care, for example, focuses primarily on care for the soil, but also, "All life forms have their own intrinsic value, and need to be respected for the functions that they perform—even if we don't see them as useful to our needs."[34]

Twelve principles flow from these ethics—from both the bottom up and the top down.[35] The first 6 start from a personal (=individual) point of view, beginning with the need to observe and interact systemically to provide design inspiration. Permaculture recognizes the importance of energy, but redirects human use to reinvest energy, especially renewable forms, rather than wasting it. It also focuses on designing a system that provides a yield as an incentive for true self-reliance—rather than a system that ends up focusing *primarily* on yield, as happens when profit-making is the goal.

The Permaculture system is designed to both apply self-regulation and accept feedback, so the system can remain in balance with the requirements of social-environmental sustainability for the long-term benefit of the community. This principle is especially important to any Rights of Nature system for food supply. Another principle, equally important in Rights of Nature, is to avoid producing waste. In Permaculture literature, the symbol for this principle is the earthworm, which converts plant litter into soil-nourishing humus.

While the first six principles of Permaculture focus on organisms and individuals from the bottom up, the second set of principles focus more on patterns and details of relationships from the top down. They embrace the ideal that systems are best designed to include an appreciation for and nurturance of the ecosystem's composition, structure, and function in both Nature and human culture, an aspect of which is agriculture. In essence, Permaculture focuses on integrating the dynamic, functional, self-reinforcing feedback loops of the system's components.

Permaculture is also built on the principle of valuing the edge effects of adjoining plant communities (including vegetated fencerows intermingled with agricultural fields) or adjoining soils of various types, as well as natural communities marginalized by human values, such as estuaries. The respect for marginal and invisible aspects of a system increases that system's functional sustainability. Finally, Permaculture seeks to use and respond to change (the underlying dynamic of every natural system's productive sustainability), which, as always, depends on human flexibility to accommodate the novelty.

Agroecology

Agroecology is another system that, similar to Permaculture, has care for the ecosystem at its heart. A working definition of it is, "The application of ecology to the design and management of sustainable agroecosystems."[36] Its design principles are based on traditional knowledge, alternative agriculture, and local food systems. Like Permaculture, it interlinks ecology, culture, economics, and society to sustain both agriculture and the environment within the culture. Although productive sustainability forms the heart of "sustainable agroecosystems," agroecologists recognize that broad principles must be designed and implemented locally because both ecosystems and their human cultures vary place to place.

Maintaining the integrity of Nature in an agricultural system is a tricky matter, especially under the recent influence of industrial agriculture, which seeks to wring profit from every aspect of the land. A whole-systems approach must, as agroecologists agree, maintain the natural resource base, and manage pests and diseases through internal regulation rather than pesticides and herbicides. Part of keeping the ecosystem resilient includes recovery from the harvest, which can take many different forms.

Agroecology, like other "alternative" agricultural movements, begins with moving away from conventional, industrial agriculture. The conversion process includes using natural inputs, adjusting the farming to the local environment, creating a sustainable design, and recycling nutrients.

The Agroecology Research Group is actively working with all these principles on the ground as the movement grows in sophistication. For example, there have been projects investigating "shade coffee" in Brazil, organic tomato farming in Ecuador, and creating revolving funds to build water tanks in Brazil. Projects in the United States and Canada have included research on sustainable sugarcane farming in the Everglades, creating urban market gardens in Toronto, the use of narrow strip cropping in Kansas, and organic strawberry farming in California. Agroecology principles are designed to adapt to local conditions, as projects far afield show, such as working with sorghum-genetic diversity in Ethiopia and soil and water conservation in Burkina Faso, among others.[37]

Seed Banks

Both of these movements fit well with local seed banks designed to maintain and stimulate use of diverse and traditional crops. Locally owned and controlled, seed banks easily create long-term storage by constant use and re-sowing of seed to protect and maintain viability in storage (which degenerates over time). Seed banks have become essential repositories of genetic diversity owing to the worldwide trend toward monocultures and genetically modified crops (Figure 13.3).

One of many highly successful seed banks is the Genetic Resource Ecology Energy Nutrition Foundation, which began work in the 1990s in Tamil Nadu and Karnataka, India. Despite the fact that women in Bengal, India, had once used more than 150 varieties of local plants to feed their families, the project had to begin in a surprising place: "… with farmers [who] had to go through an 'unlearning' process, as years of modernized agriculture had taken them very far away from a sustainable production. Many farmers did not seem aware that traditional crops and varieties had been lost.… It was even more difficult to convince them that some of the traditional varieties could yield as well as the introduced, commercial varieties that they had become used to."[38]

The Genetic Resource Ecology Energy Nutrition Foundation organized art, culture, and music festivals to celebrate local crops and raise interest in sustainable agriculture. The project then proceeded to map local seeds, and provide seeds to interested farmers. Finally, the Foundation was ready to create seed banks, where seeds are provided free to members who sowed them and returned double the amount to the seed bank after harvest. The seed banks developed activities to promote the use of local varieties of food crops, and banned the use of chemical pesticides, herbicides, and fertilizers.

FIGURE 13.3
The Svalbard Global Seed Vault provides a safe backup of seeds on food crops conserved by seed banks worldwide. This picture from inside the vault shows the shelves with the boxes holding the seed samples. Photograph by NordGen/Dag Terje Filip Endresen (https://commons.wikimedia.org/wiki/File:Storage_containers_in_Svalbard_Global_Seed_Vault_01.jpg).

At first, the Foundation met with much resistance—even disbelief—from multinationals and university scientists, but they persevered. They held seed fairs after harvest, so farmers could swap stories and practices, and learn about such things as crop diversity. The Foundation eventually became an umbrella organization that trains and serves more than 30 sustainable agriculture operations. From the outset, women were, and continue to be, big supporters of the seed bank and the return to sustainable, local agriculture.[39]

However, seed banks can also be misused, as tools of large-scale corporate agriculture to preserve genetic diversity for further manipulation—a use not in accord with a Rights of Nature system. As one research paper put it, "the conservation of old landraces was essential in order to ensure the future production of new crop varieties. Even if the newer crop varieties of the Green Revolution were very productive in the present, the long-term success of agricultural production depended on being able to access genetically diverse material."[40] Protecting such diversity uses high-technology seed-banking, the so-called "seed freezing" to fix genotypes and stabilize them through seed proxies—"a novel way of recording the genetic Constitution of crop plants so that they could be committed to memory and retrieved from cold storage as required."[41]

Wild-Caught Foods

A significant portion of the world food supply is still wild-caught, though the proportion varies by region. The largest supplier of wild-caught foods is, of course, the sea. Fish, shellfish, and other foods are vitally important to the diets of millions of people.

But the sea is overfished. Consider, for example, that overexploiting of the large, predatory, marine fishes, such as sharks and tuna, allows the populations of smaller, plankton-feeding fishes to proliferate:

> Sharks, billfish [such as sailfish, marlin, and swordfish], cod, tuna and other fish-eating fish—the sea's equivalents to lions on the Serengeti—dominated the marine world as recently as four decades ago. They culled sick, lame and old animals and kept populations of marine herbivores in check, preventing marine analogs of antelopes from overgrazing their environment.... [Moreover,] physical and chemical changes, driven by Earth's warming climate, threaten to diminish the maximum size that any species—predator or prey—can attain.[42]

This kind of exploitation not only influences community structure directly through the removal of larger-bodied fishes but also indirectly because larger-bodied fishes may exert top-down control on other species.[43] In fact, the extinction of a biophysical function brought about by overfishing takes precedence over all the other pervasive human disturbances to coastal biophysical systems, including pollution, degradation of water quality, and anthropogenic climate change.[44] Extinction of biophysical function caused by overfishing is brought about by five linked changes, which interact essentially as follows:

1. Commercial overfishing of predators.
2. Small, plankton-feeding species, such as sardines, proliferate.
3. Continual overfishing the larger individuals of any species—predator or prey—leads to a genetic shift toward progressively smaller and later maturing, less fertile individuals, which may prevent populations from ever recovering.[45]
4. The control of carbon dioxide by the ocean declines as the carbon dioxide–using plankton is reduced through consumption.
5. The end result is a change in the biophysical environment of the ocean.

This set of feedback loops is termed a "trophic cascade," where the change in one shifts all the others.[46]

American Overfishing and Aquaculture

Laws that regulate ocean fishing in American waters are mind-boggling in their complexity, a combination of state and federal policies. But in the early 21st century, the United States saw the inevitable: overfishing of many—if not most—fishing stocks was leading to ecosystem collapse. Managers responded by setting catch limits for every species the country fishes. This was done through the reauthorization of the Magnuson–Stevens Act, which governs all fishing in U.S. waters.[47] The reauthorization included new language that required each fishery to have annual catch limits in place by the end of 2011. Although they did not quite finish reorganizing fisheries management by that deadline, federal managers aimed to have catch limits in place for all fishing stocks by the end of 2012.

The problem has been that Regional Management Councils, which set the fishing limits for the more than 500 stocks under federal jurisdiction, routinely authorized more fishing than could be sustained—even when scientific evidence pointed to a need to reduce fishing. Under the new language, Regional Councils were obligated to place limits on fishing across all species.

Predictably, many fishing interests argued that the limits were arbitrary, the science was incomplete, and the restrictions were too tight, leading to destruction of small fisheries businesses. The National Oceanic and Atmospheric Administration, which regulates U.S. fisheries, itself admits that estimating population size of an ocean fishery is a very imperfect science, and this has led to endless arguments among fishermen and regulators about the health of this or that stock.[48] Especially contentious are the disagreements that erupt when an overfished stock appears on the rebound: should more fishing be allowed, or will the fishery collapse again? Where is the sustainable limit?

This is the kind of problem that can only be solved by a Rights of Nature system. Economic interests are linear and always focused on "more," thus guaranteeing that fishing will always remain right on the edge of depletion, *if* the law provides adequate safeguards. If it does not, overfishing will occur. This is the tragedy of the commons, discussed earlier. Once the Rights of Nature is fixed in the Constitutions, laws can be passed that place fishing activities inside the ocean's right to flourish. It will end the constant friction at the border between just-barely-sustainable fishing and overfishing. However, universal catch limits are a start in the right direction by recognizing that status quo cannot continue as a viable option.

Mark Spalding, president of the Ocean Foundation, said that people on both sides of the debate need to acknowledge that the United States is facing the sort of transformational moment in fishing that it did a half-century ago in forestry. Until the mid-1960s, the government allowed loggers unfettered access to public lands wherein "We had to have this wrenching, put-the-brakes-on-and-turn-the-truck-around" process, he said, adding that when it comes to setting universal catch limits, "this is a monumental achievement."[49]

One tool more and more frequently used, where appropriate, is "catch shares," authorized under the Magnuson–Stevens Act. The National Oceanic and Atmospheric Administration defines catch shares, as a general term, to describe the practice of dedicating a secure share of fish to individual fishers or fishing organizations for their exclusive use. The first American catch-share program was implemented in 1990 for the Mid-Atlantic surf clam and ocean quahog fishery; there are now 16 such programs nationwide.

Designed locally, they aim to ensure that annual catch limits are met while reducing bycatch—the catch of unwanted species. They can be designed to sustain fisheries communities by giving fishers the option to fish more if they have improved the fishing stocks with various management practices. In other words, it shifts the fisher's incentives from maximizing volume to maximizing value. Put differently, they now have a reason to collaborate as trustees of the resource upon which all generations depend—rather than overfishing as much as possible for personal benefit.[50]

The usual result of setting a catch limit can lead to many fishing boats competing with each other to catch as much as possible before the limit is reached. This often leads to unsafe practices, too many boats, and high levels of bycatch. Catch shares give each fishing group a quota, and they must stop when it is reached. Fishers can catch their quota at more leisure; over-capacity (i.e., consolidation of the fleet to economically efficient levels) allows those who exit fishing to receive compensation for their catch shares via the sale price of their shares. The problem with catch shares is setting and allocating the initial limit so the stock does not continue to be overfished. These are the often-contentious processes that the National Oceanic and Atmospheric Administration and Regional Councils are supposed to oversee, as well as regulating locally designed catch shares.[51]

Rightly designed, catch shares can be a powerful tool in overcoming the tragedy of the commons, because they give fishers an economic rationale for conserving the resource. It "makes good business sense" to conserve the stocks, reduce bycatch, and reduce harm to the ecosystem. To do this effectively, catch shares need to be community-oriented, rather than corporatized. If they are purchased and consolidated into corporate profiteering tools, catch shares easily become insensitive to the needs of communities and small fishing businesses. Catch shares can lead to such innovative management solutions as new net designs, or protecting small fisheries in coastal communities. They have been successfully implemented not only in the United States but also in other countries, such as Belize, where gillnetting and illegal fishing were depleting stocks. After catch shares were put in place, illegal fishing dropped 60 percent, and nearly three-quarters of fishers reported catching more. Fishermen working in two ecological gems along the spectacular Belize Barrier Reef—Glover's Reef and Port Honduras—have reported larger catches, and scientific surveys indicate the over-utilized reef fish populations are beginning to recover.[52]

But, as all parties recognize, catch shares is not the ultimate solution to ocean depletion, not to mention the biophysical integrity of estuaries, climate change, loss of coral reefs, ocean pollution, toxic algal blooms resulting from fertilizer runoff, and degradation of fish habitat.[53]

Catch shares are, however, greatly preferable to the other answer to dwindling fish stocks: salmon aquaculture or "salmon farming." A controversial technique used in North America, South America, and Europe, salmon farming is the industrial production of salmon from eggs to market, usually in a "sea-cage" or "ocean net-pen," which can be thought of as floating feedlots placed in an inlet from the open ocean. Most of these farms are the size of two American football fields—115,200 square feet total—containing up to a million fish per feedlot (Figure 13.4).

Because salmon are carnivorous, it takes approximately 4.5 to 11 pounds of wild prey fishes, such as sardines, to produce 2.2 pounds of farm-raised salmon. As with all agricultural practices, however, salmon farming involves environmental consequences. These include discharges of wastes and unconsumed feed from the pens that may accumulate on the ocean floor and surrounding environments. Some solutions to this problem involve locating the ocean net-pens in the vicinity of strong currents that serve to flush the area, as well as letting sites remain fallow (empty) for some period—often 6 months—to allow the ocean floor time to recover.[54]

Despite the industry's positive outlook for salmon farming, termed "Marine Harvest" in their industrial handbook,[55] these feedlots are nurseries for sea lice (tiny crustaceans with a predisposition to preying on fish) that proliferate in salmon farms and spread out

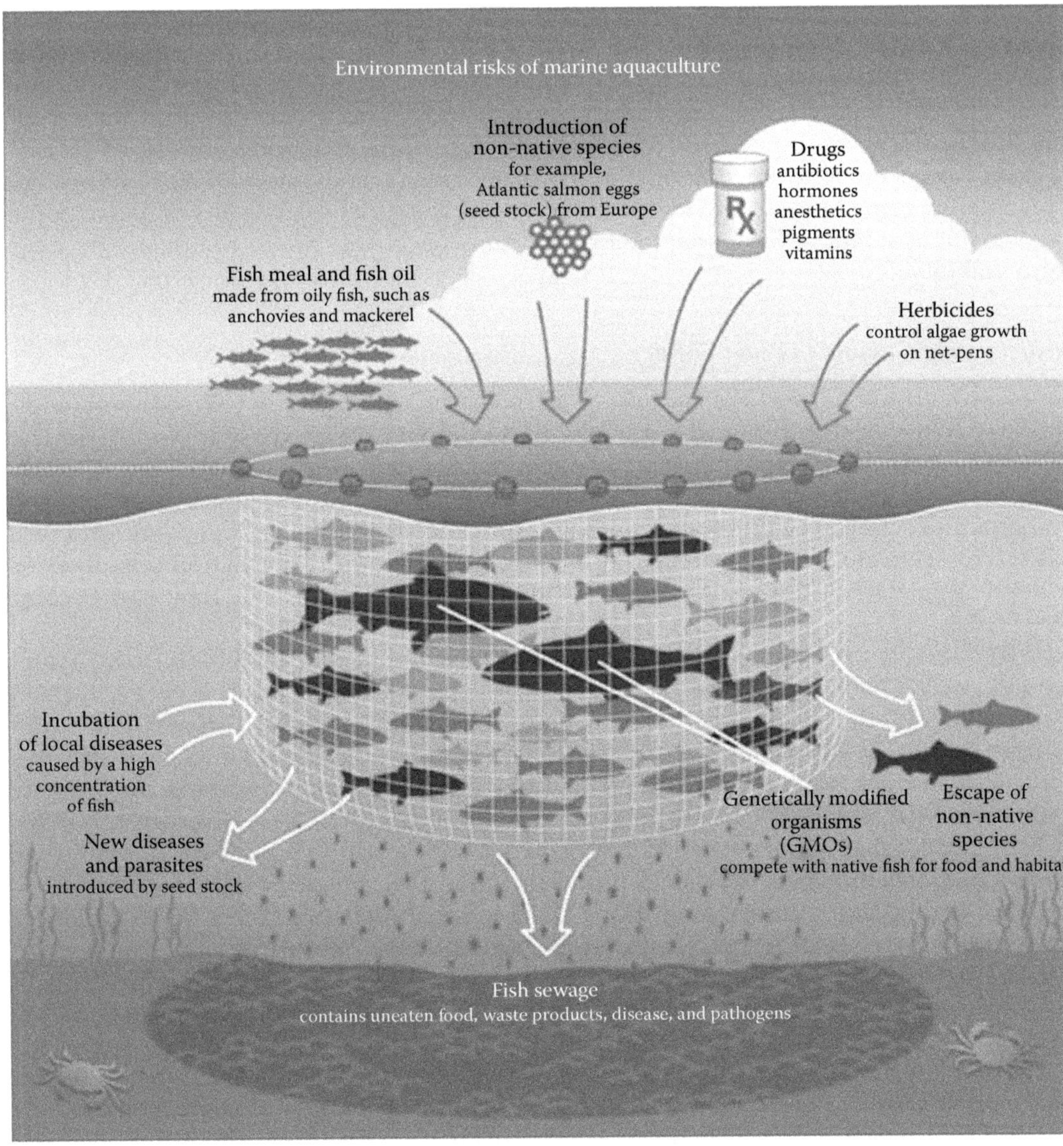

FIGURE 13.4
Aquaculture's effect on the environment. Illustration by Dr. George Pararas-Carayannis (https://commons.wikimedia.org/wiki/File:Risks_aquaculture_550.jpg).

to afflict wild fish. In one study, over a billion sea lice were produced in just 12 farms in a two-week period just prior to the out-migration of wild juvenile salmon.

While a few lice on a large salmon may not cause serious damage, large numbers on the same fish, or just a couple of lice on a juvenile salmon, can be harmful or fatal. Their feeding activity can cause serious fin damage; skin erosion; constant bleeding; and deep, open wounds (sometimes eating away the flesh of a juvenile's head down to the bone) creating a pathway for diseases, such as viral infectious salmon anemia. In addition, escaped farm fish can heighten the potential for the transfer of such diseases as infectious salmon anemia, among others.[56]

There is yet another profound problem with escaped, farm-raised salmon: namely, interbreeding with wild stock, because fish farms are typically located in the vicinity of wild fish habitats. Not surprisingly, escaped fish may disperse over large geographic areas and mix with their wild cousins, entering rivers tens to hundreds of miles from the escape site during the spawning season. Moreover, the successful spawning of escaped farmed salmon has been widely documented in rivers both inside and outside their native range. This interbreeding has the potential to not only alter the genetic variability of native populations, but also reduce local adaptability and thus negatively affect a population's long-term viability.[57]

Rejuvenating Gambia's Oyster Fishery

Innovative efforts to avert the tragedy of the commons are being developed worldwide, fitting in to the local culture and environment. In Gambia, on the coast of West Africa, for example, careful management has led to a rebounding of the local oyster fishery. Gambia is the smallest country on the African mainland, but it has a priceless jewel: Tanbi Wetland National Park. Tanbi is a bit smaller than the island of Manhattan, in New York, and well known for its lush mangrove forests. Oysters from the mangroves have traditionally been a delicacy in Banjul, the nearby capital. But, as frequently happens, they had become overfished.

"TRY," the Oyster Women's Association, was founded in 2007 to help the local oyster fishers, mostly women, reduce the overfishing that resulted from a growing human population and higher demand. (The foundation's name is derived from: "The world is your oyster—if you TRY!") Despite oystercatchers roving deeper into the mangrove forests, the oyster catch was still diminishing. TRY began by organizing the women, providing microfinance loans to help them open other businesses during the rainy season, learn other skills and handicrafts, and provide opportunities for daughters to complete their education. TRY not only lengthened the non-catch season each year to give the oysters more time to grow but also negotiated exclusive rights for its members to harvest oysters, ending the free-for-all in the wetlands. The cessation of competition has been crucial to allowing the oysters a chance to reach maturity and reproduce. In addition, parts of Tanbi are closed on rotation for extended periods to further aid the oysters.

The result was 6,300 hectares (15,568 acres) of oyster mangroves under improved management, improved oyster quality, and a rise in the price per cup of oysters, greatly to the benefit of the oystercatchers. There has also been a betterment of working conditions and education, especially among girls.[58]

The complex, multi-agency management plan, which resulted in TRY being given exclusive rights to the Tanbi oyster beds, took years to forge and welded the oyster-catching women into a web of collaborators: "...[t]he women haven't simply been handed over the rights to the wetlands to do with as they please: they are committed to looking after them. They are now the official custodians of Tanbi. ... In the past, some harvesters used machetes to chop away whole roots covered in oysters, big and small; this was wiping out juvenile oysters and damaging the forest itself. Now, as well as being much more selective and careful about taking only individual oysters of the right size, the women of TRY are also trialling an aquaculture technique similar to the one used for mussels, hanging ropes to catch young oysters from the water."[59]

Management Considerations for a Rights of Nature System

There are many excellent systems being developed worldwide to reposition agriculture as a sustainable enterprise, and make wild-caught foods sustainable. However, these programs must be placed in a Rights of Nature system to guarantee that the process of determining sustainability *begins within* the ecosystem that supports it. If this is not done, local efforts will remain patchwork and only locally effective, while the larger ecosystem processes will continue to deteriorate. Furthermore, increasing pressure on a system can, at any time, unravel such efforts as Permaculture, catch shares, or cooperative fisheries management. Here are some overall blueprint principles for converting the American food-supply system into one that is truly sustainable once the Rights of Nature are placed in state and federal Constitutions as the primary rights-holder, so the legal system ensures that Nature's right to flourish comes first.

Agriculture

1. Pass laws under a Rights of Nature framework requiring adoption of principles similar to those hammered out in Permaculture and Agroecology systems.
2. Require all agricultural producers to collaborate in mapping local ecosystems and Rights of Nature requirements for ecosystem resilience in the local region, and place laws into effect encouraging and incentivizing the strengthening of critical species in that ecosystem.
3. Provide microfinance options and technical assistance in rural and nearby urban areas to re-localize the food supply economy, so that locally grown products are the mainstay of the surrounding urban and rural populations. This includes designing a network that incentivizes imports of necessary foodstuffs from the nearest centers of production, and always prioritizes local and regional production and distribution. Community-supported agriculture, a widespread and burgeoning movement, is the best example.
4. Require all agriculture to be organic, without use of pesticides, herbicides, fertilizers, or genetically modified crops (GMOs). Provide a framework for converting the U.S. agriculture system to a sustainable one, including technical expertise and subsidies to provide help during the transition.
5. Provide federally-funded incentives for crop rotation, diverse cropping, hedgerow expansion, wetland and riparian protection, and soil nutrition, including necessary research and scientific support. Many such programs already exist in the U.S. Department of Agriculture, such as the Environmental Quality Incentives Program and Regional Conservation Partnership Program. They could be easily expanded and broadened.
6. Provide microfinance options to American farmers to locally improve the ecosystem and crop diversity, including seed banks, incentives for maintaining local livestock breeds, and expansion of local food networks between rural and nearby urban areas.
7. Reorganize agricultural laws so the basis of American agriculture is a chain or web of obligations, beginning with Rights of Nature and extending to the soil, crops, wildlife, waters, farming communities, farmers, consumers, and suppliers in the

local area. Study of feudal obligation webs between manorial tenants, farmers, and landowners provides a valuable model for designing expanded contemporary ones.

8. Provide financing and technical assistance to change both the delivery systems and methods of water use to conserve agricultural use of surface water. Depending on the part of the country, this can include drip irrigation systems, use of brackish water for irrigation, planting of drought-resistant crops, and use of instream meters and piping to eliminate instream impoundments for water storage, which negatively affect a river's ecology.
9. Provide financing and technical assistance to explore alternatives to groundwater use. Groundwater overuse, by agriculture and increasingly by municipalities as well, leads to groundwater depletion, lowering of the water table, reduction in surface water to streams and lakes, reduced riparian vegetation, land subsidence, and saltwater intrusion. The Ogallala Aquifer in the Great Plains area is notoriously depleted, for example.[60]

Wild-Caught Foods

1. Reset universal catch limits on all fished ocean species once a Rights of Nature system is in place, so that the ecosystem can maintain itself and flourish without constant fishing pressure.
2. Create large-scale marine sanctuaries to allow fisheries to recover and flourish, so that errors in setting catch limits do not prove fatal or stressful to the ecosystem. (For a discussion of such areas, see: *Interactions of Land, Ocean and Humans: A Global Perspective*.[61])
3. Implement non-corporatized catch shares or a similar management technique in every fishery that incentivizes the intrinsic value of fish stocks over the competitive, economic maximization of the catch.
4. Eliminate salmon farming in the ocean and all other water bodies, including inlets, lakes, reservoirs, estuaries, rivers and streams. Farmed salmon interfere with wild fish and have helped deplete wild salmon stocks across North America and Europe, as well as weaken them genetically.
5. Set limits, based on research of the ecosystem's needs to flourish free of human interference, to restrict overuse of any other wild-caught foods. This includes such foods as clams, deer, wild mushrooms, including truffles, wild berries, river fish, and many others. All are vulnerable to over-harvesting as they become popular and better known. Map ecosystem productivity for the relevant wild-caught foods and pass laws under a Rights of Nature system enforcing limits and/or seasonal restrictions on the wild harvest so that the ecosystem's own right to flourish is respected first and foremost.

Endnotes

1. [Ecuadorian] Constitution, full language: http://pdba.georgetown.edu/Constitutions/Ecuador/english08.html (Title II, Chapter 2, Section Two, Article 13) (accessed August 10, 2015).
2. *Ibid.* (Title II, Chapter 2, Section Two, Article 14).
3. *Ibid.* (Title II, Chapter 4, Article 57).

4. *Ibid.*
5. *Ibid.* (Title VI, Chapter 3, Article 281).
6. *Ibid.*
7. Language of the Bolivian law (compete text): http://www.worldfuturefund.org/Projects/Indicators/motherearthbolivia.html (Chapter III, Article 7). (accessed August 12, 2015).
8. Encyclical Letter **Laudato Si** of The Holy Father Francis on Care for Our Common Home [in English]. http://w2.vatican.va/content/francesco/en/encyclicals/documents/papa-francesco_20150524_enciclica-laudato-si.html (Chapter 2, II, Section 67) (accessed August 19, 2015).
9. *Ibid.* (Chapter 3, II, Section 129).
10. Agricultural Adjustment Act of 1933. http://www.legisworks.org/congress/73/publaw-10.pdf (accessed May 24, 2016).
11. Tom Philpott. A reflection on the lasting legacy of 1970s USDA Secretary Earl Butz. http://grist.org/article/the-butz-stops-here/ (accessed August 16, 2016).
12. (1) John A. Baden. Has efficient agriculture cost the farm its soul? *Corvallis Gazette-Times*, Corvallis, OR. (March 11, 1999); and (2) Andy Jones. Eating Oil. *Resurgence* 216 (2003):39–45.
13. Tobin J. Hammer, Noah Fierer, Bess Hardwick, and others. Treating cattle with antibiotics affects greenhouse gas emissions, and microbiota in dung and dung beetles. *Proceedings of the Royal Society B, Biological Sciences*, 283 (2016), doi: 10.1098/rspb.2016.0150.
14. Chris Maser. *Interactions of Land, Ocean and Humans: A Global Perspective.* CRC Press, Boca Raton, FL. (2014) 308 pp.
15. Irrigation & Water Use: Overview. http://www.ers.usda.gov/topics/farm-practices-management/irrigation-water-use.aspx (accessed August 13, 2016).
16. Irrigation & Water Use: Backgound. http://www.ers.usda.gov/topics/farm-practices-management/irrigation-water-use/background.aspx (accessed August 13, 2016).
17. (1) Stephen L. Buchmann and Gary Paul Nabhan. *The Forgotten Pollinators.* Island Press, Washington, DC. (1997) 292 pp.; and (2) Reese Halter. *The Incomparable Honeybee and the Economic of Pollination*. Rocky Mountain Books, Victoria, BC, Canada. (2011) 105 pp.
18. Chris Maser. *Interactions of Land, Ocean and Humans: A Global Perspective. op. cit.*
19. (1) Scott Newman. China Admits That One-Fifth of Its Farmland Is Contaminated. http://www.npr.org/sections/thetwo-way/2014/04/18/304528064/china-admits-that-a-fifth-of-its-farmland-is-contaminated (accessed May 25, 2016); and (2) Jennifer Duggan. One fifth of China's farmland polluted. http://www.theguardian.com/environment/chinas-choice/2014/apr/18/china-one-fifth-farmland-soil-pollution (accessed May 25, 2016).
20. Miguel A. Altieri, M. Kat Anderson, and Laura C. Merrick. Peasant agriculture and the conservation of crop and wild plant resources. *Conservation Biology*, 1 (1987):49–58.
21. This paragraph and the previous two are based on: Cathleen J. Wilson, Robin S. Reid, Nancy L. Stanton, and Brian D. Perry. Effects of land-use and tsetse fly control on bird species richness in southwestern Ethiopia. *Conservation Biology*, 11 (1997):435–447.
22. Gregory A. Jones, Kathryn E. Sieving, and Susan K. Jacobson. Avian diversity and functional insectivory on North-Central Florida farmlands. *Conservation Biology*, 19 (2005):1234–1245.
23. (1) The discussion of domestic farm animals extinction is based on: (1) *The Associated Press.* Should we save pandas AND pigs? *Corvallis Gazette-Times*, Corvallis, OR. (January 21, 1996); and (2) Consultative Group on International Agricultural Research. Rare breeds of farm animals face extinction. *ScienceDaily*, September 4, 2007. https://www.sciencedaily.com/releases/2007/09/070903094320.htm (accessed May 25, 2016).
24. Food and Agriculture Organization of the United Nations. One third of farm animal breeds face extinction. December 5, 2000. http://www.fao.org/News/2000/001201-e.htm (accessed May 25, 2016).
25. (1) *The Associated Press.* Should we save pandas AND pigs? *op. cit.*; and (2) Consultative Group on International Agricultural Research. Rare breeds of farm animals face extinction. *op. cit.*
26. Heritage Animal Breeds and Heirloom Crop Varieties. 28. Family Farm Highlights—Census of Agriculture. https://www.agcensus.usda.gov/Publications/2012/Online_Resources/Highlights/NASS%20Family%20Farmer/Family_Farms_Highlights.pdf (accessed May 25, 2016).

27. *Ibid.*
28. Family Farm Highlights—Census of Agriculture. https://www.agcensus.usda.gov/Publications/2012/Online_Resources/Highlights/NASS%20Family%20Farmer/Family_Farms_Highlights.pdf (accessed May 25, 2016).
29. The discussion of agriculture and climate is based on: (1) John F. Morton. The impact of climate change on smallholder and subsistence agriculture. *Proceedings of the National Academy of Sciences*, 104 (2007):19680–19685; and (2) S. Mark Howden, Jean-François Soussana, Francesco N. Tubiello, and others. Adapting agriculture to climate change. *Proceedings of the National Academy of Sciences*, 104 (2007):19691–19696.
30. (1) Community supported agriculture. https://www.nal.usda.gov/afsic/community-supported-agriculture (accessed August 13, 2016); and (2) Community-supported agriculture. https://en.wikipedia.org/wiki/Community-supported_agriculture (accessed August 13, 2016).
31. (1) Chris Maser. Why protect "natural" areas within our dynamic, cultural landscape? pp. 25–29. *In*: *Science and the Management of Protected Areas*. J.H.M., Willison, S. Bondrup-Nielsen, C. Drysdale, and others (eds.). Elsevier Science Publishers B.V., Amsterdam, The Netherlands (1992); and (2) Chris Maser. Patterns across the landscape. *Forum*, 10 (1995):103–106.
32. ARS Honey Bee Health and Colony Collapse Disorder. http://www.ars.usda.gov/News/docs.htm?docid=15572 (accessed May 26, 2016).
33. Permaculture ethics. https://permacultureprinciples.com/ethics/ (accessed May 27, 2016).
34. Earth Care. https://permacultureprinciples.com/ethics/earth-care/ (accessed May 27, 2016).
35. Summary of permaculture principles from: Essence of Permaculture. https://holmgren.com.au/downloads/Essence_of_Pc_EN.pdf (accessed May 27, 2016).
36. Agroecology. http://www.agroecology.org/ (accessed May 27, 2016).
37. The foregoing is based on: *Ibid.*
38. Vandana Shiva. 2000. Globalization and poverty. *Resurgence*, 202 (2000):15–19.
39. Vanaja Ramprasad. Community seed banks for maintaining genetic diversity. http://www.agriculturesnetwork.org/magazines/global/securing-seed-supply/community-seed-banks-for-maintaining-genetic (accessed May 27, 2016).
40. Sara Peres. Saving the gene pool for the future: Seed banks as archives. *Studies in History and Philosophy of Science Part C: Studies in History and Philosophy of Biological and Biomedical Sciences*, 55 (2016):96–104.
41. *Ibid.*
42. Jeremy B.C. Jackson, Michael X. Kirby, Wolfgang H. Berger, and others. Historical overfishing and the recent collapse of coastal ecosystems. *Science*, 293 (2001):629–637.
43. Serinde J. van Wijk, Martin I Taylor, Simon Creer, and others. Experimental harvesting of fish populations drives genetically based shifts in body size and maturation. *Frontiers in Ecology and the Environment*, 11 (2013):181–187.
44. (1) J.E. Carscadden, K.T. Frank, and W.C. Leggett. Ecosystem changes and the effects on capelin (*Mallotus villosus*), a major forage species. *Canadian Journal of Fisheries and Aquatic Sciences*, 58 (2001):73–85; and (2) Janet Raloff. Big fishing yields small fish: Researchers map predator loss and predict unstable oceans. *Science News*, 179 (number 8, 2011):28.
45. The discussion of shark finning is based on: (1) Shelley C. Clarke, Murdoch K McAllister, E.J. Milner-Gulland, and others. Global estimates of shark catches using trade records from commercial markets. *Ecology Letters*, 9 (2006): 1115–1126; (2) Janet Raloff. New Estimates of the Shark-Fin Trade http://www.sciencenews.org/view/generic/id/7907/title/Food_for_Thought__New_Estimates_of_the_Shark-Fin_Trade (October 20, 2012); (3) Jaymi Heimbuch. Alarming Scale of Global Shark Fin Trade Revealed in New Photos. http://www.treehugger.com/ocean-conservation/taiwan-shark-finning-ban-set-go-effect-next-year.html (accessed October 20, 1012); (4) Nadia Draske. Lopped off. *Science News*, 180 (number 9, 2011):26–29; (5) Boris Worm, E.B. Barbier, N. Beaumont, and others. Impacts of biodiversity loss on ocean ecosystem services. *Science*, 314 (2006):787–790; (6) Victoria Gill. Many sharks 'facing extinction.' http://news.bbc.co.uk/2/hi/science/nature/8117378.stm (accessed November 25, 2012); (7) Boris Worm, Brendal Davis, Lisa Kettemer, and others. Global catches, exploitation rates, and rebuilding options for

sharks. *Marine Policy*, 40 (2013):194–204; and (8) Matt McGrath. Shark Kills Number 100 Million Annually, Research Says. http://www.bbc.co.uk/news/science-environment-2162917 (accessed March 1, 2013).

46. No author given. 250 Dead Stingrays Found on Beach. http://abcnews.go.com/International/wireStory/250-dead-stingrays-found-mexican-beach-19684042#.UebKGxwU4pd (accessed July 17, 2013).
47. Magnuson–Stevens Reauthorization Act. http://www.nmfs.noaa.gov/ia/iuu/msra_page/msra.html (accessed May 28, 2016).
48. Juliet Eilperin. U.S. tightens fishing policy, setting 2012 catch limits for all managed species. https://www.washingtonpost.com/national/health-science/us-tightens-fishing-policy-setting-2012-catch-limits-for-all-managed-species/2011/12/30/gIQALLObjP_story.html (accessed May 28, 2016).
49. *Ibid.*
50. The preceding two paragraphs are based on: How catch shares work: A promising solution. https://www.edf.org/oceans/how-catch-shares-work-promising-solution (accessed May 28, 2016).
51. What Are Catch Shares? http://www.nmfs.noaa.gov/sfa/management/catch_shares/about/what_are_catch_shares.html (accessed May 28, 2016).
52. (1) Fishing rights help curb overfishing in Belize. https://www.edf.org/oceans/fishing-rights-help-curb-overfishing-belize (accessed May 28, 2016); and (2) Restoring Belizean fisheries and coral reefs. https://www.edf.org/oceans/help-us-restore-belizes-stunning-coral-reef-system (accessed May 28, 2016).
53. How catch shares work: A promising solution. *op. cit.*
54. (1) This paragraph is based in part on: What Is Salmon Farming? http://www.farmedanddangerous.org/sa lmon-farming-problems/what-is-salmon-farming/ (accessed June 2, 2016); (2) An overview of Atlantic salmon, its natural history, aquaculture, and genetic engineering. http://www.fda.gov/AdvisoryCommittees/CommitteesMeetingMaterials/VeterinaryMedicineAdvisoryCommittee/ucm222635.htm (accessed June 2, 2016); and (3) Jeffrey Hays. Salmon and Salmon Farms. http://factsanddetails.com/world/cat53/sub340/item2191.html#chapter-10 (accessed June 2, 2016).
55. Salmon Farming Industry Handbook 2015: Marine Harvest. http://www.marineharvest.com/globalassets/investors/handbook/2015-salmon-industry-handbook.pdf (accessed June 2, 2016).
56. The preceding two paragraphs are based on: (1) Encyclopedia Britannica Advocacy for Animals. The Pros and Cons of Fish Farming. http://advocacy.britannica.com/blog/advocacy/2008/08/the-pros-and-cons-of-fish-farming/ (accessed June 2, 2016); (2) Craig Orr. Estimated sea louse egg production from Marine Harvest Canada farmed Atlantic salmon in the Broughton Archipelago, British Columbia, 2003–2004. *North American Journal of Fisheries Management*, 27 (2007):187–197; (3) Alexandra Morton and Rick D. Routledge. Mortality rates for Juvenile Pink *Oncorhynchus gorbushca* and Chum *O. keta* salmon infested with Sea Lice *Lepeophtheirus salmonis* in the Broughton Archipelago. *The Alaska Fisheries Research Bulletin*, 11 (2005):146–152; and (4) Mark Crane and Alex Hyatt. Viruses of fish: An overview of significant pathogens. *Viruses*, 3 (2011):2025–2046.
57. (1) Eva B. Thorstad, Ian A. Fleming, Philip McGinnity, and others. Incidence and impacts of escaped farmed Atlantic salmon *Salmo salar* in nature. *Norwegian Institute for Nature Research*, Special Report 36 (2008):1–110; (2) Environmental consequences of escaped Atlantic salmon. http://preventescape.eu/?page_id=17 (accessed June 2, 2016); (3) Kjetil Hindar, Ian A. Fleming, Philip McGinnity, and Ola Diserud. Genetic and ecological effects of salmon farming on wild salmon: Modelling from experimental results. *ICES Journal of Marine Science*, 63 (2006):1234–1247; (4) Laura Weir and James William Grant. Effects of aquaculture on wild fish populations: A synthesis of data. *Environmental Reviews*, 13 (2005):145–168; and (5) F. Besnier, K.A. Glover, and Ø. Skaala. Investigating genetic change in wild populations: Modelling gene flow from farm escapees. *Aquaculture Environment Interactions*, 2 (2011):75–86.
58. The preceding three paragraphs are based on: (1) Helen Scales. *Spirals in Time: The Secret Life and Curious Afterlife of Seashells*. Bloomsbury Sigma, New York, NY. (2015) 304 pp.; (2) The Gambia Integrated Programming for Environmental Stewardship. https://www

.wilsoncenter.org/sites/default/files/Fatou%20Janha_TRY%20Oyster%20Women's%20Association_The%20Gambia_Integrated%20Programming%20for%20Environmental%20Stewardship_0.pdf (accessed May 29, 2016); (3) Ba Nafaa: Gambia-Senegal Sustainable Fisheries Project. http://www.crc.uri.edu/projects_page/gambia-senegal-sustainable-fisheries-project-usaidba-nafaa/ (accessed May 29, 2016); and (4) The world is your oyster—If you TRY! http://www.new-ag.info/en/developments/devItem.php?a=3162 (accessed May 29, 2016).

59. Helen Scales. *Spirals in Time: The Secret Life and Curious Afterlife of Seashells. op. cit.* (p. 117).
60. Groundwater depletion. http://water.usgs.gov/edu/gwdepletion.html (accessed August 15, 2016).
61. Chris Maser. *Interactions of Land, Ocean and Humans: A Global Perspective. op. cit.*

14

Sources of Energy

Probably the largest single problem confronting the world today is how to provide energy for private and public use in a manner that eliminates both the need for additional land, air, or water than humans currently use *and* the myriad ways the existing energy systems already pollute the global environment, such as:

1. Dams that choke free-flowing rivers and all their ecological benefits, ranging from maintaining the biophysical integrity of their deltas to providing critical habitat for anadromous fish.
2. Fossil fuels that cause massive climate change and incalculable groundwater pollution from drilling for oil and natural gas, plus habitat fragmentation from pipelines and despoliation from oil spills on land and at sea.
3. Coal mining, especially mountaintop-removal mining, which destroys entire ecosystems, and the burning of coal to produce electricity that not only is grossly air polluting but also leaves large amounts of toxic sludge called "coal ash."
4. Nuclear power that is massively damaging because it leaves dangerous, ineradicable radioactive waste that cannot be destroyed—only unsafely stored for thousands of years.
5. Wind farms that are both energy inefficient and cause massive deaths of birds and bats.
6. Wave-energy farms—the technology of which is on the cusp of commercial development—that will, if put in place, cause problems for migrating cetaceans, as well as inshore fisheries.

The Rights of Nature

The only energy solution for a Rights of Nature system is one that allows all ecosystems to flourish while providing humans the energy they require. In other words, the energy system and its infrastructure must be *completely* sustainable and *completely* recyclable.

By bitter experience, we have learned that our current sources of energy have wide-ranging, massive, detrimental environmental effects worldwide. Fortunately, however, there is now enough Rights of Nature language already in working Constitutions and laws to outline a basic blueprint for a new energy system, one based on entirely different principles from those currently employed.

The Ecuadorian Constitution, giving Nature "the right to integral respect for its existence and for the maintenance and regeneration of its life cycles, structure, functions and evolutionary processes,"[1] sets the blueprint into place. Delving further into the Constitution, we find that in Article 73, it states, "The State shall apply preventive and restrictive measures on activities that might lead to the extinction of species, the destruction of ecosystems and the permanent alteration of natural cycles."[2] Not only that, Article 72 says simply, "Nature has the right to be restored."[3]

The Bolivian Law of the Rights of Mother Earth has similar, but more detailed, language: Mother Earth has the right to "maintain the integrity of living systems and natural processes that sustain them."[4] In addition to the right of restoration, Bolivia provides Mother Earth with the right of pollution-free living, "to preservation of any of Mother Earth's components from contamination."[5] Further, Mother Earth has: "The right to maintenance or restoration of the interrelationship, interdependence, complementarity and functionality of the components of Mother Earth in a balanced way for the continuation of their cycles and reproduction of their vital processes."[6]

Bolivia devotes an entire subchapter of its law to State Obligations and Societal Duties to protect Mother Earth. These range from requirements to prevent damaging human activities to developing "balanced forms of production and patterns of consumption" that safeguard the "integrity of the cycles, processes and vital balance of Mother Earth."[7]

Clearly, an energy system based on the Rights of Nature must: (1) allow Nature to flourish and regenerate; (2) be so structured that Nature can maintain the integrity of its life cycles, functions, and productivity; (3) restrict use of processes that destroy ecosystems, pollute, contaminate, and permanently alter natural cycles; and (4) restore the biophysical interrelationships and complementarity of Nature's components—destroyed by the current energy infrastructure—so they can function sustainably.

Is it possible to have an energy system based on social-environmental sustainability? Yes, it is. The blueprint in this chapter, and the technology to realize it, is either partly in place or on the near horizon. Once the Rights of Nature are enshrined in the U.S. and state Constitutions, as well as the required laws to implement those rights, the political will to undertake the necessary steps, driven by logistics rather than technology, will be forthcoming.

The New Energy System—Solar Power

There is a bewildering plethora of laws, at both national and state levels, relating to energy production; facility locations; energy sale and distribution; infrastructure of every kind from coal trains to nuclear reactors; energy exploration; financing; environmental safeguards, ranging from curbing methane leakage in hydraulic fracturing to fish ladders for dams; and many other topics. Since the entire current energy structure needs to be completely shifted into a new paradigm, we are not going to use the reader's time to describe the massive legal structure that upholds, incentivizes, permits, and maintains the current system of energy production and delivery in the United States. Instead, we shall explore the new paradigm and draw its blueprint.

The beginning statement is the most important, and is the frame for all future discussion: the only true source of renewable energy on Earth is sunlight. In addition, the *only true investment* in the global ecosystem is energy from solar radiation (materialized sunlight). Everything else is merely the recycling of already-existing energy. In a business sense, for example, one makes money ("economic capital") and then takes a percentage of those earnings and recycles them by putting them back into the infrastructure of the enterprise for maintenance of buildings and equipment to facilitate making a profit by protecting the integrity of the initial outlay of capital over time. In a business, one recycles economic capital after the profits have been earned. "Biological capital," on the other hand, must be "recycled" *before* the profits are earned to stay even. This means forgoing some potential monetary gain by leaving enough of an ecosystem intact for it to function in a productively sustainable manner.

In order for solar power to be the principal source of future energy, it must not only be comprehensive but also simple to manufacture, to install in both new buildings and existing ones, and to maintain—using no more resources and land than humanity is currently using. These conditions require the use of solar energy to mimic natural processes as closely and humbly as humanly possible. In other words, we must create a "solar tree" that mimics the structure and function of a natural tree. But in order to describe a solar tree, we must look first at how an actual tree lives and uses solar energy for its life processes.

Nature's Tree

A tree is a highly effective collector and user of solar radiation, transforming it into the energy needed for the tree to grow and flourish—as well as nurture its surrounding environment. I (Chris) shall therefore endeavor to paint a verbal portrait of a tree in one of its myriad forms as a solar conduit. To paint such a portrait, it is necessary to examine the leaves, flowers, fruits, branches, trunk, and roots.

Although the parts are discussed as separate components, remember that a tree is a living being, an integrated living system in constant motion, and is surrounded by and infused within a large system called the biosphere, which includes all living beings. I shall discuss the tree beginning within the context of itself, beginning with the leaves and progressing to the roots, so the reader may understand how trees function as solar conduits of energy.

Leaves

Leaves come in limitless sizes and shapes, no two of them ever exactly alike. They have many functions, be they the broad leaves of a maple; the needles of a noble fir; or the small leaves and long, green thorns of a desert acacia. Despite their shape and size, they are amazingly compact energy converters, using chlorophyll to harness the sun's energy. Availing themselves of the sunlight, they convert carbon dioxide, water, and elemental nutrients from the soil into simple sugars that are distributed throughout and among ecosystems, where they are a critical part of the world's food web.

Leaves also transpire water, creating a humid microclimate around their individual surfaces in a tree's crown. Deciduous leaves change hue with their dying in autumn, when the nitrogen within is pulled down to the roots and stored over the winter. Winter is the season of leafless trees and bare-limbed shrubs. What you cannot see it that inside the frozen bud is a miniature leaf with all the means for capturing the sun's energy, just waiting for spring to release it from bondage.

Unlike deciduous trees, which lose their leaves seasonally and stand in naked slumber for part of each year, such conifers as Douglas fir and western hemlock shed about a third of their needles annually. As the needles die and turn yellowish while giving up their nitrogen, they loosen from their moorings and spin to the forest floor. There, they serve as food for a host of organisms and thus, through many circuitous routes, are eventually incorporated into the forest soil.

Thus, while a deciduous tree or forest annually produces two entirely different habitats, one in full leaf and another following leaf fall, a coniferous tree or forest produces a continuous habitat of relatively similar characteristics throughout the year. In addition, broad leaves decompose rapidly and pass into the soil within a year, whereas coniferous needles may take a decade or more to break down and recycle through the system.

Flowers and Fruits

Each seed has, hidden under its coat, not only the size and shape of the future tree and its leaves but also the size, shape, color, odor, and season of its flowers, as well as its fruits. Like all other components of life, there is an infinite variety in the composition of flowers and fruits. However, they all have one thing in common: the flowers are the materials through which the fruit, the storage units of genetic "knowledge," are created. In turn, the genetic "knowledge" is the source of the next generation of trees.

Branches and Trunk

The leaves, flowers, and fruits of a deciduous tree are nurtured, held for a time securely in place, united, and ultimately allowed to fall from branches during autumn. With the advent of spring, however, the sleeping trees begin to awaken and once again cover their branches with leaves. The health of a tree's crown determines in large measure the health of its trunk, which is a great conduit of flowing energy that unites the roots and the crown.

The trunk is the main ascending axis of a tree, a stalk or a stem. It connects and holds the branches together in their particular array, which is different in every tree. In addition, the branches—the conduits—conduct the solar energy collected in the leaves into the trunk, which in turn transmits the energy into the roots. In return, the trunk translocates water and nutrients from soil, via the roots, into the crown to nourish the leaves so they can continue to collect and process solar energy.

Roots

As a tree's leaves harvest the sun's energy, so its roots grip the soil while reaping the sun's energy stored in darkness from millennia past. The soil, where non-living and living components of the landscape join, is also where past and present flow one into the other and determine a tree's future.

Soil supports the plants and animals that in turn create and maintain the myriad hidden processes that translate into soil productivity. Into this incredibly thin band of seething activity, the very ferment of terrestrial life, a tree thrusts its roots that it may withdraw energy long stored in Nature's warehouse and replace energy derived from the present net worth of its photosynthetic exchange, the leaves of its crown.

What is a root? Most people probably think of a root as the underground portion of a plant that serves as support, draws food and water from the surrounding soil, and stores food, all the while holding the soil together. A root, however, is far more than this.

To examine the notion of a root, we will venture into a largely unknown, hidden world, with the slightly buried seed of a Douglas fir as our guide. It is spring and the seed begins to swell as it absorbs moisture from the warm soil. The seed's coat splits, and a tiny rootlet begins to penetrate the Earth, as small, green seed leaves reach toward the sun. Thus, as the seedling's roots spread through the soil, the new, non-woody tip of a tiny feeder root comes in contact with a week-old fecal pellet of a deer mouse. The deer mouse had dined on a truffle (the belowground fruiting body of a fungus) the night before it deposited the pellet. The pellet, packed full of the truffle's spores, is still soft from the moisture in the soil, and the root tip has little difficulty penetrating it.

Inside the pellet, the root tip comes in contact with the spores that have passed unscathed through the mouse's intestinal tract. Tiny, thread-like hyphae, the vegetative part of a fungus (meaning "web" in Greek), form a mantle around the tree's feeder root. As the hyphae grow into and around the seedling's root tips, they also grow out into the soil, where they join billions of miles of gossamer threads from other mycorrhizal-forming fungi. These hyphae act as extensions of the seedling's root system, as they wend their way through the soil absorbing such things as water, phosphorus, and nitrogen and sending them into the seedling's roots. As the seedling grows, it produces sugars that feed the fungus, which in turn expands through the soil, as it is nourished by and nourishes the seedling.

A tree is therefore a product of both the sun's light and soil's darkness, because the nutrients of darkness feed the leaves of the tree's crown in light and the sugars formed from the sun's radiation feed the tree's roots and their fungi in the soil.

The Solar Tree

To mimic a tree in our use of solar energy for a Rights of Nature system, we must begin to think by analogy in our technological energy delivery system. This procedure will allow us to study the tree's processes closely and adapt their effectiveness to an artificial energy system that will sustainably supply humanity's energy requirements.

Solar panels are the leaves that collect and disperse the solar energy. Solar voltaic cells are a technological device that mimics a chloroplast and its chlorophyll by collecting sunlight during the daylight hours and converting it into electricity in the cell for human use. These cells are arranged in a grid-like pattern on the surface of the solar panel, an artificial rendition of a plant's green leaf, the function of which is to mimic the photosynthetic process. The various research questions and the knowledge they produce represent the flowers and fruits that lead to the improvements in the next generation of a solar system and its components. (See Figures 8.4 through 8.6 for examples of solar panels, pages 130–132.)

The necessary technology is coming closer to the goal by leaps and bounds, as research into "artificial leaves" shows. An artificial leaf is an inexpensive fuel cell that needs water, carbon dioxide, and sunlight to create oxygen and hydrogen, which can then be used for electricity. Caltech's Joint Center for Artificial Photosynthesis has been working on artificial leaves and has designed successful prototypes. At the moment, conventional solar panels are still more efficient at converting sunlight to electricity, and some of the elements required for artificial leaves are rare earth metals—costly and environmentally damaging to mine and process.

But research is underway to improve artificial leaves so that humans could convert sunlight into hydrogen to create electricity and store it for future use. Hydrogen is a more benign energy-storage alternative in several ways, when compared to batteries. The ability to efficiently turn sunlight into hydrogen on a residential scale could change the world's energy economy.[8]

The wires connecting the solar panels (the "leaves") are the branches. Where the wires from one or more solar array(s) collect and form the main conduit into a home or business is analogous to the trunk, while the electrical system inside the building is a bio-simulation of the root system in its function, and the storage batteries of whatever kind represent the reciprocal energy interchange of the mycorrhizal-forming fungi. Like the real tree, no energy is "lost."

Management Considerations for a Rights of Nature System

However, to make a solar energy system even remotely as effective as a tree, especially on a large scale, the new Rights of Nature energy system must be very different from the current energy paradigm. As the conversion gets underway, more details will become apparent—the flowers and fruits of the solar tree that lead to the next generation of the solar-energy system. The Rights of Nature paradigm includes both the switch to small-scale solar power *and* removal and recycling of existing energy infrastructure, as well as restoration of damage to the devastated ecosystems used for energy production.

Creating the Solar Tree Energy System

Here is an initial outline of the basic requirements for shifting to small-scale solar power nationwide:

1. Ensure that all energy/power required by human beings—for whatever purposes—can be provided via rooftops or other nearby solar panels in such locations as window sills, alongside driveways, or perhaps even in sewage ponds.
2. Design solar energy units to power enormous skyscrapers, as well as residences of every kind—from single-family homes to large apartment complexes. They must produce electricity, hot water, and heat or cooling throughout the structure.
3. Ensure via legislation, grants, incentives, funded research, and market intervention that solar energy becomes the sole source of power for industrial processes, except insofar as secondary sources, such as gravity flow of wastewater in pipes, electricity harvested from vehicle vibrations as cars drive over piezoelectric crystals inserted under roads, or the friction from machinery can produce a secondary energy source without use of additional resources.
4. Design the solar energy system so it mimics a tree, in that we devise the ability to convert sunlight into energy for our use, as leaves do; store the energy when not using it, like roots do; and then use the energy again when needed.
5. Design all local solar energy systems to provide capacity not only for night but also for a longer time in case there is little sunlight for extended periods (e.g., during winter) and/or a larger amount is needed for some purpose. Focus design to achieve *deep sustainability*—that is, it must be systemic and stay within or reduce the current, global footprint created by human behavior, and not use new lands, resources, or waters anywhere in the world. Massive solar farms with power lines to carry and distribute the energy are an example of shallow sustainability that symptomatically increases the human footprint in the name of "renewables." The recently announced expansion of Dubai's Mohammed bin Rashid Al Maktoum Solar Park is a good example of this problem. At 13 megawatts, the Park is already the largest independent power producer, single-site, solar energy project in the world. The second phase, due to begin production in 2017, will bring 200 megawatts more online.[9]
6. Ensure solar panels or artificial leaves are produced using non-toxic, easily available ingredients or recycled components, rather than resources requiring mining or other damaging production techniques. Solar panels currently require cadmium chloride, which is highly toxic, expensive to produce, and dangerous to

dispose of. But cadmium chloride can be replaced with a nontoxic substance such as magnesium chloride, which is also much cheaper.[10]

7. Ensure solar panels are energy-efficient to produce. For example, crystalline solar panels are currently extremely energy intensive to make, though the base is merely silicon. Amorphous panels are cheaper because they use less semiconductor material, but also are less efficient and effective, so more panels are required to produce a given amount of power. However, new amorphous cells are being developed, which generate electricity from different wavelengths of sunlight by stacking the semiconductor layers on top of one another.[11]
8. Incentivize research via government-funded intensive think tanks to devise a way to create and store usable solar energy with *no additional harm* to the environment in the extraction, manufacturing, or recycling of the materials used to build and/or update current solar energy systems. Artificial leaves or similar on-the-horizon technologies provide the route map for reaching this goal. Auxiliary energy sources that require no additional resources must be able to link into the solar system, such as "LucidEnergy"—an innovation for generating electricity from turbines installed in city water pipes—which can be coupled with solar energy and further reduce carbon dioxide emissions from the burning of fossil fuels.[12]
9. Design the solar energy system to provide the energy for all the world's existing buildings, from family housing to commercial skyscrapers and industrial structures, because tearing down and rebuilding even a fraction of the existing structures would incur a massive environmental cost.
10. Design the solar energy system so it does not require expensive or extensive upgrades of electrical equipment in buildings (e.g., the wiring). It must be easy to install, and must be used with minimal disruption and without new equipment.
11. Eliminate the need for large-scale energy grids. Design and install individual solar systems on a building-by-building basis to render large energy grids and storage facilities unnecessary. If grids for extra energy are needed, they must be small, local, interconnected nodes, not requiring separate, specialized structures or power lines of any size to store power independently from the source.

    Re-localizing the energy system to the community level is an essential part of changing the current massive grid-based system, which uses unsustainable amounts of resources worldwide, is fossil-fuel dependent, and vulnerable to disruptions. The Transition Town movement is perhaps currently the most well known and successful effort to re-localize energy and other needs to the community or regional level.[13]
12. Incentivize, research, and design solar technology and engines that fully power all vehicles, including airplanes and ships, via panels and/or solar-storage units, so that no hybrids that use fossil fuels, or any other kind of fuel, will be needed. This will necessitate creation of a new kind of engine beyond the traditional internal combustion engine.
13. Pass laws both at the federal and state levels to mandate government intervention in markets in order to *require* solar power, and provide upfront financial assistance for the conversion. This intervention includes placing a mandatory requirement in every municipal building code that all new buildings, whether residential, commercial, or industrial, be fully solar. Below a certain financial level of income,

people who cannot afford solar panels (even with financial incentives) and are living in established dwellings get a minimum amount of solar energy per month, free of charge. Beyond that, they would have to pay for it. Persons with income above the financial baseline must pay for all the energy they use. Moreover, beyond a given number of kilowatt-hours, the price begins to go progressively upward per kilowatt-hour of use. Although such an energy network will take time to reach fruition, the beginning is already taking shape.[14]

14. End all fossil fuel subsidies. This currently amounts to about 5.6 trillion dollars annually (direct payments, tax breaks, and unpaid environmental costs).[15] Direct these funds instead to creation of an equitable energy network, and to distributing small-scale solar energy technology. Sunlight is part of the global commons, and thus everyone's birthright. In that sense, solar energy belongs to everyone and ought to be available *and* affordable to all citizens of every community.

The Transition to Solar Power

Unlike the biological tree, which nourishes the environment in both life and death, today's solar power arrays come with the detrimental cost of destructive mining to obtain the cadmium that is added to other environmentally unfriendly materials used in their construction, such as plastics.[16] The challenge, therefore, is to imitate a natural tree in such a way that the construction and function of each iteration of a solar system is increasingly friendly to the environment, as a whole, for the benefit of all life through the generations.[17]

It is not enough, however, for the solar tree to merely replace the existing energy system. A major part of the solar conversion in a sustainable Rights of Nature system must include the removal and recycling of existing energy infrastructure. The vast array of pipelines, dams, energy plants, and mining debris must be carefully recycled for other uses, and damaged ecosystems must be repaired.

If this seems like an overwhelming—or even impossible—task, remember that the government, businesses, community groups, and individuals assembled the logistics, human power, organization, and legal structure to create the existing system. It is no different marshaling a similar organizational network to remove and recycle it. Both the human power and the resources are there, as well as the cultural structures needed to successfully undertake the task of recycling and repair, which has already been successfully done on a smaller scale in many places. The requirements and the means to undertake this endeavor include:

1. Remove and recycle all energy infrastructure after a full conversion to solar power. The unneeded infrastructure includes: (1) power poles and lines, (2) power towers, (3) hydroelectric dams, (4) wind farm equipment, (5) wave farm equipment, (6) mountaintop-removal mining infrastructure, (7) coal ash ponds, (8) underground coal mining infrastructure, (9) coal power plants, (10) nuclear power plants, (11) nuclear waste dumps, (12) oil and gas plants, (13) oil and gas refineries, (14) oil and gas pipelines, (15) liquefied natural gas facilities, (16) hydraulic fracturing and all its equipment, (17) large-scale solar arrays and their associated infrastructure, and (18) all the subsidiary industrial and manufacturing facilities and materials needed to provide the world's existing energy infrastructure.

2. Retool industrial plants to manufacture sustainably that which is truly needed, or close them and redistribute work to sectors where new capacity is required. This is, again, a question of logistics. Community planning will be necessary, because leaving such transitions to the "free market" simply guarantees that industry will reorganize around the greatest opportunities for profit, combined with the greatest opportunity to externalize environmental and social costs to the public through all generations.
3. Create the detailed logistical maps for removing energy infrastructure. The larger the infrastructure to be removed, the more the logistics must be studied and thoroughly planned in advance—just as they were for the construction.
4. Removal of any large dam requires, at the very least: (1) an extensive review of engineering details to determine the best way to remove the dam, whether by breaching, partial removal from the side, or some other approach; (2) building or re-engineering roads and other infrastructure as necessary for the removal; (3) if necessary, creating and maintaining of a well-managed market to ensure the complete recycling and reuse of the materials; (4) having on hand the equipment and industrial processes necessary to decontaminate everything that requires it; (5) ensuring, before any removal begins, that the benefits the dam provided—irrigation, municipal water, water storage, electricity—are more efficiently provided with less environmental cost, or the need reduced via recycling and conservation; and (6) conducting the studies and developing the protocols necessary to help the river regain its functional, biophysical integrity, such as re-growing a vital riparian zone to helping control water-borne sediments, create shade, add food for aquatic organisms through vegetative debris, and in general revitalize the river as needed. Colossal dams that destroy vast ecosystems require the most study and preplanning to result in successful removal, sustainable replacement of the dam's benefits or successful conservation that eliminates the needs met by the dam, and ecosystem restoration.

   Since 1999, more than 400 dams have been removed nationwide, both those providing electricity and those providing irrigation or municipal water. The rivers rebounded rapidly, pushing the clogged sediment downstream. For example, after removal of Gold Ray Dam on Oregon's iconic Rogue River, the river rebounded quickly to providing much better habitat for salmon in the previously dammed segment.[18]
5. Decommission and demolish all nuclear power plants. Though nuclear power is not a major component of the U.S. energy system, there are a few plants in use, and they are highly dangerous, especially as the structures age and the radioactive materials are less effectively protected. For example, the Trojan nuclear plant in Oregon was demolished. Its cooling tower was the first nuclear cooling tower dismantled in the United States. The radioactive portions of Trojan are stored at Hanford, while the spent fuel is stored onsite[19] (Figure 14.1).
6. Make available substantial and sustained federal funding for the necessary research to find successful ways of fully decontaminating radioactive waste of all kinds, especially nuclear waste. So far, this has not happened. "Permanent" storage of nuclear and other contaminated radioactive waste is not a viable choice, because no storage option is, in fact, safely permanent.

FIGURE 14.1
Portland General Electric's now-closed Trojan Nuclear Power Plant, located on the Columbia River near Prescott, Oregon. Circa 1974. Photograph by the U.S. Department of Energy (https://commons.wikimedia.org/wiki/File:HD.6B.328_(11842583975).jpg).

7. Fund via government grants to private research firms and support for government research stations full-scale analysis into the best methods to render harmless the toxic materials in contaminated energy sites or equipment, such as coal ash ponds and electrical transformers. *No removal must take place until that is done.* For example, many transformers contain polychlorinated biphenyls—a group of toxic compounds that have been widely used, primarily in electrical equipment. Since this electrical equipment has been widely distributed in estuaries and lakes, their bottom sediments must not be disturbed until methods of demobilizing the contaminants are found. Coal ash contains highly toxic arsenic, mercury, lead, and chromium, among other ingredients. Coal ash ponds cannot be drained and the ecosystems cannot be restored until the toxins are made nontoxic and innocuous.
8. Dismantle and recycle all wind turbines and wind farms, and provide federal funding to states, local organizations, and restoration nonprofits to aid in the recovery of the sites. The turbines not only kill millions of birds and bats, as discussed earlier, but also are a classic, symptomatic example of shallow sustainability that actually increases the biophysical costs of the power and enlarges the environmentally destructive human footprint.

9. Remove buried oil and gas pipelines with an overall goal of protecting the environment, using equipment that can "pull" the pipelines with minimal disturbance to the Earth's surface.
10. Restore to the greatest extent possible the vast areas of the United States that have been stripped bare in the hunger for coal and other resources necessary to create and maintain the energy infrastructure. Mountaintop-removal mining is the most egregious example, but there are many others, especially including hydraulic fracturing. "Restoration" means committing massive federal research and financial resources on learning how to repair the biophysical integrity of the environmental processes in areas devastated by energy-related mining and drilling.
11. Create the kind of intensive think tanks needed to tackle these removal, restoration, and recycling problems. It is no different in principle from the research facilities the government created to develop the atom bomb—the finest scientific minds were put to the task, and they succeeded. The same can *and must* be done to solve the decontamination and restoration objectives outlined here. Research and understanding of organic processes, including the humblest and least visible, such as soil buildup, is advanced and sophisticated. The tools are there. The portal to success is the political will to see the paradigm shift to completion.

To create a society in which the Rights of Nature can truly flourish will require massive changes in the current energy system worldwide. This is not sliding backward, nor is it wishful thinking. It is the only way to move forward to cultivate and advance social-environmental sustainability. In addition, this paradigm change will simultaneously provide vast stockpiles of reusable metal, concrete, and equipment. It will stimulate new research and understanding of the natural world, the human world, and ways in which they not only coexist but also strengthen resilience and thus each other.

Many and varied are the proposals for a future free of energy derived from fossil fuels, ranging from massive investment in nuclear power to large-scale wind farms. But energy requirements must be met through small-scale solar and auxiliary means that do not increase the human footprint. This kind of systemic, deep sustainability is the only one consonant with the Nature's Laws of Reciprocity and thus the Rights of Nature.

Other Recycling: From Waste to Zero Waste

Recycling of the existing energy infrastructure is only one part of the larger picture of changing to a Rights of Nature system. The economy must shift from creating new products from raw materials to recycling and reusing existing products. This simultaneously reduces stress on the natural environment, where resource extraction takes place, and slims the country's energy budget. Such is the case because it is far more energy-intensive to extract resources and manufacture products than it is to reuse existing products in new forms as needed.

"Waste"—the contrived, economic concept for something that fails to produce a visible profit—does not exist in Nature. Everything is part of Nature's interactive feedback loops that we experience as the varying degrees of ecological resilience and productivity. With this in mind, the Bolivian Law of Mother Earth, as mentioned above, provides the

Earth both the rights to pollution-free living and to resilience from the integrity of its feedback loops—the interrelationships of which form the continuation of Nature's cycles. However, the massive aggregates of human-generated "waste," much of it toxic, seriously skews these Earth-oriented rights, which are necessary for the existence of all life, including humans.

The human equivalent of Nature's Laws of Reciprocity, that nothing is wasted, is the goal of *zero waste*. Changing from the current throwaway society to a zero-waste society requires many components. Again, like the energy transition, zero waste has been implemented on a small scale in some communities, and is a policy goal in others. Consider Maryland's Plan to Reduce, Reuse, and Recycle Nearly All Waste Generated in Maryland by 2040:

> Zero waste is an ambitious, long-term goal to nearly eliminate the need for disposal of solid waste and to maximize the amount of treated wastewater that is beneficially reused. It involves rethinking the way products are designed in order to prevent or reduce waste before it ever occurs. Discards that cannot be avoided should be designed for efficient recovery through recycling. Throughout their lifecycles, materials should be used and managed in ways that preserve their value, minimize their environmental impacts, and conserve natural resources. Ultimately, products that cannot be redesigned or recycled should be replaced with alternatives. Zero waste goals are intended to be challenging and to require comprehensive action.[20]

And Maryland is not alone. The leaders of Japan's Subaru Corporation have achieved zero waste in the manufacture of their automobiles in the United States because, as they said:

> **Who We Are Is What We Leave Behind**. It's been said our lives, our legacies, are simply the sum total of all the choices we make. Theodore Roosevelt certainly understood this when, in 1906, he fought the conventional wisdom of his time and set aside millions of acres of land to be preserved for future generations. And it's something Subaru understood when, over a decade ago, we became the first U.S. auto manufacturer to achieve zero landfill, with all waste recycled or turned into electricity. It wasn't easy. Doing the right thing rarely is. But like President Roosevelt, we made a commitment to something we believe in: the future. It's this promise that now leads us to share our expertise with the National Park Service and the National Parks Conservation Association as we work together toward the goal of making our irreplaceable national treasures zero landfill as well. Because love the earth means understanding you can't throw anything away, because there simply is no "away."[21]

Although much of what is discarded by our society may break down or be consumed by microorganisms over time, there is much that will undoubtedly be around for centuries. As the world's population continues to grow and towns and cities continue to expand ever farther into the countryside, the human-generated garbage will increase proportionately, while the available land area in which to hide it will continually shrink. This situation is untenable in design and sustainability. Author and businessman Paul Hawken puts it succinctly:

> Industry has transformed civilization and created material wealth for many people. But it only succeeds by generating massive amounts of waste. Our production systems function by taking resources, changing them to products and discarding the detritus back into the environment. This process is overwhelming the capacity of the environment to metabolize our waste, and, as a result, the health of our living systems is slowly grinding down. There are simply too many manufacturers making too much too fast. … By the time you get to the consumer, 98 percent of the problem has already occurred[22] (Figure 14.2).

FIGURE 14.2
A Caterpillar 826C landfill compactor being used at an Australian landfill site. Photograph by Ropable (https://commons.wikimedia.org/wiki/File:Landfill_compactor.jpg).

According to Hawken, there is a growing movement (the Factor Ten "Club") in both Europe and the United States to radically reduce our requirements from resources. The Factor Ten "Club" consists of scientists, economists, and experts in public policy who are calling for a 90 percent reduction (Factor Ten) in the use of materials and energy over the coming 40 to 50 years.

Factor Ten states that, to achieve dematerialization over the next 30 to 50 years (one generation), a decrease in the use of energy and the flow of materials by a factor of 10 is required, along with an increase in resource productivity and efficiency by a factor of 10. In other words, to attain social-environmental sustainability, we need to reduce resource turnover by 90 percent on a global scale, within the next 50 years.[23]

The goal of Factor Ten is to invent products, technologies, and systems that deliver the services people require without 90 percent of the fuel, metal, wood, packaging, and waste that is currently used. Keep in mind, admonishes Hawken, that we want *the service* a product provides, not the thing itself. For example, we want clean clothes, not necessarily the detergent that cleans them.[24]

To fulfill the goal of Factor Ten, specific questions must be asked (the flower on the previously discussed solar tree): How much human-generated waste can we convert into food for microorganisms so that it can continually cycle throughout the environment, as a renewable source of energy? In addition, we must ask of each new item that technology proposes to produce—while it is still in the design stage: Will this item be biodegradable?

If the answer is "yes," then we must ask: By what mechanism will it be biodegradable? We must then further inquire to what degree, under what conditions, and in what time frame the item will biodegrade.

If the answer is "no," then we must ask why the product is not biodegradable, and if it can be re-designed to become biodegradable. If not, then we must determine whether there is a biodegradable substitute that can serve the same purpose, or if one can be made. If not—*do not make that product!*[25]

The upshot is that society can no longer afford to produce and use things that are wasteful and are not rapidly biodegradable. But unless these questions are specifically asked and answered, new technology will most likely multiply the things that are non-biodegradable *to increasing degrees*. This does not mean, however, that planned obsolescence, as a corporate marketing strategy, will disappear. It means that whatever is discarded will take up more and more space in an unusable form for ever-longer periods, like industrial chemicals, metals like aluminum, some plastics, and nuclear waste already do.

Nonbiodegradable materials, especially those that are toxic and capable of spreading, can make planning for social-environmental sustainability exceedingly difficult, even at the scale of a local community. The negative effects of such materials are often more poignant at the bioregional scale, however, where they may affect an entire water supply, not to mention the growing accumulations of plastic and other garbage in the oceans of the world.[26] (For a thorough discussion of plastics and other garbage in the oceans, see *Interactions of Land, Ocean and Humans: A Global Perspective*.[27])

Electronic waste ("e-waste") is a perfect example. It is the fastest-growing waste stream in the world. The country producing the largest amount of e-waste? The United States. Something like 50,000 dump trucks' worth of e-waste goes to recycling centers every year, much of it containing toxic materials like mercury, cadmium, and lead. But though recyclers often deny it, e-waste is frequently shipped abroad.

The Basel Action Network of Seattle has been monitoring (via tracking devices) e-waste shipments from the United States to other countries since 2015, and their initial investigations show how weak American e-waste recycling oversight is. No one knows how much e-waste is shipped to poorer countries like Mexico, China, Pakistan—and especially, to rural Hong Kong—for disassembly. Estimates vary widely. The United Nations estimates that 10 percent to 40 percent of American e-waste is exported. The foreign recycling scrapyards, such as those in Guiyu, China, tend to use few or no environmental safeguards, and studies show high levels of worker, soil, and water contamination by toxic metals.

The 1989 Basel Convention is an international treaty signed by most developed countries to stop industrialized nations from dumping hazardous waste in poorer countries. The United States is the only industrialized nation in the world that has refused to ratify the Basel Convention. Nor does the United States have any federal laws banning export of e-waste, and the Basel Action Network's recent investigations show very reputable American e-waste recycling firms export electronics. Without federal laws, e-waste recycling is left to a patchwork of state laws that vary in effectiveness and have no control over export.

E-waste recycling is one of the most important areas for federal market intervention to create and ensure a robust market for the recycled elements of e-waste, so that companies have incentive to recycle safely and find markets for resale of the recycled commodities ranging from plastics to copper. Zero waste strategies must also research ways to retool electronic devices rather than manufacture new ones every few months. Currently, tens of thousands of new electronic devices are sold to U.S. consumers each year.[28]

Management Considerations for a Rights of Nature System

A partial list of policy changes required to move society to zero waste necessary in a Rights of Nature system includes the following. However, as society initiates these changes, other restructuring necessities will become apparent. With Nature as the primary rights holder, sensitivity to Nature's Laws of Reciprocity will make the additional steps easily discernible.

1. Redesign markets to accept and prefer products made from reused materials. This requires government intervention in financial incentives, legal prohibitions on "greenfield" products,[29] protection of ecosystems so they cannot be exploited by the industrial system, and similar adjustments. (The term "greenfield" was originally used in construction and development to reference land that has never been used [e.g., green or new], where there was no need to demolish or rebuild any existing structures. Today, the term *greenfield* project is used in many industries, including software development, where it means to start a project without the need to consider any prior work.)
2. Retool the legal system to require all new products be made from reused materials, which will require creating markets both for the raw resources and the subsequent products. Products must be completely biodegradable and organic (e.g., food waste) or completely recyclable.
3. Mine existing landfills for reuse of materials, and recalibrate technology for this purpose. This will take government intervention to incentivize, and perhaps subsidize, the industries that take the risks to enter this new frontier. Some communities, such as the Berkshire County Council in England, are already eyeing old landfills for their glass, metals, and plastics to enrich recycling opportunities.[30]
4. Close industries that cannot make their products without additional raw resources. Alternatively and as appropriate, aid industries via federal tax incentives to research and develop new products made of recycled or biodegradable components, especially based on appropriate human-scale biomimicry.
5. Prioritize zero waste in federal, state and local laws. The goal is always "deep sustainability"—not increasing the human footprint on the earth.
6. Provide federal incentives and grants for new research on recycling and reuse of materials not thought to be recyclable. On the flipside, write laws so the legal system restructures subsidies and incentives to heavily penalize greenfield technologies and manufacturing that require raw materials and further extraction of natural resources.
7. Restore landfills and dumps whose products cannot be used for recycling and render the materials biodegradable. This will also require incentivized research to ensure that no toxic waste, contamination plumes, and groundwater-poisoning leachates remain onsite. Many products have been thrown away in massive quantities on the assumption that they could not be recycled, or at least only with high cost and low return, such as Styrofoam or waxed cardboard. But there are always other uses. Styrofoam, for example, can be mixed with cement to make building blocks that are insulating, flexible, and strong. Though with current technology the cement still requires a lot of energy to make, Styrofoam blocks are an important and obvious example of the right kind of innovation.[31]

8. Tackle the burgeoning electronic waste problem via federal regulation of incentives in order to prohibit shipment abroad, research the best technologies for recycling the hazardous components, create robust recycling markets for components, and provide incentives for retooling devices for further use rather than manufacture of new devices.

The shift from addictive consumerism to zero waste may seem wrenching to contemplate. But, in fact, all the pieces are in place for the new Rights of Nature paradigm to emerge. The technologies are available—or nearly so. However, without a recognized and acknowledged crisis to force a major reorganization, it is rare for deep-seated change to arise, because the average person changes only when absolutely forced to.

Now that environmental signs unmistakably show the hour has come for deep, society-wide change, it is heartening to realize that the shifts are already occurring. They just have not yet become the central paradigm. But when the Rights of Nature are enshrined in the U.S. federal and state Constitutions, the necessary legal mandates will be much easier to enact. Once the new paradigm's legal requirements are accepted and implemented, the blueprint of social-environmental sustainability will be established through which human society can flourish without ravaging the legacy of Nature's productive capacity on which all life depends.

Endnotes

1. [Ecuadorian] Constitution, full language: http://pdba.georgetown.edu/Constitutions/Ecuador/english08.html (Title II, Chapter VII, Article 71) (accessed August 10, 2015).
2. *Ibid.*
3. *Ibid.*
4. Language of the Bolivian law (compete text): http://www.worldfuturefund.org/Projects/Indicators/motherearthbolivia.html (Chapter III, Article 7, I(1)) (accessed August 12, 2015).
5. *Ibid.* (Article 7, I(7)).
6. *Ibid.* (Article 7, I, (5)).
7. *Ibid.* (Chapter IV, Article 8, Section 2).
8. (1) Evan Ackerman. Artificial Leaf Is 10 Times Better at Generating Hydrogen from Sunlight. http://spectrum.ieee.org/energywise/energy/renewables/artificial-leaf-is-ten-times-better-at-generating-hydrogen-from-sunlight (accessed August 16, 2015); and (2) Erik Verlage, Shu Hu, Rui Liu, and others. A Monolithically Integrated, Intrinsically Safe, 10% Efficient, Solar-Driven Water-Splitting System Based on Active, Stable Earth-Abundant Electrocatalysts in Conjunction with Tandem III-V Light Absorbers Protected by Amorphous TiO_2 Films. http://authors.library.caltech.edu/59897/ (accessed August 16, 2015).
9. Lorraine Chow. Dubai to Build World's Largest Concentrated Solar Power Plant. http://www.ecowatch.com/dubai-to-build-worlds-largest-concentrated-solar-power-plant-1891164329.html (accessed August 16, 2015).
10. Tofu ingredient could revolutionise solar panel manufacture. http://phys.org/news/2014-06-tofu-ingredient-revolutionise-solar-panel.html (accessed March 15, 2016).
11. How Are Solar Panels Made? http://www.reuk.co.uk/How-are-Solar-Panels-Made.htm (accessed March 15, 2016).
12. (1) LucidEnergy. http://www.lucidenergy.com/lucid-pipe/ (accessed January 15, 2016); and (2) LucidEnergy. http://www.lucidenergy.com/wp-content/uploads/2012/10/ProductInfo_Oct2012.pdf (accessed January 15, 2016).

13. Burns H. Weston and David Bollier. *Green Governance: Ecological Survival, Human Rights and the Law of the Commons*. Cambridge University Press, New York, NY. (2013) 390 pp.
14. Concept Note for Workshop on Equitable Energy Access. http://www.undp.org/content/dam/uspc/docs/SPC%20SE4ALL%20Concept%20Note.pdf (accessed January 16, 2016).
15. Pete Dolack. IMF reports: Fossil fuel subsidies worth $5.6 trillion per year. http://www.theecologist.org/News/news_analysis/2884881/imf_reports_fossil_fuel_subsidies_worth_56_trillion_per_year.html (accessed January 16, 2016).
16. Agency for Toxic Substances and Disease Registry. http://www.atsdr.cdc.gov/ (accessed March 15, 2016).
17. Elizabeth Palermo. New Solar Battery Could Generate Cheaper Clean Energy. http://www.livescience.com/48335-solar-battery-renewable-energy.html (accessed March 15, 2016).
18. Scott Learn. After dam removals, Oregon's Rogue River shows promising signs for salmon. http://www.oregonlive.com/environment/index.ssf/2010/10/early_signs_good_for_dam_remov.html (accessed March 15, 2016).
19. Trojan Nuclear Power Plant. https://en.wikipedia.org/wiki/Trojan_Nuclear_Power_Plant (accessed March 15, 2016).
20. Maryland Department of the Environment. Zero Waste Maryland: Maryland's Plan to Reduce, Reuse and Recycle Nearly All Waste Generated in Maryland by 2040. http://www.mde.state.md.us/programs/Marylander/Documents/Zero_Waste_Plan_Draft_12.15.14.pdf (accessed January 17, 2016).
21. Subaru. Who we are is what we leave behind. *Sunset Magazine*, March (2016):33.
22. Paul Hawken. Undoing the damage. *Vegetarian Times*, September (1996):73–79.
23. The preceding two paragraphs are based on: *Ibid.*
24. Sustainability Concepts: Factor 10. http://www.gdrc.org/sustdev/concepts/11-f10.html (accessed January 5, 2016).
25. Paul Hawken. Undoing the damage. *op. cit.*
26. (1) Kimberly Amaral. Plastics in Our Oceans. http://www.whoi.edu/science/B/people/kamaral/plasticsarticle.html (accessed January 7, 2016); (2) Sailors for the Sea. Plastic Pollution and Its Solution. http://sailorsforthesea.org/resources/ocean-watch/plastic-pollution-and-its-solution (accessed January 7, 2016); and (3) Great Pacific Garbage Patch. *National Geographic*. http://education.nationalgeographic.org/media/reference/assets/great-pacific-garbage-patch-1.pdf (accessed January 7, 2016).
27. Chris Maser. *Interactions of Land, Ocean and Humans: A Global Perspective*. CRC Press, Boca Raton, FL. (2014) 308 pp.
28. (1) Katie Campbell and Ken Christensen. On the Trail of America's Dangerous, Dead Electronics. http://www.opb.org/news/series/circuit/tracking-dangerous-dead-electronics/ (accessed August 15, 2016); and (2) Tony Schick and Cassandra Profita. NW E-Cycle Programs Tested by Electronic Waste Exports. http://www.opb.org/news/series/circuit/northwest-states-reckon-with-undetected-e-waste-exports/ (accessed August 15, 2016).
29. (1) Greenfield Products. http://www.greenfieldpi.com/ (accessed April 5, 2016); and (2) Greenfield Products Invests $6.3 Million to Expand Its Union City, Tennessee, Manufacturing Center. http://www.areadevelopment.com/newsItems/3-17-2014/greenfield-products-production-center-expansion-union-city-tennessee782378.shtml (accessed April 5, 2016).
30. Lorna Howarth. Wealth in waste. *Resurgence*, 180 (1997):23.
31. Dawn Killough. New Insulated Concrete Blocks Made from Recycled Styrofoam. http://greenbuildingelements.com/2015/06/23/new-insulated-concrete-blocks-made-from-recycled-styrofoam/ (accessed April 5, 2016).

15

Mining and Drilling

Mining of all kinds is one of the most destructive activities modern society undertakes. In this chapter, we look at the four major kinds of mining in the United States: coal mining, hardrock mining, oil and gas drilling, and deep-sea mining. We give a historical overview of each, look at the legal framework that governs them, and discuss the colossal environmental damage caused by each. Finally, we lay out a program for ending mining and implementing a restoration program to aid in the recovery of these areas under a Rights of Nature framework.

The Rights of Nature

Under a Rights of Nature system, what is to be done with mining? It is perhaps the largest, single question we must answer—outside of how to reinvent the energy sector in the country.

Let us look first at what a Rights of Nature system requires. Title II, Chapter 7, Article 71 of the Ecuadorian Constitution affords Nature "the right to integral respect for its existence and for the maintenance and regeneration of its life cycles, structure, functions and evolutionary processes,"[1] and thereby gives the initial compass reading for mining operations. Delving further into the Constitution, we find that in Article 73 it states, "The State shall apply preventive and restrictive measures on activities that might lead to the extinction of species, the destruction of ecosystems and the permanent alteration of natural cycles."[2] Not only that, but Article 72 states simply, "Nature has the right to be restored."[3]

The Bolivian Law of the Rights of Mother Earth has similar, but more detailed, language: Mother Earth has the right to "maintain the integrity of living systems and natural processes that sustain them."[4] In addition to the right of restoration, Bolivia provides Mother Earth not only with the right of pollution-free living, "to preservation of any of Mother Earth's components from contamination,"[5] but also with equilibrium, that is, "The right to maintenance or restoration of the interrelationship, interdependence, complementarity and functionality of the components of Mother Earth in a balanced way for the continuation of their cycles and reproduction of their vital processes."[6]

Bolivia devotes an entire subchapter of its law (Chapter IV, Article 8) to State Obligations and Societal Duties to protect Mother Earth. These range from requirements to prevent damaging human activities to developing "balanced forms of production and patterns of consumption" that safeguard the "integrity of the cycles, processes and vital balance of Mother Earth."[7]

The Universal Declaration of the Rights of Mother Earth, growing out of the Bolivian and Ecuadorian Rights of Nature language, is even more detailed. It begins in Article 1 with the statement, "Mother Earth is a living being." The definition of that is wide-ranging: "Mother Earth is a unique, indivisible, self-regulating community of interrelated beings that sustains, contains and reproduces all beings."[8] The Declaration extends Mother Earth's inherent rights to "all the beings of which she is composed" and sets the framework for dealing with conflict: "The rights of each being are limited by the rights of other beings and any conflict between their rights must be resolved in a way that maintains the integrity, balance and health of Mother Earth."[9]

Mother Earth's rights, like those of any other being, include first and foremost the right to life and to exist; to be respected; "to continue … vital cycles and processes free from human disruptions"; to water and clean air; to be free from contamination, pollution, and toxic waste; the right to "full and prompt" restoration if the rights are violated by human beings.[10]

Human beings have obligations as well as rights, the basic one being to respect and live in harmony with Mother Earth. This has many aspects, but includes the obligation to respect, conserve, and restore the integrity of vital ecological processes and cycles, as well as guaranteeing that violations of Mother Earth's inherent rights are rectified, and those held responsible are held accountable for restoring Mother Earth's integrity.[11] To put this set of obligations into other language, the Declaration's vision includes the obligation to have those who despoil the Earth—for whatever reason—pay for the restoration.

Types of Mining: Past and Present

Let us look at the three major different kinds of mining, from their earliest days in America to their current methods of production and their environmental consequences, as well as a summary of laws that regulate them. That sets the stage for a discussion of how best, under a Rights of Nature system, to end mining and drilling in the United States.

Coal

Early Coal Mining

In colonial days, America's abundant coal reserves were infrequently mined because Great Britain had little use for coal. After independence, however, interest turned to America's own coal, especially the bituminous coal reserves in the Richmond basin of east Virginia. Despite the interest, growth potential in the coalfields was limited for several reasons, including transportation bottlenecks and labor shortages (Figure 15.1).

By the 1830s, anthracite coal—which has a higher carbon content and is much harder than bituminous coal—became commonplace for domestic and industrial use, far outstripping British imports or bituminous coal mining. This increase was due mainly to the breakthrough that allowed anthracite to be use in the making of iron. As a result, coal extraction spread beyond Virginia to Ohio, Maryland, and at least 20 other states by the onset of the Civil War. Most coal operations were small and labor intensive, partly because coal seams, as in Pennsylvania, were often near the surface. Nevertheless, in 1860, coal production topped 20 million tons for the first time in American history.

The need for coal during the Civil War kept prices high. After the war, railroad expansions to the West aided in the opening of new coalfields and provided easy transportation to market. Technological innovation allowed ever-deeper shafts and the use of steam power in mines, while electric cutting machines eliminated the need for miners' traditional skills. Coal firms became larger and better capitalized to take advantage of these technological improvements.

FIGURE 15.1
Child coal miners—drivers and mules, in a Gary, West Virginia, mine. Photograph by Lewis Wickes Hine (1874–1940) (https://commons.wikimedia.org/wiki/File:Child_coal_miners_(1908)_crop.jpg).

In the late 19th century, the rise of coke (processed bituminous coal whose impurities are burned away) fueled a massive increase in the iron and steel industries; coke coal became the leading type of American coal, and America became more and more dependent on it. By 1880, coal had surpassed wood as the primary fuel. Although the ensuing cutthroat competition and boom-and-bust cycles were devastating to miners, they stimulated the organization of labor unions, which began in the 1860s in Pennsylvania, culminating in the successfully organized United Mine Workers of America. The union subsequently gained full recognition by the turn of the 20th century, and had upward of 250,000 members by 1903.

Mine safety was a primary concern, especially as larger corporate concerns took over the coalfields. Corporations also paid miners less and squeezed them further by making them pay to lease mining equipment and/or by paying them in scrip so they could buy necessities only at company stores. There was no federal regulation, and state regulation was extremely lax. West Virginia, for example, had the highest mine-death rate, and was also the site of the notorious Monongah mining disaster in 1907. The methane gas ("firedamp") in the mine ignited, killing at least 362 men and leaving more than 1,000 children fatherless. As a result, Congress created the Bureau of Mines to begin federal oversight.

Coal was the premier American fuel in 1900. Coal-mine output was about 267 million tons that year. Pennsylvania was the top coal-producing state, followed distantly by Illinois and other states with much smaller percentages of the production.[12]

Current Coal Mining

Coal mining now bears little relation to the early, almost artisanal coal mining of the early 19th century, which, in addition to being small scale, was usually underground. Coal mining is now a corporate enterprise, highly technological and consisting of vast strip mines. The United States has the largest coal deposits of any country in the world. Though there are nearly 1,500 coalmines in the United States, the largest coal-producing state has been Wyoming since 1988, when extraction of the Powder River Basin's coal began. The nation's seven largest coal mines are all in Wyoming. Around 37 percent of U.S. coal comes from the Powder River basin, more than all Appalachian coal production combined. West Virginia is America's second place coal producer.[13]

How big and colossally destructive are coal strip mines? Just to take a single example: The Kayenta coal strip mine, on Navajo Nation lands in Arizona, covers 40,000 acres, contaminates watersheds (in the desert, where water is already scarce), destroys all ecological functions and leaves a dust-swept moonscape behind, obliterates sacred lands, and ruins the health of hundreds of workers. The 250 coal-waste ponds at the mine site are laced with toxic heavy metals.[14]

The form of coal mining that has garnered the most attention, however, is mountaintop-removal mining, now common in Appalachia, especially in West Virginia, Virginia, Kentucky, and Tennessee. Though practiced since the 1960s, it has become widespread and is expanding—despite severe environmental effects (see Figure 9.3, page 149, for an image of mountaintop-removal mining).

Mountaintop-removal mining for coal in Appalachia has destroyed at least 470 mountains, burying something like 2,000 miles of streams under the rubble, and destroyed 800,000 acres of biologically rich forest. Coal slurry, which remains after the coal is washed, contains impurities, coal dust, and chemical agents. It is injected into abandoned underground mines or held in vast slurry ponds near the mine sites. There are hundreds of these slurry impoundments throughout Appalachia. Ash from the combustion of coal is dumped in rural areas, despite the fact that it contains arsenic, mercury, cadmium, and similar heavy metals, which are extremely toxic. But Congress exempted coal ash from hazardous waste rules in 1980, and state laws are often lax and inconsistent.[15]

Though coal provides less than half of America's energy, and coal consumption is declining more dramatically than any other energy sector, the mining still continues. In 2014, coal production increased 1.5 percent, though productive capacity decreased and the number of employees continues to decline nationwide. The production uptick came despite the fact that steam-coal, used to generate electricity, fell 29 percent between 2007 and 2015. Moreover, the price and availability of other fuels, including renewable fuels and natural gas, have continued the decline in coal use. In some states, such as Ohio and Pennsylvania, the decline has been dramatic, where the use of coal has fallen 49 percent and 44 percent, respectively.[16]

Nevertheless, ten states are still heavily dependent on coal for power production, with West Virginia (95 percent), Kentucky (93 percent), and Wyoming (88 percent) leading the pack. Coal-dependent utilities and states have sought to protect the coal industry, despite its overwhelming environmental effects. President Barack Obama's Clean Power Plan, which aimed to cut U.S. emissions by 32 percent as of 2030, focuses on a gargantuan shift to renewable sources of energy. Each state had to prepare proposals for how to meet the Plan's goals. But a coalition of 27 states, utilities, and coal miners resisted the Clean Power Plan, and brought suit to halt it. The U.S. Supreme Court, in early 2016, ruled that all legal challenges must be resolved before the Clean Power Plan could be implemented, delivering a key defeat to federal efforts to move the nation from coal and other fossil fuels.[17]

Environmental Concerns and Environmental Laws

The principal law regulating this devastating activity is the Surface Mining Control and Reclamation Act of 1977,[18] which is enforced through the U.S. Office of Surface Mining Reclamation and Enforcement. The law grew out of state failures to regulate either environmental damage or health and safety problems at mines. As surface coal mining jumped from 33 percent in 1963 to 60 percent of the total ten years later, the concern burgeoned.

As might be expected, trying to reclaim abandoned coalmines is one of the Office of Surface Mining Reclamation and Enforcement's main functions. The agency has collected more than $10 billion in fees since 1977 and distributed more than $8 billion; but there are still more than $4 billion worth of abandoned coal sites awaiting reclamation in the United States. The Office is responsible for distributing funds to states and Tribes for high-priority cleanups and reclamation activities.

Regulating coal mining is the agency's other function. The Office sets basic environmental standards, and requires permits before mining can begin. In addition, regulators may inspect mines, and fine companies that violate the Surface Mining Control and Reclamation Act or equivalent state acts.

Though technically the Office regulates coal mining, it often plays a simple oversight role, which allows 24 states—the majority of the coal-producing states—to regulate their own coal production. The federal oversight can include delegating administration of most coal mining requirements on federal lands to states, which has occurred in 14 cases. Not surprisingly, state programs are often very weak and poorly enforced.[19]

Most coal mining is subject to the federal Clean Water Act,[20] enforced by the U.S. Army Corps of Engineers, which is responsible for permits relating to mining waste in streams and wetlands. However, the actual permitting process for coal mining is delegated to the Office of Surface Mining Reclamation and Enforcement or state-regulated programs. Other environmental laws that apply include the Clean Air Act,[21] Endangered Species Act,[22] and National Environmental Policy Act.[23]

Nevertheless, the regulatory framework to protect the environment from coal mining must be considered a colossal failure. Between 1996 and 2005, only 5.7 percent of coal-mined lands were considered fully restored, with pre-mining surface and restored groundwater. Coal ash ponds, which hold the slurry remaining after production, are toxic and pose dangers to communities. Coal ash cannot be used beneficially, though Pennsylvania especially has sought to use it as part of mine reclamation.

Coal companies frequently do not appreciate shouldering the costs necessary to fully reclaim a site, all the more because true, full reclamation is exceedingly rare and takes tremendous amounts of money, time, and expertise. Even to restore agricultural fields or a basic monoculture of grassland is difficult, never mind full biological complexity. Coal companies have stated that reclaimed areas are for "economic development," but something like 89 percent of sites are not used for any such purpose, according to 2009 studies by the watchdog organization Appalachian Voices.[24]

Companies have been known to hire small contractors to mine the coal, which then fold when the reclamation costs stack up. Other times, companies create shell corporations to mine and dissolve the corporation once mining is complete. The losers are the land itself, and the shattered communities that must depend on degraded lands and waters. If any real restoration is going to occur, it requires massive commitment of time, resources, and expertise.[25]

Hardrock Mining

Early Hardrock Mining

Hardrock mining includes extraction of several different minerals from rock, the most high profile of which are gold, copper, and nickel. The United States has produced copper since the 1840s, when mining began in Michigan. The Michigan State Geologist's announcement of copper deposits in 1841 triggered a mining boom from 1845 through 1881, during which Michigan produced more than 75 percent of American copper most years, and more than 95 percent of the copper in peak years.

Early mining of fissure veins required chiseling by hand a single giant mass of copper into small enough pieces to haul it out of the mine. Later mines, having lower-grade ores, were blasted to obtain the ore, which was mechanically hoisted to the surface. Michigan copper companies began to consolidate in the early 20th century, ultimately leaving only two major companies. Production peaked in 1916, and declined thereafter into the 1960s, when the copper industry fizzled out. Michigan copper country not only has large piles of sterile sand in waterways from early mining but also was denuded of trees, as timber was essential for mine support and for generating the steam needed to power mining equipment.[26]

Commercial gold mining started in the United States around 1799, when gold was discovered in North Carolina, but large-scale production of gold began with the 1848 California gold rush. Early American hardrock mining was primarily for gold, and was done laboriously by hand. There were two main methods: winnowing gold from the sun-dried pay dirt with the aid of the wind, if no water was available, or by panning, which used water to wash away the dirt and concentrate the gold. A few simple technological innovations allowed more gravel to be shoveled by hand into sluice boxes, where flowing water separated the gold ("placer mining"). Simple placer mining peaked around 1852.

Large-scale hydraulic mining for gold began about the same time, using high-pressure nozzles to tear riverbanks apart and separate out the gold in riffle sluices. Mercury was added to aid in recovery. But even in the 19th century, the practice was so injurious to streams needed by agriculture that it was banned in 1884. Increasing industrialization and deep shaft mining technology—including electricity, ventilation pumps, and stream-powered drillers with diamond bits—made gold mining attractive to the next generation of miners, starting in the late 19th century.

Dredge mining for gold also began right at the end of the 19th century, with chains of buckets digging up the river mud and depositing it in sluice boxes on a boat. These gold dredges were a common sight in early 20th-century California, where it was big business and controlled by three main companies. As the 20th century rolled on, so did the corporate control of dredge gold mining. Large-scale dredge mining for gold continued into the 1960s.

Cyanide was used to extract gold from crushed ores beginning in 1896, and quickly became the most common method, especially as the majority of gold mines began processing lower-grade ore. By 1936, the United States had something like 3,200 gold mines scattered across California. Gold mines did, and still do, frequently produce other ores in addition to gold, most notably copper.[27]

Current Hardrock Mining

Hardrock mining in the United States today is big business, led by huge corporations at unimaginable scales. Large pits and mountaintop-removal mines are common methods of hardrock mining. Open-pit mining requires developing a pit with concentric rings or ridges so the excavation equipment can dislodge the ore after blasting it loose. Large loading shovels place the ore in enormous hauling trucks. The gravel is then hauled to refining and extracting facilities.

Refining the ore is the most dangerous process of all. Especially common is "heap leaching," in which cyanide is sprinkled over heaps of ore, thereby dissolving the gold and other metals into large, waiting sluices. It is then further refined and purified.

Gold is sometimes a by-product of a copper mine, many of which are profitable even with extremely low amounts of gold in the ore, such as 0.073 ounces per ton, if the price of gold is high enough. U.S. production soared in the 1980s, along with the price of gold and the increased use of the heap-leach processing to recover low-grade ores. The United States is the fourth-largest gold producer in the world (after China, Australia, and Russia), with the two largest gold-producing states being Nevada and Alaska.

Some of the biggest open-pit mines are in the United States—the largest in the world being the Hull Rust open pit iron mine near Hibbing, Minnesota. The pit is 5 miles long, 2 miles wide, and 535 feet deep. It covers more than 2,000 acres. The deepest mining pit is the Bingham Canyon Mine near Salt Lake City, which is nearly 3 miles across, three-quarters of a mile deep and removes 450,000 tons of rock every day (Figure 15.2).[28] Although Bingham Canyon is mainly a copper mine, it also produces gold.

FIGURE 15.2
Bingham Canyon copper mine in Bingham Canyon, Utah, is owned by Kennecott Utah Copper Corporation, first formed as the Boston Consolidated Mining Co. in 1898. Photograph by Spencer Musick (https://commons.wikimedia.org/wiki/File:Bingham_mine_5-10-03.jpg).

The United States is the world's fourth-largest copper producer, accounting for a third of the world's copper—behind Chile, the largest, followed by China. Arizona has been the leading American copper state since 2007, producing 60 percent of the U.S. total. The first open-pit mine was located at Ajo in 1917. Before that, all copper mining was done underground. One of the largest open pits in the United States is the Morenci copper mine in Arizona.[29]

Copper in the United States is mainly used for construction and electric equipment. As previously stated, other minerals are often found in conjunction with copper, including molybdenum, gold, and silver. Thus, open-pit copper mines can produce multiple minerals, and usually do.[30]

As prices increase and demand swells, new, devastating copper mines continue to be developed and proposed in Arizona. The Safford mine, which consists of two pits, began full production in 2008—the first new industrial copper mine in Arizona in 30 years.[31]

New proposals include the Superior Mine south of Tucson, which would destroy the Apache's sacred lands under a copper pit two miles wide and 1,000 feet deep. Although Congress has approved the federal land exchange required for the project, opponents remain undaunted.[32]

Another proposal lurking in the wings is the Rosemont Copper Mine, also south of Tucson, which would destroy a large swath of the Santa Rita Mountains in another pit a mile wide and half a mile deep. Moreover, this mine would be excavated in the middle of an aquifer crucial for Tucson's water supply.[33]

Hardrock mines are often proposed for the most pristine areas.[34] I (Cameron) have, for several years, been part of an environmental and community coalition fighting a nickel strip mine complex proposed by Red Flat Nickel Corporation on National Forest lands in the North Fork Smith River watershed of southern Curry County, Oregon. This region is one of the most unspoiled in the entire Untied States; the North Fork Smith River and its tributaries are completely pristine as they flow into the Smith River. The Smith River in turn, which empties into the Pacific Ocean in northern California, is one of the largest free-flowing rivers in California. In addition, the Smith River and its tributaries provide the water supply for Crescent City, Gasquet, and many rural residents, and the Smith River is a critically important river for wild salmon. The effects of one or more large surface mines in this area would be obscenely devastating of the region's unique ecology, hydrology, and water quality. Oregonians know what a nickel mine looks like: Oregon had the only nickel mine ever to operate in the United States, located at Riddle in Douglas County. It was a large-scale mountaintop-removal operation. Closed since the 1990s, the Glenbrook Mine is a vast moonscape of scraped earth and piles of toxic tailings.

Environmental Concerns

The effects of the heavily industrialized, open-pit mines that characterize modern mining are among the most devastating human activities on the planet. They are biological deserts and profoundly disturbed, polluted aquifers—the latter because of toxic metals leaching into streams and groundwater for decades, both from the pits and from the massive piles of overburden (rock or soil overlying a mineral deposit, archaeological site, or other underground feature).

Abandoned or exhausted mines produce some of the deadliest pollution and ecosystem destruction. An example is Berkeley Pit, outside Butte, Montana. It is a gigantic mining pit a mile and a half wide—all that is left of a colossal copper mine that was active until the 1980s. The groundwater that had been pumped out during the mine's active years slowly filled the pit, creating a toxic brine from leached minerals and heavy metal poisons, such as arsenic, lead, and zinc. Water in the pit continues to rise, and will eventually contaminate Butte's groundwater. Even now, the fog rising from pit's water is poisonous.[35]

The Gold King Mine along Cement Creek, a tributary of the Animas River near Silverton, Colorado, is a notorious recent failure to clean up an abandoned mine. It turned ghastly orange in 2015 after a spill of contaminated water replete with such toxic heavy metals as cadmium and lead and other toxic elements, such as arsenic, beryllium, zinc, iron, and copper. The spill caused Colorado, New Mexico, and the Navajo Nation to declare disaster emergencies for stretches of the Animas and San Juan Rivers, as the toxic waste spread downstream toward Lake Powell in Utah.[36]

There are an estimated half million abandoned mines in the United States. In addition, mining has contaminated the headwaters of more than 40 percent of watersheds in the West and will cost tens of billions to clean up.[37]

Remediation and reclamation of hardrock mines, whether smaller and underground or the vast pits, are difficult if not essentially impossible. The situation is greatly exacerbated by lack of funding for reclamation. Congress did not provide any remediation funding for mines, nor did it provide any compensatory royalty payments from mining operations that could be used for cleanup of abandoned or exhausted mines.

Laws of Hardrock Mining: Permits and Reclamation

The central problem for the regulation of mining in the United States is the 1872 General Mining Law,[38] passed by Congress to promote mining of hardrock minerals, which remains in effect as enacted nearly a century and a half ago. This law granted miners broad rights to work the public lands, and restricted the government's ability to limit access, require fees or royalties, or prevent the physical devastation that mining always entails.[39]

Hardrock mines now require various permits from air- and water-quality agencies, both state and federal, as well as from federal land managing agencies, if the mine is proposed for federal land. In addition, federal law, under the Federal Land Policy and Management Act of 1976,[40] requires the recording and maintenance of claims. Nevertheless, these requirements are secondary to the 1872 General Mining Law.[41]

As a result of this antiquated law's preeminence, the American West is home to a large body of contaminated mine sites. How many? Nobody knows for certain, thanks to spotty and inconsistent recordkeeping. The Western Governors Association in 1998 estimated there were more than 250,000 abandoned hardrock mining sites in 13 states, the majority of which are in Arizona, California, Nevada, and New Mexico. Teasing out how many of the total are situated on public lands is even more difficult due to lack of inventories and standardized terminology, but it is likely a substantial proportion.

Clearly, the reclamation problem is staggering. "For decades, state and federal regulators have struggled to understand the nature and scope of the environmental threats associated with hardrock mining and to develop strategies to address them. … Many hardrock mining sites, however, have been abandoned or have been inactive for long periods and thus cannot easily be regulated under the public mining laws or the pollution control laws directed at currently operating facilities."[42]

Regulators have been partially successful in bringing other environmental statutes to bear on contemporary mining, despite the 1872 law's preeminence. Statutes such as the

Clean Water Act apply to hardrock mining and require a permit before pollutants may be discharged to U.S. waters. Stormwater runoff from current mines is also regulated.[43]

Potentially, the best regulatory tool is the federal Comprehensive Environmental Response, Compensation and Liability Act of 1980, the "Superfund" law.[44] Superfund has a strong liability framework that allows the Environmental Protection Agency to recoup its cleanup costs from responsible parties. The Environmental Protection Agency can also require such responsible parties to clean up sites if there is imminent danger to public health or welfare. The Superfund law is not limited to protection of surface waters, but covers damage to land, soil, subsurface land, and groundwater as well.[45]

The problem is that the costs of reclamation are immense, long-term, and frequently lead to miniscule actual results because the damage is insuperable and resistant to any reclamation: "[e]nforcement at mining sites presents immense challenges, however. The costs of cleaning up hardrock mining sites can be enormous. … At many contaminated hardrock mining sites, massive quantities of waste rock, tailings and other materials often must be covered with impermeable soil, or even excavated, transported and disposed of in carefully engineered repositories. Acid mine drainage or other surface water contamination at mining sites may require the installation and perpetual operation of sophisticated storm water and wastewater treatment systems. At some sites, full restoration of the environment may be unattainable at any cost."[46]

Cleanup of such "mega-sites" on private lands can be intimidating, since they are often abandoned, financing is difficult, and costs run from the hundreds of millions into the billions of dollars. But the situation is worse for mines on public lands, where the Environmental Protection Agency is not the primary responder, but rather such federal land-managing agencies as the U.S. Forest Service or Bureau of Land Management.[47]

The Environmental Protection Agency, under the Superfund law, can require the federal agency to accept liability the same as a private operator, but this is uncommon. However, when federal land managers seek to make mining companies clean up a site, the companies often file a so-called "contribution action," seeking to make the agencies partially liable for hardrock mining contamination on grounds that they are co-owners or operators. They justify this action by claiming the agency is managing the public lands, where the mining is taking place, and/or encouraged production of the minerals. This is despite the fact that the United States has no choice but to allow unregulated hardrock mining on federal public lands—thanks to the 1872 Mining Law. Courts, however, have generally refused to impose liability on the United States, despite the claims of mining companies.[48]

Threats to public health from abandoned mining sites are largely unknown, thanks to inconsistent recordkeeping. However, it is clear from surveys by the Environmental Protection Agency that at least tens of thousands of mines are in desperate need of cleanup to avoid becoming serious, long-term, public health hazards.

Various agencies are working on the problem, but funding is inconsistent, and the science of reclamation is simply not sophisticated enough to recover mined lands and contaminated waters, if it is possible at all. Problems include subsurface cleanup methodologies in uranium mines; the difficulties of detecting mercury and other metals in streams, where they fluctuate on a daily or seasonal basis; evaluating the effects of toxic metals in food chains—and a staggering host of other problems.[49]

The bottom line, as with coal mining, is that any real restoration will require commitment of massive resources, both to reclamation science and to actual reclamation on the ground. Many imperfections will remain, even if reclamation becomes a national priority, but at least some of the problems will be mitigated.

Oil and Gas Drilling

Early Oil and Gas Drilling

Oil seeps had been noticed, and locally used, in the United States by Native tribes for centuries. Early European explorers noted the seeps of oil and natural gas in western Pennsylvania and New York. Interest grew substantially in the mid-1850s, as scientists reported on the potential to manufacture kerosene from crude oil, provided a sufficiently large supply could be found.[50]

Whereas early wells drilled for salt brine sometimes—but only incidentally—produced oil and gas, the first rig designed to drill for oil was the successful Drake Well in Titusville, Pennsylvania, in 1859. It was drilled by Colonel Edwin L. Drake for the newly formed Seneca Oil Company. Drake struck oil just as the company had withdrawn his funding, and his personal finances were exhausted. The subsequent oil boom did not benefit Drake, and he died a poor pensioner despite his fame as the first person in North America to drill commercially for oil.

The Drake well attracted a rush of investment, leading to the Pennsylvania oil boom, the first in the United States. It lasted until the early 1870s. In the first year of the Pennsylvania oil rush, wells produced something like 4,500 barrels of oil. Ten years later, in 1869, the annual output nationwide swelled to 4 million barrels. Pennsylvania oil production peaked in 1891, when the state produced 58 percent of the national total.[51]

A rash of oil drilling in western Appalachia followed this first oil boom (Figure 15.3). Shortly thereafter, under the impetus of the Civil War, oil drilling spread through western Pennsylvania and down the Ohio River Valley.[52]

FIGURE 15.3
Often used for drilling brine wells, a "spring-pole" well discovered oil in Appalachia. Steam-powered cable-tools drilled faster and deeper. Photograph from "The World Struggle for Oil," a 1924 film by the U.S. Department of the Interior (http://aoghs.org/technology/oil-well-drilling-technology/).

The U.S. oil industry subsequently expanded to wherever oil was found nationwide, ranging from the Gulf Coast to California.[53] By the Natural Gas Act of 1938, the federal government stepped in to regulate prices on natural gas in interstate commerce, by setting "just and reasonable" rates.[54] The Natural Gas Policy Act of 1978 enforced price controls to all natural gas in the country, whether sold through interstate commerce or not.[55] These were finally phased out under the Natural Gas Wellhead Decontrol Act of 1989.[56]

Landowner Rights to Oil and Gas

Oil and gas drilling falls under a different set of laws compared to coal or hardrock mining, both for historical reasons and because the technology and the resources extracted differ so much. The surface landowner in the United States owns oil and gas rights, unless they have been specifically severed from the surface ownership. In such "split estates," the oil and gas rights may be bought, sold, or transferred like any other real estate.[57]

Private Lands

Oil and gas drilling on private lands is regulated by the individual states. However, the degree to which local communities can regulate or ban oil and gas drilling is a highly controversial and ongoing battle, especially in states with large amounts of hydraulic fracturing. Most states allow the oil and gas owner to use as much of the surface land as needed to extract the hydrocarbons. Courts have generally upheld this right of "subsurface supremacy," as it is called, holding that ownership of the rights to oil and gas would be meaningless otherwise.

There are two conflicting legal doctrines covering oil and gas extraction. The "rule of capture" allows a driller to drain the oil out from under their own land, even if it originated on neighboring land and flowed elsewhere as a result of geological forces or drainage. This doctrine gives landowners incentive to pump oil as quickly as possible to capture the neighbors' oil, and can lead to reducing the overall pressure in an oilfield.

The other doctrine, which many states use to curb the rule of capture, is the "correlative rights doctrine." This protects the rights of neighboring owners (and prevents waste) by regulating extraction.

Oil and gas companies frequently lease the mineral rights from the owner. Leases last for a certain period ("primary term"), as long as the lessee pays the rent. The primary term is usually one to ten years, but if oil and gas are produced in paying quantities, the lease is typically extended. Leases can include bonuses, rents, and royalties. The bonus is an upfront payment at the beginning of the lease; the rent is an annual payment, made until the property begins producing commercial quantities. The royalty is a portion of the gross value of the oil and gas, paid before any cost deductions.[58]

Federal Lands

The U.S. government regulates oil and gas leasing on federal lands and on private lands, where the federal government owns the oil and gas rights. There may be up to 57 million acres of land in the United States with private surface owners, but federally owned minerals. Between 1982 and 2004 (just before the boom in hydraulic fracturing), the federal government had leased, or offered for lease, 229 million acres of public land in 12 Western states. More than double that amount, about 570 million acres, of federal lands are available for leasing. Federal leases are usually for a primary term of 10 years, but they can be extended for as long as oil and gas are produced in commercial quantities. Leases allow the lessee to use as much of the surface land as needed to develop the oil and gas.

Under the Mineral Leasing Act of 1920, as amended,[59] and the Mineral Leasing Act for Acquired Lands of 1947, as amended,[60] the Bureau of Land Management is the federal agency responsible for oil and gas leasing on federal lands, including National Forests and private lands, where the federal government owns the rights. The Bureau only offers leases for those lands evaluated through its planning process, and can deny leases in sensitive areas. In practice, however, the Bureau often waives environmental and wildlife protections at the request of oil and gas companies.

The maximum size for a Bureau lease is roughly 10,200 acres, with different size limits for competitive or non-competitive leases. The Bureau's rules allow one person or entity to control a maximum of 246,000 acres of oil and gas leases in any one state at the same time.

There is controversy over the size of the required bonding, on grounds that actual cleanup costs far outstrip the modest fee requirements, especially in the case of abandoned wells. Controversy also rages over the requirements that drilling operators must abide by environmental protections prohibiting "undue damage" to surface or subsurface lands. Whether the Bureau of Land Management has always enforced the environmental safeguards, as required, is in dispute.[61]

Environmental Concerns

Reclaiming old oil wells is a special problem, especially as many abandoned wells are now near new fracked wells. Methane can leak into the old well, and homes and water supplies, from the new well. This is a major problem in Pennsylvania, for example, the legacy of the oil boom in the early 20th century. Oil wells were often just carelessly abandoned, or perhaps plugged with whatever junk lay to hand. Oil and other toxins seep from these old wells, contaminating water, farmland, forests, and people's backyards. There are a staggering number of abandoned oil and gas wells in the United States—at least 2.6 million. State agencies only know the location of about 10 percent of them and don't have the resources to locate the rest. Citizen researchers focusing on public lands and homeowners who report leaking wells on their property are filling some of the gap, but old well location is a very patchy process at best.[62]

And there is more to come. A precipitous drop in oil prices since 2014 has caused more than 60 oil producers to declare bankruptcy, often abandoning their wells in the process. This puts oil-producing states like Texas in line to wrestle with environmental problems from thousands of newly abandoned oil wells. Texas alone may end up with a portfolio of more than 12,000 such wells, far more than regulators can handle now. It would cost something like $165 million to plug them all properly. This is far more than Texas regulators can cope with, already struggling with a backlog of up to 25,000 earlier untended wells.

Paying for reclaiming abandoned wells has become problematic, as location and capping costs far exceed required producer fees. In 2013, for example, Texas received bonding fees of $4 million from drillers, who abandoned 1,500 wells. The fees covered approximately a fifth of the cost of plugging them all. Capping a single well can cost more than $17,000. States like Texas are looking to increased driller fees, higher bonding requirements on drillers, and perhaps increased general funding. But when Louisiana required a new $7 bond for every foot drilled to raise money for plugging the 3,000 abandoned wells in the state, the drillers caused a political fracas, and regulators were forced to back down.[63]

Hydraulic Fracturing

The technology for hydraulic fracturing, or "fracking," as it is known, has been available since the 1940s, but was little used until 2003. Fracking allows drillers to tap into previously unavailable natural gas reserves up to 8,000 feet below the ground, especially in shale formations (Figures 15.4 and 15.5).

In fracking, the driller pumps water, chemicals, and sand into a well under high pressure. This fractures the surrounding rock formations and opens passages that allow natural gas to flow through the fractures to the production well. The mixture of chemicals involved in fracking is long, but includes such things as benzene and diesel fuel. Once the fracturing has occurred, the "carrying fluid" can flow back to the production well with the gas, but 60 percent to 80 percent of the fluid usually remains underground rather than being drawn back to the surface.

Fracking uses tremendous amounts of water per well. Early techniques required 20,000 to 80,000 gallons per well, but advanced technologies use 1.2 to 4 or even 5 million gallons per well. Despite such a dramatic use of water, fracking received a great boost by being exempted from key environmental protection statutes in the Energy Policy Act of 2005[64] under the George W. Bush Administration. This law ushered in sweeping protections for fracking by removing the fledgling industry from many provisions of the Clean Air Act, the Clean Water Act, the Safe Drinking Water Act,[65] and the Superfund law. The Energy Policy Act not only exempted the fluids used in fracking from these key environmental laws but also created a loophole exempting companies drilling for natural gas from disclosing the chemicals they used. Although clean water laws routinely require such disclosures, bills to remove this loophole for fracking have died in Congress.[66]

FIGURE 15.4
Seneca Resources drill pad on land leased from Pennsylvania Department of Conservation and Natural Resources in Loyalsock State Forest, Lycoming County, PA. Photograph courtesy of Richard A. Martin, PaForest Coalition.org.

FIGURE 15.5
Hydraulic fracturing rig on private land in Lycoming County, PA. Photograph courtesy of Richard A. Martin, PaForest Coaliton.org.

As of 2013, there were at least two million fracked wells in the United States, and the vast majority of new wells use hydraulic fracturing technology. Sixty-seven percent of American natural gas production came from fracked wells by 2013, and more than 43 percent of oil production.[67]

Environmental Concerns

As the number of fracking wells in the United States has grown, so have the environmental concerns, which range across the board: releases of ozone; unregulated large-scale emissions of methane (an extremely potent greenhouse gas); contamination of groundwater and drinking water; massive and unsustainable uses of water, especially in dry areas; and fracking-induced earthquakes from reinjection of fracking fluids. In response, Vermont completely banned fracking as of 2012, and New York, after a two-year moratorium while studies were conducted, also banned fracking in 2015.[68] Maryland placed it under moratorium until 2017. On the other hand, Texas and Oklahoma, both states with very high rates of fracking, signed legislation in 2015 prohibiting local municipalities from enacting prohibitions on fracking.[69] Taken as a whole, there are three primary, critical environmental concerns in fracking: (1) air quality, (2) water, and (3) earthquakes.

Air Quality: Methane Emissions

Not until 2012 did the Environmental Protection Agency issue the first federal air standards for fracked natural gas wells, which require capturing gas that escapes into the air, and making it available for sale. As of 2015, the Agency still had sketchy data about the extent of spills from fracked wells and associated fracturing fluids, their environmental effects, groundwater contamination, and reasons for the accidents.

There was no federal regulation, and almost no state regulation, until 2016, when the Environmental Protection Agency issued rules for controlling emission from new oil and gas wells, focusing on leakage of methane and the venting of air in the natural gas production process. There are still no federal rules for *existing* wells—resulting in unregulated methane emissions from 75 percent of oil and gas equipment. Moreover, well-substantiated complaints against the Environmental Protection Agency allege that some agency-linked scientists purposely undercalculate methane emissions in studies designed to estimate the problem as part of the regulation process.

Allegations have especially centered on the one-time Chair of the Agency's Science Advisory Board, whose influential studies in 2013 and 2014 showed lower-than-expected emission rates. Actual rates of methane emission are likely much higher: "Because EPA, academic and industry personnel have ignored or misrepresented these problems instead of addressing them ... methane emissions remain poorly quantified, although it seems very likely that they are much higher than estimated by EPA. For example, a 2016 EDF study (Lyon 2016[70]) found methane leakage rates 90% higher than original estimates."[71]

The risks of cancer from methane emissions are very troubling, despite the picture from sketchy data. Earthworks, using data from the Environmental Protection Agency's National Emissions Inventory, released a map in 2016 showing that more than 9 million people in six states face a higher level of cancer risk due to toxic air emissions, especially methane, from active oil and gas installations—most of them fracking wells. A total of 12.4 million people live within a half-mile radius of active oil and gas wells or equipment, and there are 11,543 schools and 639 medical facilities in that same danger zone nationwide.[72]

Water: Drinking and Agriculture

The incredibly high use of water in fracking has already come under fire, as farmers in arid regions, such as Colorado and New Mexico, find themselves competing with the fracking industry for water.[73] As for Texas, "Fracking boom sucks away precious water from beneath the ground, leaving cattle dead, farms bone-dry and people thirsty."[74]

Increasingly, urban areas in these regions are facing the same problems. Opportunities for recycling water are limited, since most of the water used stays in the ground or is reinjected, as brine waste, after the fracking is completed. Moreover, most water used in fracking is fresh, despite the fact that wells are often in water-stressed regions, where more than 80 percent of the water is already allocated. A 2013 report found that nearly half of almost 26,000 wells evaluated are in high or severely water-stressed areas. Massive water use by the fracking industry is not sustainable.[75]

It is estimated that up to 30,000 new wells were fracked between 2011 and 2014—often located near households and locally used sources of drinking water. Around 9.4 million people lived within a mile of a fracked well between 2000 and 2013, and at least 6,800 sources of drinking water were located within the same distance—serving more than 8.5 million people.[76]

As the Environmental Protection Agency conservatively noted in its draft report, "Although proximity of hydraulic fracturing activities to a drinking water resource is not in of itself sufficient for an impact to occur, it does increase the potential for impacts. Residents and drinking water resources in areas experiencing hydraulic fracturing activities are most likely to be affected by any potential impacts, should they occur."[77]

The Agency's study found that fracking can affect water supplies by excessive withdrawals, spills of fracking fluid, fracking directly into underground sources of drinking water, underground migration of liquids and gases, as well as poor treatment of wastewater. Although the Environmental Protection Agency found only "specific instances" of impacts on drinking water rather than systemic affects, the draft report admits possible causes for these sketchy findings include: insufficient data, lack of long-term studies, and the "inaccessibility of some information on hydraulic fracturing and potential impacts."[78]

Injecting the waste fluid from fracking back underground is causing a ghastly array of problems. The chemicals in fracking fluids number at least 2,500, of which more than 650 are known or suspected carcinogens. There have been numerous instances of groundwater contamination from the fracking process, and increasing problems of surface-water spills from accidents or discharge of fracking brine through conventional wastewater treatment facilities. By 2012 and 2013, even reluctant federal regulators were able to document contamination at various sites, both of surface water and groundwater. But the exemptions to federal drinking water laws have hampered the Environmental Protection Agency's investigations. Consequently, the magnitude of the problems have not been well researched.[79]

Earthquakes: Growing in Intensity

Perhaps the most alarming effect of fracking is earthquakes. The initial fracturing process causes tiny earthquakes, detectable only to sensitive instruments. The U.S. Geological Survey does not consider these to be of concern.

But earthquakes are common and becoming stronger as wastewater is reinjected back underground at the end of the hydraulic fracturing process that starts with injections of water and chemicals at high pressure deep into underground shale formations.[80]

By 2012 and 2013, fracking-related earthquakes were intensifying in several locations: there was a 5.3 magnitude quake in southern Colorado in 2013 and a 5.6 magnitude quake east of Oklahoma City that destroyed 14 homes. In the 10 years leading up to 2013, there had been a 600 percent increase of wastewater injection-related earthquakes above 3.0 magnitude—strong enough to be felt aboveground. Arkansas and Ohio have both recognized the connection with the fracking wastewater disposal process, and Arkansas implemented a local injection ban.[81]

Oklahoma is suffering from a frightening swarm of earthquakes: they began increasing in 2009, when there were 20 quakes people could feel aboveground. In 2015, there were more than 900 such quakes, and 30 of them were above 4.0 on the Richter scale. The quakes continued into 2016.[82]

Oklahoma regulators finally took initial steps after their own homes were rattled by quakes. The Corporation Commission ordered operators of 250 injection wells to reduce the amount of wastewater they inject, and began beefing up state monitoring staff. Scientists are reporting that Oklahoma's natural earthquake faults are stimulated by the fracking wastewater disposal, which numbers in the billions of gallons, leading to temblors of increasing magnitude and frequency. There has been an 81 percent leap in the volume of reinjected wastewater between 2009 and 2014 in Oklahoma—exactly the period in which seismicity increased dramatically.[83]

In response to this devastating situation, the Sierra Club has filed suit against three major energy companies whose fracking and wastewater injection are highly correlated with the Oklahoma quakes.[84]

Oil and Gas Pipelines

The other aspect of oil and gas drilling, and fossil fuel use in the United States, is the pipelines necessary to carry the oil and gas from production areas through distribution hubs to urban and industrial sites, where it is used. There is no other feasible way to transport massive volumes of oil and gas around the country, though fossil fuels are also transported by truck and rail. Pipelines crisscross the United States, which has the largest network of energy pipelines in the world, with more than 2.5 million miles of pipe, including 190,000 miles of petroleum-transmission pipelines and a massive web of natural gas pipelines.[85]

The use of pipes for the transportation of oil started shortly after the first commercial oil well drilled in Pennsylvania by Edwin Drake in 1859. The first pipes were iron, but changed to steel as early as the 1860s.[86]

The Federal Energy Regulatory Commission has sole authority to permit interstate pipelines, but regulating pipeline safety falls under the jurisdiction of other agencies. Building new pipelines has become fiercely controversial in the decade from 2005 through 2015, as the fracking boom has necessitated more natural gas facilities and enlarged pipeline networks to bring natural gas to ports for shipping overseas. I (Cameron) have been part of the environmental coalition opposing the Jordan Cove Energy Project on the southern Oregon coast, which proposes a liquefied natural gas facility and a 232-mile pipeline from Coos Bay to Klamath Falls, passing through private lands, public lands, over the Coast Range, and requiring more than 400 crossing of rivers and streams. Landowner opposition has been sustained, fierce, and successful. Pipeline companies cannot use the power of eminent domain to purchase private lands for a pipeline without a certificate from Federal Energy Regulatory Commission for "public convenience and necessity." In March 2016, the Federal Energy Regulatory Commission denied the permit for the Jordan Cove Energy Project pipeline, citing lack of need for the liquefied natural gas facility and landowner opposition. In December 2016, the Commission reiterated its denial, after the company petitioned for a rehearing. Jordan Cove subsequently announced it would refile its application and seek approval again.

Similar scenarios of fierce landowner opposition have been playing out nationwide, leading to recent cancellations of some pipeline projects. Most notably, Kinder Morgan in 2016 canceled its proposed Northeast Energy Direct Pipeline from New York through Massachusetts and New Hampshire.[87]

The principal pipeline regulatory agency is the U.S. Department of Transportation Pipeline and Hazardous Materials Safety Administration, under the Hazardous Liquid Pipeline Safety Act of 1979,[88] which added transportation of liquids to the initial pipeline regulation statute of 1968. Pipeline safety laws have been continuously amended since then, most recently in 2002. Congress created the Office of Pipeline Safety in 1968 to oversee interstate pipeline safety, but it has had a poor record as a regulator since at least 1978, which has been substantiated by a General Accounting Office report in 2000. Even the petroleum industry implemented more recommendations from the National Transportation Safety Board (87 percent) than did the Office of Pipeline Safety (69 percent).[89]

Are pipelines safe? It depends what data one looks at, who collected it, which category of data one investigates, and a host of similar questions. The U.S. Department of Transportation Pipeline and Hazardous Materials Safety Administration collects this data, which shows 5,667 "significant pipeline incidents" between 1996 and 2015, with 346 deaths and 1,347 injuries. The annual number of incidents has remained fairly constant during this period, with the largest spike in 2005, but fatalities peaked in 1996, with a secondary surge in 2010.[90]

The primary federal environmental law involved in pipeline permitting is the National Environmental Policy Act, which requires federal agencies to evaluate environmental effects of federal undertakings, and present an array of alternatives. Many other federal and local environmental laws also apply, including the Endangered Species Act.[91] However, as with other regulatory systems, the one overseeing pipelines must be reckoned a massive failure, as its primary purpose is to permit pipelines whenever possible, not encourage and incentivize alternative energy systems.

In the United States, the companies who transport the oil and gas own the pipelines; federal or state regulation only oversees transportation, storage, and safety once the pipeline is built.

Offshore Drilling

The United States also maintains a risky offshore oil and gas energy program, which allows leasing on the outer continental shelf, as well as other American waters like the Gulf of Mexico and Alaskan coastal waters. The Bureau of Ocean Energy Management regulates offshore leasing, overseeing leases in the neighborhood of 26 million acres in American waters. Offshore drilling accounts for just about 5 percent of natural gas production used in this country. Despite the extreme risks to the ocean and ocean life, the Bureau estimates that 90 billion barrels of oil and 327 trillion cubic feet of gas lie as yet undiscovered on America's outer continental shelf.[92]

The Submerged Lands Act of 1953[93] granted states rights to the natural resources of submerged lands up to 3 miles from their coastlines, and also affirmed federal claims to the outer continental shelf—those lands further at sea than state jurisdiction. The Outer Continental Shelf Lands Act of 1953,[94] and later amendments, set up the framework by which the federal government maintains responsibility over submerged lands on the Continental Shelf. By a presidential proclamation in 1982, President Ronald Reagan set up the Exclusive Economic Zone, which extends up to 200 nautical miles seaward from the American coastline.

Under the Oil Pollution Act of 1990,[95] the Bureau of Ocean Energy Management also has authority over offshore oil spills: determining responsibility and liability for spills, bonding requirements, claims, and penalties. Under the Energy Policy Act of 2005, the Bureau took responsibility for marine renewable energy projects as well.[96]

Environmental Damage

Oil and gas spills are a constant worry in the offshore program, for good reason: Storms can damage drilling rigs and oil spills greatly affect marine life. Hurricane Ike destroyed oil platforms and pipelines across the Gulf of Mexico, leading to at least half a million gallons of crude oil spilled into the ocean. Hurricanes Katrina and Rita caused 125 spills from rigs on the Outer Continental Shelf, leading to nearly 685,000 gallons of petroleum products polluting the ocean. In 2006, British Petroleum Exploration Alaska (BPXA) pipelines in Prudhoe Bay, Alaska, spilled up to 267,000 gallons into the Bay over a five-day leak. The disaster was caused by British Petroleum's negligence in controlling pipe corrosion, despite warnings from employees and concern over cost-cutting measures. And there was a subsequent spill in 2009 in British Petroleum's Prudhoe Bay operations due to frozen pipes.[97]

This sampling of damages pales before the worst offshore oil spill in history: the Deepwater Horizon explosion and spill in the Gulf of Mexico in 2010, on British Petroleum's Macondo oil drilling platform. An estimated 4.9 million barrels of oil spilled from a seafloor gusher that poured oil into the sea for 87 days after the rig exploded, burned, and sank due to a methane leak (Figure 15.6). Much of the oil rose to the surface through the water column, poisoning marine life as it did so—then spreading out to the Gulf coast, its islands, barrier islands, and shorelines. The spill had an impact on 68,000 square miles of the ocean, according to satellite photos.

Two years later, oil was still found on 200 miles of Louisiana beaches, where 4.6 million pounds of oily material was removed from the state's beaches in 2013 alone. Marine life in the Gulf has continued to die in record numbers 3 and 4 years after the spill. As much as a third of the oil may have mixed with deep-ocean sediments, poised to cause extensive damage to ecosystems and commercial fisheries.[98]

Moreover, farmers and fishermen along the Gulf of Mexico had already watched for decades as their sensitive ecosystem's waters slowly got dirtier and islands eroded. Adding insult to injury, the country largely ignored the destruction caused by British Petroleum's drilling operation.[99]

In response to this spill, the U.S. government in 2011 created a new Bureau of Safety and Environmental Enforcement to enforce environmental and safety regulations. The point of this reorganization was to separate resource management from safety oversight, so that regulators would have more independence. It also sought to strengthen the role of environmental review and analysis in the offshore program.[100]

FIGURE 15.6
Fire boat response crews battle the blazing remnants of the offshore oil rig *Deepwater Horizon* in the Gulf of Mexico, off the coast of New Orleans, April 21, 2010. Photograph by the U.S. Coast Guard (https://commons.wikimedia.org/wiki/File:Defense.gov_photo_essay_100421-G-0000L-003.jpg).

Only in 2016 did the courts finalize the settlement with British Petroleum over natural resource injuries, for a total of some $8.8 billion. It is the largest natural resource damage assessment ever undertaken. The National Oceanic and Atmospheric Administration, as the lead restoration agency, will oversee the $7.1 billion earmarked for restoration projects over a 15-year period. The Administration has studied the types of restoration needed for a comprehensive, integrated, restoration alternative to provide the best benefit possible in the restoration program.[101]

Once again, as with coal mining, hardrock mining, and onshore oil and gas operations, it is clear that the regulatory framework is an enormous and very dangerous failure—failing to prevent colossal ecosystem damage or to curb excessive and dangerous activity, especially in sensitive areas like the ocean.

The common denominator to these offshore disasters is that current methods of cleanup and restoration only remove a small percentage of the oil spilled. Most of the oil continues to wreak havoc on marine and coastal ecosystems over time. Not only that, but offshore drilling, like its land-based counterpart, is in itself toxic. Each well produces tens of thousands of waste-drilling muds, laced with mercury, lead, and cadmium that bioaccumulate in the marine ecosystem, poisoning marine life, including seafood eaten by humans.

Another waste product is the so-called "produced water," which is mixed with the oil and gas in an offshore well. It usually contains arsenic, lead, and radioactive pollutants. Wells discharge hundreds of thousands of gallons of produced water daily back into the ocean, contaminating both coastal and deep waters. It must also be remembered that each offshore well requires new onshore infrastructure, including roads, pipelines, and processing facilities, built on the most fragile and sensitive coastlines.[102]

Based in part on these kinds of dramatic environmental concerns, the U.S. government, in late 2015, canceled potential lease sales off the Alaskan coast in the Beaufort and Chukchi seas. The government also denied requests to extend leases of two other companies. Offshore oil drilling will be suspended in these areas for the next few years, under current plans for drilling rights sales.[103] And now, despite all the other anthropogenic disruption of the world's oceans, people want to mine the ocean floor in search of commercially extractable minerals.

Mining the Ocean's Floor

The prospect of a deep-sea "gold rush" opens a new, controversial Pandora's box for mining on the ocean floor to extract so-called "nodules" containing gold, copper, manganese, cobalt, and other metals from the seabed. Although the yen for such mining has been around for decades, a technical study by the United Nations' International Seabed Authority (the body overseeing deep-sea mining), as well as high commodity prices and new technology, has now brought it closer to realization.

The monetary interest has been fueled by an assessment of the eastern Pacific, a 3.1 million-square mile area known as the Clarion–Clipperton Zone, which concluded that more than 29 billion tons of nodules could be lying on the seafloor. Moreover, the rocks are projected to contain more than 7.7 billion tons of manganese, 375 million tons of nickel, 319 million tons of copper, and 86 million tons of cobalt. However, no one knows how much of that mineral wealth is accessible. (For further discussion, see *Interactions of Land, Ocean and Humans: A Global Perspective.*)

As it now stands, 17 licenses have been issued to prospect for the minerals in vast areas of the Pacific Ocean, Atlantic Ocean, and the Indian Ocean, with more likely to follow.

One of the most recent licenses granted was to UK Seabed Resources, which is a subsidiary of the British arm of Lockheed Martin, an American defense corporation. In addition, the longer, ice-free summers the Arctic Ocean is now experiencing will most likely open it to the oil and mining industries, as well as various other forms of commercial exploitation.[104]

Here it must be reiterated that mining of any sort is already known to be highly disruptive to the terrestrial biophysical system. How much more biophysical destruction to the seafloor topography would deep-sea mining cause when coupled with the already destructive practices of deep-sea fishing discussed earlier?

Moreover, University of Southampton Professor Paul Tyler, a member of the National Oceanography Centre Southampton, warns that unique species would be at risk.

> If you wipe out that area by mining, those animals have to do one of two things: they disperse and colonise another hydrothermal vent somewhere or they die.
>
> And what happens when they die is that the vent will become biologically extinct.[105]

Professor Rachel Mills, a marine chemist at the University of Southampton who has carried out research for Nautilus Minerals, a Canadian firm planning to mine hydrothermal vents off Papua New Guinea, calls for a wider debate about mining in general.

> Everything we are surrounded by, the way we live, relies on mineral resources and we don't often ask where they come from.
>
> We need to ask whether there is sustainable mining on land and whether there is sustainable mining in the seas.
>
> I actually think it is the same moral questions we ask whether it's from the Andes or down in the Bismarck Sea.[106]

There is yet another consideration with mining the ocean floor, which began with the "industrialization" of the world's oceans—"ocean sprawl," otherwise known as the proliferation of artificial structures associated with the exponential growth in shipping, sunken war material, aquaculture, and other coastal industries, all of which provide habitat for the attachment of marine organisms. Jellyfish are a prime example because the collective ocean sprawl provides habitat for their polyps and may be an important driver of the global increase in jellyfish blooms. Support for the role of ocean sprawl in promoting jellyfish blooms is based on observations and experimental evidence, which demonstrates that jellyfish larvae settle in large numbers on artificial structures in coastal waters, where they develop into dense concentrations of jellyfish-producing polyps.[107]

Management Considerations for a Rights of Nature System

It is clear from these brief tours of mining in the United States—coal mining; hardrock mining; oil and gas drilling, both onshore and offshore; and deep-sea mining—that there is no way to undertake any of these forms of mining without extensive environmental repercussions, which are completely at odds with any Rights of Nature system. This situation is especially true of a robust Rights of Nature paradigm, such as we propose, in which Nature is the primary rights-holder.

A cursory glance at the Rights of Nature documents underscores the obvious: mining in any modern form is completely incompatible with a culture that respects the Earth, Nature's Laws of Reciprocity, and maintains a commitment to ecological integrity and resilience. Based on past and present evidence, mining corporations—which perform the vast majority of U.S. mining—can reliably be expected to put corporate profits above environmental requirements, costs of restoration, or any other societal benefit.

How can modern, large-scale mining be squared with a Rights of Nature system? The answer: It cannot be. Mining, especially large-scale industrial mining, must stop. It is too destructive of the world's ecosystems and human cultures. However, ending mining does not have to be an instantaneous and chaotic process. It is well within American technical and political expertise to design a mining slowdown, beginning with a program to extract metals found in slag heaps, rubble piles, abandoned pits, and landfills. This first phase would then lead to a final end to new mining operations, and closure of existing operations, including retraining workers for other jobs.

The second, concurrent, phase is restoration to the best extent possible. That does not mean restoration to the best extent *as currently practiced,* because that falls far below the possible. It requires a deep commitment to restoration research, funding, training, and implementation—again, all well within current capabilities and funding horizons.

The third prong of ending mining is finding new ways to recycle metals, create recycling markets, and reduce need. Already half of the U.S. use of copper comes from recycled copper, even with little or no focus on recycling and efficient use.[108]

In brief, here are several interrelated programs that, under a Rights of Nature system, where Nature itself is the primary rights holder, the United States must design, put in place, and implement:

1. Pass federal laws that design a program and a framework to end all mining, phasing it out on an accelerated, orderly schedule. The statutes must begin with a determination of which small-scale, low-technology mining, if any, is compatible with the Rights of Nature and could be allowed in limited quantities in restricted locations, with controls, monitoring, and requirements for immediate cessation if ecosystem harm results that degrades biological functioning.

   Under this standard, "recreational" suction-dredge gold mining would be prohibited, for example. It uses what is essentially a gas-powered vacuum on a raft to pull up gravels from the river floor to sluice for gold. It is small-scale but very destructive of riverine habitat, especially salmon habitat, on fragile rivers and streams in California, Oregon, Washington, Idaho, and other Western states.[109]

2. Begin with a secondary-use program to extract metals or other resources from slag heaps, waste pits, slurry ponds, landfills and the like for minerals that could be processed from them but which were disregarded in the initial mining cycle. This project will only be allowed in regions where doing so will not increase the pollution, contamination, and devastation of the mined area.

3. Pass federal statutes setting up an integrated, nationwide system of recycling mining infrastructure, including creation of and stimulus for, reuse and resale of recycled materials. Some metals, such as copper, are already fairly intensively recycled.

4. Pass federal statutes authorizing and designing a massive, focused, restoration program for all mining sites. This must begin with dedicating immense amounts of funding to government, semi-government, and university reclamation and restoration

research. These funds must be distributed nationwide, to be used locally in programs tailored to study both the needs of ecosystems suffering from mining damage and the most effective methods of undertaking restoration. Require state laws that implement these statutes in every state with mining and mining infrastructure.

5. Paying for abandoned oil and gas well reclamation is a special problem. This may require federal legislation and a generous grant program for support of state reclamation efforts. Funding could come from stiff driller/producer fees and bonding requirements, matched by an equal amount from general fund federal tax monies, and an oil and gas user tax dedicated to restoration. This three-way mix of funds would ensure the monies are not too closely tied to the oil and gas industry and thus liable to corrupt the reclamation and regulation processes.
6. Earmark restoration research funding from general, federal sources, or a broad-based user tax, in addition to fees from mining companies or oil and gas producers. Creating a broad-based funding source will avoid corruption in scientific findings.
7. Legally structure restoration programs that, although carefully tailored to local areas, follow a relatively similar pattern. The following must be required of states using federal monies for their restoration programs:
    a. Restore the outer areas of a mining or drilling operation first: the staging areas, roads and equipment, storage areas, slurry ponds, coal ash heaps, and slag heaps.
    b. Restore the most damaged areas in a "ring" fashion, with the outer edges of the most damaged areas first, progressively moving inward to the most devastated regions, where the mining was both intensive and extensive, and where biological activity was the most completely disrupted.
    c. Prioritize, via the restoration research and programs, especially sensitive areas such as coastal regions and their estuaries, as well as mountains supplying the headwaters for regional watercourses that provide potable water critical to human well-being. All sites are to be fully restored, but priority needs to be given to the most sensitive areas.
    d. Require permanent, legal protections for all reclaimed sites to ensure they become special focus areas devoted to nurturing and strengthening local keystone species and their habitats. This will aid in recovery of full ecological processes, function, and resilience.
8. Create a public–private liability structure through a federal statute in order to ensure restoration research and groundwork is extensive, thorough, and fairly distributed. This could be modeled, in part, on the Comprehensive Environmental Response, Compensation and Liability Act (the Superfund law). However, given the frequent problem of abandoned mines and wells, and the often-repeated history of corporations abandoning contaminated sites to protect their profit structure, government monies and initiative must be the centerpiece of this statutory framework.
9. Pass federal laws allowing and encouraging citizen lawsuits to monitor and enforce all restoration plans and activities. To make an effective program of citizen oversight and enforcement, new laws must:

a. Shield the conservation groups from attorneys' fees in lawsuits to protect, expand, create, or enforce activities in restoration programs.
b. Provide a federal grant program, with monies from corporate fines levied on mining companies to cover true costs of restoration, as well as those who have abandoned mines but are still solvent. These fines should be mixed with broad user fees for mineral consumption to avoid corruption in providing grants to help fund the watchdog process. Watchdog groups can apply to this program for research, restoration monitoring, community rebuilding, and lawsuit preparation costs.
c. Allow such groups to sue in the name of the devastated watershed, mountain, or other mined areas for recovery of its own ecosystem function. This is much more direct and sensible than forcing groups to sue based on their uses of the area.

As all these programs are put into place, the United States will have finally taken the first steps under the Rights of Nature system to end the colossal degradation of land, waterways, and sea with which mining has blanketed the country during the last century and a half.

Endnotes

1. [Ecuadorian] Constitution, full language: http://pdba.georgetown.edu/Constitutions/Ecuador/english08.html 2 (accessed June 13, 2016).
2. *Ibid.*
3. *Ibid.*
4. Language of the Bolivian law (complete text): http://www.worldfuturefund.org/Projects/Indicators/motherearthbolivia.html (Chapter III, Article 7, I (1)) (accessed June 13, 2016).
5. *Ibid.* (Article 7, I(7)).
6. *Ibid.* (Article 7, I(5)).
7. *Ibid.* (Article 8, Section 2).
8. Universal Declaration of Rights of Mother Earth: Global Alliance for the Rights of Nature. https://therightsofnature.org/universal-declaration/ (Article 2 (1)) (accessed June 13, 2016).
9. *Ibid.* (Article 1, (7)).
10. *Ibid.* (Article 2).
11. *Ibid.* (Article 3 (1) and (2) (a–g).
12. The foregoing discussion is based on: (1) Sean Patrick Adams. The US Coal Industry in the Nineteenth Century. https://eh.net/encyclopedia/the-us-coal-industry-in-the-nineteenth-century-2/ (accessed June 13, 2016); and (2) History of coal mining in the United States. https://en.wikipedia.org/wiki/History_of_coal_mining_in_the_United_States (accessed June 13, 2016).
13. (1) Powder River Basin. http://www.sourcewatch.org/index.php/Powder_River_Basin (accessed June 13, 2016); and (2) Existing U.S. Coal Mines. http://www.sourcewatch.org/index.php/Existing_U.S._Coal_Mines (accessed June 13, 2016).
14. Jeff Biggers. Black Cross Movement Calls Out Dangerous Navajo Coal Mines (Photos). http://www.huffingtonpost.com/jeff-biggers/pattern-of-violations-bla_b_778754.html (accessed June 13, 2016).
15. Plundering Appalachia: The Tragedy of Mountaintop-Removal Coal Mining. http://www.plunderingappalachia.org/theissue.htm (accessed June 13, 2016).

16. (1) Total U.S. Energy Production Increases for Sixth Consecutive Year. http://www.eia.gov/todayinenergy/detail.cfm?id=25852 (accessed June 13, 2016); and (2) Annual Coal Report. http://www.eia.gov/coal/annual/ (accessed June 13, 2016).
17. Matt McGrath. Obama Climate Initiative: Supreme Court Calls Halt. http://www.bbc.com/news/science-environment-35538350 (accessed February 9, 2016).
18. Surface Mining Control and Reclamation Act of 1977 (Public Law 95-87). http://www.osmre.gov/lrg/docs/SMCRA.pdf (accessed August 1, 2016).
19. The preceding four paragraphs are based on: (1) Office of Surface Mining Reclamation and Enforcement. http://www.osmre.gov/ (accessed June 13, 2016).
20. Federal Water Pollution Control Act, as Amended by the Clean Water Act of 1977. http://www.google.com/search?q=clean+water+act+&hl=en&gbv=2&oq=clean+water+act+&gs_l=heirloom-serp.12..0l10.367275.374458.0.376495.7.7.0.0.0.0.152.731.3j4.7.0....0...1ac.1.34.heirloom-serp..0.7.731.TgZo7OVGLhI (accessed June 13, 2016).
21. 1990 Clean Air Act Amendment Summary. https://www.epa.gov/clean-air-act-overview/1990-clean-air-act-amendment-summary (accessed June 13, 2016).
22. Endangered Species Act of 1973. https://www.fws.gov/laws/lawsdigest/esact.html (accessed June 13, 2016).
23. National Environmental Policy Act of 1969. http://www.epw.senate.gov/nepa69.pdf (accessed June 13, 2016).
24. The foregoing two paragraphs are based on: Appalachian Voices. Mountaintop Removal Reclamation Fail. http://ilovemountains.org/reclamation-fail/ (accessed June 13, 2016).
25. Coal mine reclamation. http://www.sourcewatch.org/index.php/Coal_mine_reclamation (accessed June 13, 2016).
26. The preceding two paragraphs are based on: Copper mining in Michigan. https://en.wikipedia.org/wiki/Copper_mining_in_Michigan (accessed June 13, 2016).
27. The preceding four paragraphs are based on: Summary of Gold Mining Techniques in Western United States 1842–1996. https://www.tchistory.org/tchistory/more_gold.htm (accessed June 13, 2016).
28. The foregoing five paragraphs are based on: The world's deepest, biggest and deadliest open pit mines. http://www.losapos.com/openpitmines (accessed June 15, 2016).
29. Copper mining in Arizona. https://en.wikipedia.org/wiki/Copper_mining_in_Arizona (accessed June 15, 2016).
30. Copper mining in the United States. https://en.wikipedia.org/wiki/Copper_mining_in_the_United_States (accessed June 15, 2016).
31. Safford Mine. http://www.fcx.com/operations/USA_Safford.htm (accessed June 15, 2016).
32. (1) Rayan Randazzo. Rio Tinto sees potential in Resolution Copper mine near Superior despite low prices. http://www.azcentral.com/story/money/business/energy/2015/12/10/rio-tinto-sees-potential-resolution-copper-mine/77052144/ (accessed June 15, 2016); and (2) Emily Bregel. Superior mine opponents say they will continue to protest. http://tucson.com/news/local/superior-mine-opponents-say-they-will-continue-to-protest/article_edc84b51-dfab-5932-b4f7-8525df9518fc.html (accessed June 15, 2016).
33. Save the Scenic Santa Ritas. http://www.scenicsantaritas.org/key-facts (accessed June 15, 2016).
34. John Myers. Range lawmakers want Twin Metals access approved. http://www.bemidjipioneer.com/news/region/3940811-range-lawmakers-want-twin-metals-access-approved (accessed June 15, 2016).
35. The world's deepest, biggest and deadliest open pit mines. *op. cit.*
36. (1) Susan Montoya Bryan and Ellen Knickmeyer. Animas River disaster: Colorado mine waste spill called low-risk; downstream towns still concerned. http://www.mercurynews.com/nation-world/ci_28621087/animas-river-disaster-epa-chief-takes-responsibility-colorado (accessed June 15, 2016); and (2) 2015 Gold King Mine waste water spill. https://en.wikipedia.org/wiki/2015_Gold_King_Mine_waste_water_spill (accessed June 15, 2016).

37. Liquid Assets 2000: America's Water Resources at a Turning Point. https://nepis.epa.gov/Exe/ZyPDF.cgi/20004GRW.PDF?Dockey=20004GRW.PDF (accessed June 15, 2016).
38. General Mining Act of 1872. https://en.wikipedia.org/wiki/General_Mining_Act_of_1872 (accessed June 15, 2016).
39. The 1872 General Mining Law. http://www.nevermined.org/reform.shtml (accessed June 15, 2016).
40. Federal Land Policy and Management Act of 1976: Public Law 94-579, 94th Congress. http://www.blm.gov/flpma/FLPMA.pdf (accessed June 15, 2016).
41. Lacey McCormick. 8 Hard Truths about Hardrock Mining. http://blog.nwf.org/2015/08/8-hard-truths-about-hardrock-mining/ (accessed June 15, 2016).
42. John F. Seymour. Hardrock mining and the environment: Issues of federal enforcement and liability. *Ecology Law Quarterly*, 31 (2004):795–956.
43. Federal Water Pollution Control Act, as Amended by the Clean Water Act of 1977. https://www3.epa.gov/npdes/pubs/cwatxt.txt (accessed June 15, 2016).
44. Comprehensive Environmental Response, Compensation, and Liability Act of 1980 "Superfund." http://www.epw.senate.gov/cercla.pdf (accessed June 15, 2016).
45. *Ibid.*
46. John F. Seymour. Hardrock mining and the environment: Issues of federal enforcement and liability. *op. cit.* (pp. 802–803).
47. Act of 1980 "Superfund." http://www.epw.senate.gov/cercla.pdf (accessed June 15, 2016).
48. John F. Seymour. Hardrock mining and the environment: Issues of federal enforcement and liability. *op. cit.*
49. (1) Philip L. Verplanck (ed.). Understanding contaminants associated with mineral deposits. *U.S. Geological Survey Circular* 1328 (2008):1–96 (accessed June 15, 2016); and (2) Cleaning up Contamination in Hard Rock Mine Lands http://toxics.usgs.gov/highlights/amli.html (accessed June 15, 2016).
50. History of the petroleum industry in the United States. https://en.wikipedia.org/wiki/History_of_the_petroleum_industry_in_the_United_States (accessed June 15, 2016).
51. The preceding two paragraphs are based on: Pennsylvania oil rush. https://en.wikipedia.org/wiki/Pennsylvania_oil_rush (accessed June 15, 2016).
52. History of the petroleum industry in the United States. *op. cit.*
53. *Ibid.*
54. (1) Natural Gas Act of 1938. https://en.wikipedia.org/wiki/Natural_Gas_Act_of_1938 (accessed June 15, 2016); and (2) The History of Regulation. http://naturalgas.org/regulation/history/ (accessed June 15, 2016).
55. The Natural Gas Policy Act of 1978. https://ballotpedia.org/Natural_Gas_Policy_Act_of_1978 (accessed June 15, 2016).
56. Natural Gas Wellhead Decontrol Act of 1989. http://www.eia.gov/oil_gas/natural_gas/analysis_publications/ngmajorleg/ngact1989.html (accessed June 15, 2016).
57. The preceding two sentences are based on: The History of Regulation. *op. cit.*
58. The above discussion is based on: (1) Oil and gas law in the United States. https://en.wikipedia.org/wiki/Oil_and_gas_law_in_the_United_States (accessed June 15, 2016); and (2) Oil and gas regulation in the United States: Overview. http://us.practicallaw.com/9-525-1545?source=relatedcontent (accessed June 15, 2016).
59. Mineral Leasing Act of 1920 as Amended. http://www.blm.gov/style/medialib/blm/ut/vernal_fo/lands___minerals.Par.6287.File.dat/MineralLeasingAct1920.pdf (accessed June 16, 2016).
60. *Ibid.* p. 50 (accessed June 16, 2016).
61. The foregoing discussion of the Bureau of Land Management is based on: (1) Who Owns The West? Oil and Gas Leases. http://www.ewg.org/oil_and_gas/part2.php (accessed June 16, 2016); and (2) Oil and gas regulation in the United States: Overview. http://us.practicallaw.com/9-525-1545?source=relatedcontent (accessed June 16, 2016).

62. Jennifer Oldham. In the Birthplace of U.S. Oil, Methane Gas is Leaking Everywhere. http://www.bloomberg.com/news/articles/2016-06-20/in-the-birthplace-of-u-s-oil-methane-gas-is-leaking-everywhere?utm_content=buffer6ed02&utm_medium=social&utm_source=twitter.com&utm_campaign=buffer (accessed August 15, 2016).
63. The preceding two paragraphs are based on: Oil Bust Leaves States with Massive Well Cleanup. http://abcnews.go.com/Politics/wireStory/oil-bust-leaves-states-massive-cleanup-39969993 (accessed August 1, 2016).
64. Energy Policy Act of 2005. http://energy.gov/sites/prod/files/2013/10/f3/epact_2005.pdf (accessed June 16, 2016).
65. Title XIV of the Public Health Service Act Safety of Public Water Systems (Safe Drinking Water Act). http://www.epw.senate.gov/sdwa.pdf (accessed June 15, 2016).
66. This paragraph is based on: (1) A Brief History of Hydraulic Fracturing. http://www.eecworld.com/services/258-a-brief-history-of-hydraulic-fracturing (accessed June 16, 2016); and (2) Fracturing Responsibility and Awareness of Chemicals Act. https://en.wikipedia.org/wiki/Fracturing_Responsibility_and_Awareness_of_Chemicals_Act (accessed June 16, 2016).
67. How Is Shale Gas Produced? http://energy.gov/sites/prod/files/2013/04/f0/how_is_shale_gas_produced.pdf (accessed June 16, 2016).
68. It's official: New York bans fracking. https://www.rt.com/usa/270562-new-york-fracking-ban/ (accessed June 16, 2016).
69. This overall paragraph is based on: Oil and gas regulation in the United States: Overview. *op. cit.* (accessed June 16, 2016).
70. David R. Lyon, Ramón A. Alvarez, Daniel Zavala-Araiza, and others. Aerial surveys of elevated hydrocarbon emissions from oil and gas production sites. *Environmental Science & Technology*, 50 (2016):4877–4886.
71. COMPLAINT and request for investigation of fraud, waste and abuse by a high-ranking EPA official leading to severe underreporting and lack of correction of methane venting and leakage throughout the US natural gas industry. http://www.ncwarn.org/wp-content/uploads/EPA-OIG_NCWARN_Complaint_6-8-16.pdf (p. 5) (accessed June 16, 2016).
72. Earthworks. Interactive Map Shows Where Toxic Air Pollution from Oil and Gas Industry Is Threatening 12.4 Million Americans. http://www.ecowatch.com/interactive-map-shows-where-toxic-air-pollution-from-oil-and-gas-indus-1891173774.htmls (accessed June 16, 2016).
73. (1) New Study: Hydraulic Fracturing Faces Growing Competition for Water Supplies in Water-Stressed Regions. http://www.ceres.org/press/press-releases/new-study-hydraulic-fracturing-faces-growing-competition-for-water-supplies-in-water-stressed-regions (accessed June 16, 2016); and (2) Hydraulic Fracturing & Water Stress: Water Demand by the Numbers. http://www.ceres.org/resources/reports/hydraulic-fracturing-water-stress-water-demand-by-the-numbers (accessed June 16, 2016).
74. Suzanne Goldenberg. A Texan tragedy: Ample oil, no water. https://www.theguardian.com/environment/2013/aug/11/texas-tragedy-ample-oil-no-water (accessed June 16, 2016).
75. New Study: Hydraulic Fracturing Faces Growing Competition for Water Supplies in Water-Stressed Regions. Supplies in Water-Stressed Regions. *op. cit.*
76. Assessment of the Potential Impacts of Hydraulic Fracturing for Oil and Gas on Drinking Water Resources. https://www.epa.gov/sites/production/files/2015-07/documents/hf_es_erd_jun2015.pdf (accessed June 16, 2016).
77. *Ibid.* (p. ES-6).
78. *Ibid.* (p. ES-6).
79. Environmental impact of hydraulic fracturing in the United States. https://en.wikipedia.org/wiki/Environmental_impact_of_hydraulic_fracturing_in_the_United_States (accessed June 16, 2016).
80. How is hydraulic fracturing related to earthquakes and tremors? https://www2.usgs.gov/faq/categories/10132/3830 (accessed June 16, 2016).
81. Mike Soraghan. Earthquakes: Disconnects in Public Discourse around 'Frackin' Cloud Earthquake Issue. http://www.eenews.net/stories/1059963291 (accessed June 16, 2016).

82. William Yardley. Oklahoma takes action on fracking-related earthquakes—But too late, critics say. http://touch.latimes.com/#section/-1/article/p2p-86080332/ (accessed June 16, 2016).
83. Kyle Murray. Class II Saltwater Disposal for 2009–2014 at the Annual-, State-, and County-Scales by Geologic Zones of Completion, Oklahoma. https://www.researchgate.net/publication/288833796_Class_II_Saltwater_Disposal_for_2009-2014_at_the_Annual-_State-_and_County-_Scales_by_Geologic_Zones_of_Completion_Oklahoma?channel=doi&linkId=5685967f08ae1975839520d4&showFulltext=true (accessed June 16, 2016).
84. Lorraine Chow. Lawsuit Filed over Oklahoma's 'Fracking' Earthquakes as Its Third Largest Quake Is Felt in 7 Other States. http://ecowatch.com/2016/02/17/lawsuit-filed-oklahoma-earthquakes/ (accessed June 16, 2016).
85. Where Are Liquids Pipelines Located? http://www.pipeline101.com/where-are-pipelines-located (accessed June 16, 2016).
86. The History of Pipelines. http://www.pipeline101.com/the-history-of-pipelines (accessed June 16, 2016).
87. Joe Carroll. Kinder Cancels Gas Pipeline, Demands Collateral on Slump. http://www.bloomberg.com/news/articles/2016-04-20/kinder-morgan-profit-drops-under-weight-of-energy-market-rout (accessed June 16, 2016).
88. Hazardous Liquid Pipeline Safety Act of 1979. http://www.phmsa.dot.gov/staticfiles/PHMSA/DownloadableFiles/Files/Pipeline/Hazardous%20Liquid%20Pipeline%20Safety%20Act%20of%201979.pdf (accessed June 16, 2016).
89. The foregoing paragraph is based on: A brief history of federal pipeline safety laws. http://pstrust.org/about-pipelines1/regulators-regulations/a-brief-history-of-federal-pipeline-safety-laws/ (accessed June 16, 2016).
90. Regulatory Development Support Services Pipeline Safety: Safety of Hazardous Liquid Pipelines Notice of Proposed Rulemaking (NPRM) http://www.phmsa.dot.gov/staticfiles/PHMSA/DownloadableFiles/Files/2137_AE66_NPRM_Hazardous_Liquids_NPRM_RIA.pdf (accessed June 16, 2016).
91. Oil and gas regulation in the United States: Overview. http://us.practicallaw.com/9-525-1545?source=relatedcontent (accessed June 16, 2016).
92. Oil & Gas Energy Program. http://www.boem.gov/Oil-and-Gas-Energy-Program/ (accessed June 17, 2016).
93. Submerged Lands Act. http://www.boem.gov/uploadedfiles/submergedla.pdf (accessed June 17, 2016).
94. Title 43 > Chapter 29 > Subchapter III—Outer Continental Shelf Lands. http://www.boem.gov/Outer-Continental-Shelf-Lands-Act/ (accessed June 17, 2016).
95. Oil Pollution Act of 1990. https://en.wikipedia.org/wiki/Oil_Pollution_Act_of_1990 (accessed June 17, 2016).
96. The preceding two paragraphs are based on: BOEM Governing Statutes. http://www.boem.gov/Governing-Statutes/ (accessed June 17, 2016).
97. (1) Protecting Our Ocean and Coastal Economies: Avoid Unnecessary Risks from Offshore Drilling. https://www.nrdc.org/sites/default/files/offshore.pdf (accessed June 17, 2016); and (2) Prudhoe Bay oil spill. https://en.wikipedia.org/wiki/Prudhoe_Bay_oil_spill (accessed June 17, 2016).
98. The preceding two paragraphs are based on: (1) Affected Gulf Resources. http://www.gulfspillrestoration.noaa.gov/affected-gulf-resources/ (accessed June 17, 2016); and (2) *Deepwater Horizon* oil spill. https://en.wikipedia.org/wiki/Deepwater_Horizon_oil_spill (accessed June 17, 2016).
99. Alon Harish. BP Oil Spill: Two Years Later, Dispersants' Effects Still a Mystery. http://abcnews.go.com/US/bp-oil-spill-years-dispersants-effects-mystery/story?id=16727991#.T_sApRwU64A (accessed July 9, 2012).
100. The Reorganization of the Former MMS [Minerals Management Service]. http://www.bsee.gov/About-BSEE/BSEE-History/Reorganization/Reorganization/ (accessed June 17, 2016).

101. Trustees Settle with BP for Natural Resource Injuries to the Gulf of Mexico. http://www.gulfspillrestoration.noaa.gov/2016/04/trustees-settle-with-bp-for-natural-resource-injuries-to-the-gulf-of-mexico/ (accessed June 17, 2016).
102. The preceding two paragraphs are based on: Protecting Our Ocean and Coastal Economies: Avoid Unnecessary Risks from Offshore Drilling. *op. cit.*
103. U.S. Backs Away from Offshore Arctic Drilling. http://www.npr.org/sections/thetwoway/2015/10/16/449286988/u-s-backs-away-from-offshore-arctic-drilling (accessed June 17, 2016).
104. This paragraph is based on: (1) David Shukman. Deep Sea Mining 'Gold Rush' Moves Closer. http://www.bbc.co.uk/news/science-environment-22546875 (accessed June, 17, 2016); and (2) John Vidal. Arctic sea ice levels to reach record low within days. http://www.guardian.co.uk/environment/2012/aug/23/arctic-sea-ice-record-low (accessed June 17, 2016).
105. *Ibid.*
106. *Ibid.*
107. Carlos M. Duarte, Kylie A. Pitt, Cathy H. Lucas, and others. Is global ocean sprawl a cause of jellyfish blooms? *Frontiers in Ecology and the Environment*, 11 (2013):91–97.
108. Mining Arizona. http://arizonaexperience.org/land/mining-arizona (accessed June 17, 2016).
109. California Department of Fish and Game: Suction Dredging Permitting Program. http://www.icmj.com/UserFiles/file/recent-news/Review-of-Available-Suction-Dredging-Studies.pdf (accessed August 13, 2016).

16

International Trade: Moving Goods and People

Moving goods and people across land and sea and trading goods one culture has in exchange of those of another are activities as old as human culture itself. Trade has taken place in every form of transportation available to humans, from walking to carrying goods on the backs of animals in caravans, to boats and ships, bicycles, cars, trucks, and airplanes.

Why is a chapter on international trade necessary in a book on environmental sustainability and the Rights of Nature? It is essential because international trade has now burgeoned so dramatically that the trade itself jeopardizes ocean ecosystems, land ecosystems, and human communities in a variety of ways worldwide. Dominated by corporations seeking ever more intensive manufacturing and controlled patterns of consumption, combined with the greatest use of natural resources at the lowest costs, international trade now risks many important environmental and cultural values. It must, therefore, be returned to a more personal exchange of goods among cultures, rather than wholesale, international systems devised solely for the economic benefit of profit-oriented corporate networks.

Rights of Nature and Community

Ecuador

Rights of Nature must guide any discussion of the future of international trade. The Ecuadorian Constitution, giving Nature "the right to integral respect for its existence and for the maintenance and regeneration of its life cycles, structure, functions and evolutionary processes,"[1] describes the overall right. Article 73 then states, "The State shall apply preventive and restrictive measures on activities that might lead to the extinction of species, the destruction of ecosystems and the permanent alteration of natural cycles."[2] Not only that, but Article 72 states simply, "Nature has the right to be restored."[3]

The rights of community sections of Ecuador's Constitution contains critically important language about local or national control, which is being ever more insidiously undermined by expanding international trade. It behooves us to examine the language of community rights closely, keeping the serious problems of corporate-oriented international trade in mind. Among the greatest changes wrought by adherence to a 19th-century "free trade" principle is the weakening of nations' ability to protect ecosystem sustainability, resilience, and productivity; natural resources; local farming; and the cultural methods by which goods are produced and marketed.

The Ecuadorian Constitution contains a chapter entitled "Rights of the good way of living."[4] Article 13 states, "Persons and community groups have the right to safe and permanent access to healthy, sufficient and nutritional food, preferably produced locally and in keeping with their various identities and cultural traditions. The Ecuadorian State shall promote food sovereignty."[5] These community rights cut directly across the goals of modern, corporatized international trade.

In the "Healthy environment" section, the Constitution guarantees "The right of the population to live in a healthy and ecologically balanced environment that guarantees sustainability and the good way of living is recognized."[6] This includes both biodiversity and protection of Ecuador's genetic assets, as well as preventing environmental damage, as "matters of public interest."

The Constitution also has a lengthy chapter guaranteeing the rights of communities, peoples, and nations. These range from the right to maintain and develop ancestral traditions and forms of social organization to keeping ownership of their community and ancestral lands. Communities and peoples also have the right to "keep and promote their practices of managing biodiversity and their natural environment."[7] Most importantly, the Constitution upholds their right to protect and develop collective knowledge, sciences, ancestral wisdom, "genetic resources that contain biological diversity and agricultural biodiversity,"[8] and the right to restore plants, animals, minerals, and ecosystems.

The Constitution does not stop there. In laying out the "Development of Systems," the Constitution's language makes food sovereignty "a strategic objective and an obligation of the State," so that people "achieve self-sufficiency with respect to healthy and culturally appropriate good on a permanent basis."[9] This includes bolstering diversification and introducing techniques of organic farming. The overall thrust of this section is to support small and medium-size producers and set up equitable, rural–urban supply networks so urban residents have access to locally grown foods. This includes creating and promoting policies of redistribution that will enable small farmers access to land, water, and materials needed for farming. All these goals, objectives, and rights are jeopardized by the enormous international trade and resource extraction system that drastically affects most countries of the world.

Bolivia

The Bolivian Rights of Nature law contains direct language on Nature's right to the diversity of life, to "functionality of the water cycle" and its protection from pollution. Nature's rights include the "right to maintenance or restoration of the interrelationship, interdependence and functionality of the components of Mother Earth in a balanced way for the continuation of their cycles and reproduction of their vital processes."[10]

Mother Earth also has the right to "maintain the integrity of living systems and natural processes that sustain them."[11] In addition to the right of restoration, Bolivia provides Mother Earth with the right of pollution-free living, "to preservation of any of Mother Earth's components from contamination."[12] That is, "The right to maintenance or restoration of the interrelationship, interdependence, complementarity and functionality of the components of Mother Earth in a balanced way for the continuation of their cycles and reproduction of their vital processes."[13] This is an essential provision against which to balance the current overuse of international trade, which is degrading both ecosystems and human communities worldwide.

Bolivia devotes an entire subchapter of its law to State Obligations and Societal Duties to protect Mother Earth.[14] These range from requirements to prevent damaging human activities to developing "balanced forms of production and patterns of consumption" that safeguard the "integrity of the cycles, processes and vital balance of Mother Earth."[15] One of the most obvious aspects of today's international trade is the extent to which it is dominated by corporate exploitation of natural resources, manufacturing of consumer products with the sole purpose of producing profits—of which they can never have enough. As such, the current system is clearly set up to maximize international consumption, which is the polar opposite of this Rights of Nature provision.

Global Alliance for the Rights of Nature

The Universal Declaration of the Rights of Mother Earth, growing out of the Bolivian and Ecuadorian laws, is even more detailed. It begins in Article 1 with the statement, "Mother Earth is a living [sentient] being."[16] The definition of that is wide-ranging: "Mother Earth is a unique, indivisible, self-regulating community of interrelated beings that sustains, contains and reproduces all beings."[17] The Declaration extends Mother Earth's inherent rights to "all the beings of which she is composed,"[18] and also sets the framework for dealing with conflict: "The rights of each being are limited by the rights of other beings and any conflict between their rights must be resolved in a way that maintains the integrity, balance and health of Mother Earth."[19]

Mother Earth's rights, like those of any other being, include first and foremost the right to life and to exist; to be respected; "to continue their vital cycles and processes free from human disruptions;" to water and clean air; to be free from contamination, pollution, and toxic waste; and the right to "full and prompt" restoration if the rights are violated by human beings.[20]

The Declaration's language is especially important to the discussion of international trade because the Global Alliance of the Rights of Nature, which produced it, also set up the first International Rights of Nature Tribunal. Established to begin hearing cases concerning ecological degradation and collapse, the Tribunal is currently outside the existing international legal framework. It is, nevertheless, a highly important template for making international trade sustainable and nature-centered by (in part) setting up a tribunal with enforceable laws and the power to render verdicts, which must be obeyed by all countries in the international community, protecting the Earth's ecosystems and their right to flourish.

Early Trading Networks

Until the early Modern era, all trade networks were regional, separated by impenetrable continents or vast seas. Nevertheless, some ancient networks were exceedingly impressive. Frequently, though not always, trade and conquest—colonization—went hand in hand. This was especially true if the trade networks spanned great distances.

Ancient Trade Networks

Ancient trade networks created by Polynesian explorers crossed wide and otherwise unknown seas. Recent research confirms that a Polynesian trading network existed between Hawaii, Tahiti, the Tuamotus, and other Polynesian islands colonized by the Polynesian people. These staggeringly long voyages of 2,500 miles or more across the Pacific between Hawaii, Tahiti, and other islands probably began, or at least were standardized, by about 900 BCE and lasted until the 15th century.[21]

Not only that, but the Polynesians almost certainly traded with peoples on the west coast of South America, and perhaps colonized it as well. Discoveries of Polynesian-originated chicken bones in Chile make this a highly likely scenario. The bones date from the early 14th century. In addition, archeological evidence shows the Polynesians grew the sweet potato in the Cook Islands a thousand years ago; but the sweet potato originates in the

Americas. The likelihood is that Polynesians traded, and perhaps lived, among the west coast peoples of South America, especially Peru and Chile.[22]

The Viking peoples of Scandinavia, who traveled both east and west from their homelands, created two other long-distance trade networks. Swedish Vikings, beginning in the 9th and 10th centuries, sailed east to establish trade networks with Baghdad and then with far-flung Constantinople (now Istanbul). As part of their trade network, the Swedes founded or expanded Kiev and Novgorod, the capital of Ukraine. They carried furs from Russian trading partners to exchange for Arab silver from Baghdad. Once the Baghdad trade declined as Arab silver mines began to fail, the Vikings turned to Constantinople. They formed the backbone of the Byzantine Varangian Guard in Constantinople, the personal guard of the Emperor, and traded Baltic amber, furs, and slaves for Byzantine silk and other goods. Their trading networks spanned more than 2,000 miles.[23]

Vikings from Norway and Denmark sailed west, founding the colony of Iceland in the late 10th century. Less than a hundred years later, they founded the colony of Greenland, and lived there for nearly 500 years before returning to Iceland and Norway. Trade between all these regions consisted of many goods, ranging from furs, walrus ivory, and timber to Baltic amber, wool, and wine.[24]

Ever restless, western Vikings continued raiding, trading, and settling in Ireland, northeastern England, and Scotland. Their (recorded) presence in the British Isles may have begun in 789 CE, but more famously with the raid of the Lindisfarne monastery on England's east coast in 793 CE. The Vikings raided along the Irish and Scottish coasts as well, plundering Iona Abbey three times in the early 9th century. Finally, in the mid-9th century, they began to settle in the British Isles and Ireland, founding Dublin and Limerick, among other cities. They also expanded local villages and rural hamlets, as well as the city of York, which they made their English headquarters. In addition, they ruled the Orkney and Shetland Islands of northern Scotland for nearly five centuries.[25]

Launching from Greenland, the Vikings also founded a three-year colony in North America, at L'Anse aux Meadows, Nova Scotia, around the year 1000 CE. Even after abandoning the colony and retreating to Greenland, due probably to a small number of colonists and large indigenous populations with whom they clashed, the Vikings continued to trade with the Native peoples for fur, timber, and grapes.[26]

The First International Trade Route: The Spanish Empire

It was, however, the Spanish who founded the first truly international trade network in the 16th century. The Spanish conquered the Aztec peoples of Mexico, beginning in 1519. Subsequent conquests in Mexico against the Maya people and down into Guatemala expanded Spanish control in Central America, but the real leap in the Spanish empire came with the conquest of Peru and the Inca Empire. This began in 1532, but took decades to complete. The Spanish claimed control over the greater Andes region by 1542, and continued spreading in South America, exploring and conquering Paraguay and parts of what is now Argentina in the 16th century.[27]

Conquering and settling part of the Philippines allowed the Spanish to expand the long-standing, local-trade networks between native Philippine chieftains, China, Japan, and other southeast Asian countries. But their subsequent international trade network relied on a critically important discovery by Andres de Urdaneta, who sailed as part of the Philippines voyage by Miguel Lopez de Legazpi in 1565. Urdaneta discovered the route, via favorable trade winds, which would allow ships to return from Manila to New Spain, crossing the entire vast Pacific Ocean—a voyage of 12,000 miles between Manila and

Acapulco.[28] (New Spain was a Spanish colony north of the Isthmus of Panama, established after the conquest of the Aztec Empire in 1521. After 1535, the capital of New Spain was modern-day Mexico City.[29])

Legazpi's expedition in 1565 resulted in the first Spanish settlement in the Philippines.[30] By 1570, Legazpi's men had begun trading with, and ultimately conquering, the peoples of the Manila area—an event, combined with the discovery of Urdaneta's route, which would have momentous consequences for international trade. The 250-year Manila galleon trade was about to begin. It ran from 1565 to 1815, usually consisting of one gigantic galleon per year running in each direction (Figure 16.1).

The trade was fueled by silver. The Chinese had little interest in Spanish goods, but tremendous desire for Spanish silver. And Spain had silver. The Cerro Rico ("rich hill") mine, in what is now Potosi, Bolivia, discovered in the mid-16th century, turned out to be one of the richest silver mines in the world. It, and other South American mines under Spanish control, made the Manila trade what it became: the first international trade system. Potosi, founded as a mining town in 1545, was at the base of Cerro Rico; the population expanded to 200,000 at the peak of mining. About 60 percent of all the silver mined in the world in the second half of the 16th century came from Cerro Rico.[31]

Spain sent the silver to Manila via the galleon to pay for the Asian trade goods. Manila was the hub of the trade route because of its location close to China:

> …the Pacific route to China was Spain's exclusive means of direct access to the Chinese marketplace … Manila was the commercial linchpin of this global commercial network…. By 1576 the trade was firmly established, and Chinese junks made Manila a customary port of call … colonial merchants acted as intermediaries between Asian producers and American viceroyalties … Manila collected commodities from a multitude of places. From India came cottons, which in the eighteenth century were an important item of export, second only to Chinese silks. Earthen jars and a wide variety of porcelain arrived from China, alongside rugs from Persia and goods from Japan. Spices like clove, mace, pepper, and cinnamon were brought from the Moluccas, Java and Ceylon…. The galleon also carried gold, jewelry, precious jewels and uncut gems…. Silks in every stage of manufacture and of every variety of weave and pattern formed the most valuable part of their cargoes.[32]

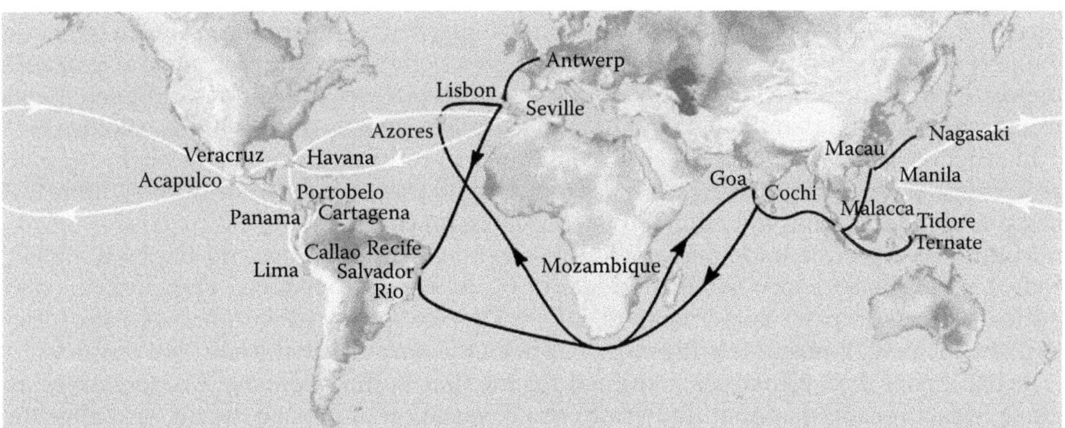

FIGURE 16.1
Map showing main Portuguese (black) and Spanish (white) oceanic trade routes in the 16th century (https://commons.wikimedia.org/wiki/File:16th_century_Portuguese_Spanish_trade_routes.pngs).

Once the galleon docked in Acapulco, the Asian trade goods were carried or shipped through Mexico, Peru, and other South American points. Manila trade goods were also carried over the Isthmus of Panama on mules and, once in Portobelo, were loaded on ships bound for Spain, making the Manila trade truly international.

The Manila voyage was unpredictable, to say the least. It was about 12,000 miles between Manila and Acapulco, and usually took six to nine months, sometimes longer. The galleons were the largest ships of their time, and could carry up to a thousand men—but instead were customarily packed with trade goods. The best and most expensive cabins for passengers were about five feet square, the conditions appalling, overcrowding constant, and rats numerous. Deaths from scurvy (a deficiency of vitamin C) and beriberi (a deficiency of vitamin B_1) were common; the food was poor and dwindled as months went by, even for the wealthiest. Storms in the North Pacific were often relentless. Disastrous shipwrecks with total or great loss of life and cargo were common, especially in the early years; the waters around Guam, the Marianas Islands, and the Philippine archipelago contain several galleon wrecks, and others are known off the coasts of California and Baja.

Nevertheless, the trade flourished, with very lucrative profits for merchants, and the Asian porcelains, silks, gems, and spices highly sought after. The Manila galleon trade was also the sole support of the Spanish colony in the Philippines during the entire time the trade was active.[33]

Modern International Trade

As other trade networks expanded worldwide in the early modern era, trade organizations and compacts made their appearance as part of the mercantilist philosophy. This led to governments imposing wage and price controls, protecting national industries, and promoting the export of finished goods while restricting imports, all in order to protect the nation's interests against that of others. The Dutch East India Company was formed in 1602 to expand and control Dutch trading activities, and several countries formed similar organizations. The engine of the mercantilist philosophy was a desire for wider markets on the part of merchants, fused with a desire to protect one's own country against excessive imports. The English Navigation Act of 1651, for example, prohibited non-European goods from import into England and gave England exclusive right to trade with its colonies.[34]

In reaction to mercantilism, ideals of liberty in production and trade began to gain acceptance, with corresponding pressure to negotiate trade agreements and end trade tariffs, prohibitive customs duties, and similar hostile practices.

In 1860, the first free-trade agreement was signed between France and England, widely considered to be a victory for liberal ideas of trade. It greatly reduced French tariffs and duties. Other European agreements followed. Despite resurgences in occasional protectionism, an ideal of liberal trading rules has been the norm since the late 19th century.

In 1927, more than 20 nations convened for the first World Economic Conference in an effort to recover international trade from the devastation of Europe during and after the First World War. It was the most comprehensive trade agreement written up to that time, and a precursor to the General Agreement on Tariffs and Trade of 1947, which "proved to be the most effective instrument of world trade liberalization, playing a major role in the massive expansion of world trade..."[35]

The Biggest Breakthrough: The Panama Canal

The most ardently sought aid for the international movement of goods since the 16th century was a canal across the Isthmus of Panama. King Charles V of Spain ordered a survey of the area in 1534, searching for a route through the Americas to reduce travel for ships sailing between Spain and Peru. Nothing came of schemes to build a canal or overland route, until the United States built the Panama Railway in 1855. It was a vital trade link and determined the route of the Canal years later. The French began canal construction in 1881, led by Ferdinand de Lesseps, who used the vast fortune he accumulated from building the Suez Canal to finance it. But the project went bankrupt due to poor management, poor design, and a horrific death rate of more than 200 workers per month from yellow fever and malaria, both common in the Panamanian jungles.

American foreign policy maneuvering under President Theodore Roosevelt led to the United States supporting Panamanian rebels seeking independence. Panama declared independence from Colombia in 1903. The United States quickly recognized the new nation and blocked sea lanes against Colombian troops seeking to dissolve the rebellion. The Panamanian ambassador signed the Hay–Bunau–Varilla Treaty with

FIGURE 16.2
The famous Culebra Cut, Panama Canal, 1907. Photograph by H.C. White Co. (https://commons.wikimedia.org/wiki/File:Panama_Canal_under_construction,_1907.jpg).

FIGURE 16.3
SS Kroonland transits the Panama Canal on February 2, 1915. She was the largest passenger ship to that date to transit the canal. Photograph by Underwood & Underwood (https://commons.wikimedia.org/wiki/File:Kroonland_in_Panama_Canal,_1915.jpg).

the United States three days after the new nation gained independence, giving the United States an indefinite right to build and administer the Panama Canal Zone and its defenses.

It took the United States 10 years to construct the Canal, from 1904 to 1914 (Figure 16.2). American engineers took over the dilapidated French equipment, put massive sanitary measures in place to lower the death rate, and hired thousands of workers. Even so, some 5,600 workers died during the Canal construction. The Canal was opened to ship traffic in August 1914. At current prices, the Canal cost the United States about $8.6 billion. It is 48 miles long, and consists of a series of locks and lakes (Figure 16.3). Over the years, the United States built various improvements as use of the canal burgeoned. In 1999, after extensive negotiations, the Panama Canal Zone was handed over to the country of Panama and its newly created Panama Canal Authority.[36]

Free Trade Expands: North American Free Trade Agreement

The North American Free Trade Agreement (commonly referred to as "NAFTA") was a trilateral free trade agreement between Canada, the United States, and Mexico, which went into effect in 1994. Building on the 19th-century philosophy of liberal free trade, the central purpose of the Agreement was to phase out most tariffs on goods traded among the three countries. It was one of many Free Trade Agreements around the world that have been signed in the last 25 years or so between various countries.

Economists generally agree that the North American Free Trade Agreement has worked, more or less, as planned: "NAFTA was designed to promote economic growth by spurring competition in domestic markets and promoting investment from both domestic and foreign sources. It has worked," wrote Gary Clyde Hufbauer and Jeffrey J. Schott, experts at the Peterson Institute for International Economics.[37]

In 2010, the United States had $918 billion in two-way trade with Canada and Mexico, and boosters have sought ways to further bolster economic output by tweaking the Agreement to create a truly seamless international marketplace. Proponents of the North American Free Trade Agreement defy critics by saying domestic problems are not the fault of NAFTA, and the real success lies in an increase in each country's Gross Domestic Product since the pact was implemented.[38]

The Dark Side of the North American Free Trade Agreement

The American balance of trade plunged from a $1.7 billion U.S. surplus in 1993 to a $16.4 billion deficit in 2012—a drastic change that would seem to indicate numerous American economic dislocations due to increased globalization. Many in the field of labor economics criticize NAFTA for placing workers in different countries in direct competition with one another, as companies lengthen supply chains by spreading production across countries to reduce costs—that is, companies seek to manufacture in whichever country has the lowest wages and lowest environmental costs in order to get a leg up in global competitiveness. This is termed "offshoring."[39]

U.S. manufacturing and shipping jobs have declined dramatically, and the North American Free Trade Agreement has also displaced millions of rural Mexican agricultural workers. A central problem is that globalization increases competition, and the losers tend to be the less educated and those non-industrialized countries with less domestic investment in their own economic structures, such as Mexico. Although foreign investment in Mexico tripled subsequent to NAFTA, it did not aid the local economy.[40]

Environmental issues (along with labor concerns) were the largest controversies at the time of NAFTA's implementation. This led to an ancillary agreement—the North American Agreement on Environmental Cooperation. By this agreement, the three countries agreed to study environmental problems, improve protections, and strictly enforce their own environmental laws. The problem, of course, is that it asked nothing new of any country—only reiterated that they comply with their existing environmental protections, which they ought to be doing anyway.

The North American Agreement on Environmental Cooperation led to setting up a new Commission for Environmental Cooperation, to rule on disputes in case one country is not adequately enforcing its own environmental laws. Unfortunately, the Commission's rules were so complex that it did not function as planned in dispute resolution for corporate accountability. Its efforts to support scientific research through grants to nonprofit organizations for solving environmental problems were more fruitful. In general, however, the North American Free Trade Agreement's environmental protections are weak and do not help much in stemming the degradation that occurs alongside exploding economic development and the ransacking of natural resources by corporations investing in projects under NAFTA in the partner countries.[41]

Some of the most horrific effects have been in Mexico, where environmental degradations caused by foreign multinationals have surged. Matamoros, for example, was once a quiet agricultural community across the border from Brownsville, Texas. But Mexico, under a border industrialization policy predating the North American Free Trade Agreement, offered incentives to U.S. manufacturing companies to operate there, and more than 90 such factories are clustered in the town. The result? "Mexico's woefully underfunded and politically weak environmental regulatory program has not kept up with the rapid industrialization. *Maquiladora* [assembly] plants in Matamoros operate without the scrutiny and environmental controls they would have faced had they stayed in the United States."[42] (Maquiladoras assembly plants import materials and equipment on a duty-free, tariff-free basis for sole purpose of exporting finished products.)

Pollution in the form of orange and purple slime ran in ditches; toxic solvents, metal-plating solutions, and other chemicals contaminated the water supplies, and severe birth defects became more common. Informal cooperation between the two countries did not solve the problems, and the North American Free Trade Agreement did not improve the situation. Corporations continued to migrate to Mexico in search of lower labor and environmental costs. A General Accounting Office Survey in 1992, just before NAFTA was implemented, found that *none* of the U.S. companies operating in the Mexican border region were in compliance with Mexican environmental laws.

As critics pointed out at the time, NAFTA simply promotes plunder, since each country must give access to its natural resources as much to the trade partners as it does to its own citizens. Foreign access to such resources was guaranteed, as long as supplies lasted, regardless of domestic shortages. In other words, NAFTA created, and has led to, unsustainable patterns of resource exploitation.

From the beginning, the North American Free Trade Agreement privileged trade over environment, with environmental laws often seen as cleverly crafted trade barriers rather than critical environmental protections. Indeed, environmental issues were an afterthought to the negotiators, forced on them by outside groups, as was also true with labor issues.[43]

Corporate Maneuvering Outside National Laws

But corporate international trade has an even darker side, not limited to the North American Free Trade Agreement. There is a hidden "Investor-State Dispute Settlement" (ISDS) process expanding alongside increasing international trade. ISDS is an instrument of public international law that grants an investor the right to use a tribunal proceeding to settle a dispute against a foreign government. Essentially, it has created international tribunals, where corporations can sue to protect their investments. The ISDS trade regime has accompanied the explosion of Free Trade Agreements and Bilateral Investment Treaties in the last 25 years. (A "Bilateral Investment Treaty" is an agreement establishing the terms and conditions for private investment by nationals and companies of one state in another state.) These international tribunals are a serious and pernicious problem, creating a set of rules that operate outside any nation's judicial system. Corporations are allowed to sue governments directly. Not only that, but corporations can sue for "indirect" expropriation; that is, they can target regulations that reduce the value of a foreign investment, such as a country's health and environmental laws.

The ISDS system gives corporate investors from one country the right to file a claim against any policy, law, or regulation in another country that the investor believes has violated investment protections. They can seek a range of damages, including unrealized profits. ISDS trade regimes are well known for intentionally vague standards and inconsistent rulings. International trade rules also ban governments from restricting flows of capital, and must treat foreign investors at least as favorably as domestic investors. This strips a government's power to pursue strategies for national development.[44] Even if a country wins such a dispute against a corporation, it faces huge legal costs; if it loses, the penalties for curbing corporate-capital excesses are in the millions of dollars.

Since the North American Free Trade Agreement deregulated the movement of corporate capital, corporations moved to protect their investments through the ISDS legal system. Chapter 11 of NAFTA set up one of these arbitration regimes, which contained all the vagueness and secrecy for which ISDS tribunals are already notorious.[45]

As of 2014, Mexico had lost at least five disputes under Chapter 11 of NAFTA, facing $200 million in penalties. Such suits have doubled since 2005, and Canada has become a frequent target as well.[46]

The United States has faced 12 North American Free Trade Agreement challenges, all from Canadian companies, and won them all. But the biggest one is just beginning: President Obama, in 2015, denied the Keystone XL Pipeline, an $8 billion TransCanada project seeking to connect oil from Canada's oil sands region to U.S. Gulf Coast refineries. TransCanada announced in early 2016 that it would seek, via the NAFTA trade tribunal, $15 billion in damages from the United States, because the company "has been unjustly deprived of the value of its multibillion-dollar investment by the U.S. administration's action."[47] TransCanada described the permit denial as "arbitrary and unjustified." In addition, TransCanada is suing the United States in U.S. federal court, arguing that U.S. rejection of the pipeline is unconstitutional.[48]

The World Bank: Example of International Corporate "Legal Rights"

The International Center for Settlement of Investment Disputes, affiliated with the World Bank, was created in 1966. The International Center is one of the secretive, extrajudicial, trade tribunals. For the first 30 years of its existence, the International Center was essentially dormant. What gave it life was the explosion of Free Trade Agreements and Bilateral Investment Treaties, of which there were nearly 3,000 by 2013. The treaties' purpose was to protect overseas investments in foreign countries, and they have worked excruciatingly well. Cases at the International Center for Settlement of Investment Disputes, and probably other international tribunals whose workings are less transparent, have burgeoned, from three cases in 2000 to 169 cases as of March 2013. More than 35 percent of them are disputes over mining or oil and gas.[49]

Across the board, Latin American countries are frequently targets of these bilateral trade lawsuits—46 percent as of 2013. Nearly a third of the mining/oil and gas lawsuits at the International Center for Settlement of Investment Disputes are also against Latin American countries.

To take just two examples: In 2010, Ecuador lost a case brought by Chevron through the International Center, and had to pay the company $700 million. In 2012, the Center ordered Ecuador to pay $1.7 billion to Occidental Petroleum for canceling its 2006 operating contract, the largest such award to date. Combined, the two cases total more than 3 percent of the country's Gross Domestic Product.

Even more notoriously, the Canadian/Australian mining firm "Pacific Rim" (now OceanaGold) sued El Salvador in 2009 under the Central American Free Trade Agreement to force the government to grant permits for a vast gold-mining project in the headwaters of the Lempi River. The Lempi provides drinking water for at least 50 percent of El Salvador's population, due in part to massive pollution of other watersheds. International, corporate-financed gold mining has contaminated more than 90 percent of El Salvador's surface water. El Salvador is the most water-stressed country in Central America. Three successive El Salvadoran governments have denied the requested permit, and sought to end new mining projects. OceanaGold wanted up to $315 million as compensation for lost profits—5 percent of the country's Gross Domestic Product, an amount equivalent to three years' worth of Salvadoran public spending on health, education, and public security combined. This case has emerged as one of the best examples of corporate overreach in the secretive, international, "free trade" legal system.

However, in a rare ruling upholding a country's needs over corporate profits, the World Bank tribunal in late 2016 denied OceanaGold any lost profits from El Salvador's refusal to grant the gold-mining permit. The tribunal also awarded El Salvador $8 million to cover its legal fees and costs in defending the case. The ruling was unanimous, based in part on the finding that OceanaGold had never acquired the rights to all the land it wanted for the mining concession. Though this particular ruling is protective of human and environmental rights, the entire secret tribunal process continues to have a chilling effect on public policies protecting environmental and human rights worldwide.[50]

Countries targeted by international resource extraction companies are fighting back. Bolivia was the first country to do so, withdrawing from the International Center for Settlement of Investment Disputes in 2007, and then amending its Constitution to prevent international tribunals from judging disputes over investments in the oil and gas sector. Ecuador withdrew from the International Center in 2009, followed by Venezuela in 2012.[51]

The Next Horizon: The Trans-Pacific Partnership

The Trans-Pacific Partnership is an even more ambitious free trade agreement—one of the largest in the world. It would bring together 12 countries of the Pacific Rim, including Australia, Canada, Chile, Japan, Malaysia, Mexico, New Zealand, Peru, the United States, and Vietnam. The goal, as always with Free Trade Agreements, would reduce or eliminate national tariffs on imported goods and increase the national economic ties. The Partnership nations currently comprise about 40 percent of international trading activity.

Tariffs (taxes on imports) would fall on a vast range of agricultural and industrial goods; at least 18,000 tariffs would be eliminated. Criticisms, now based on hard experience with other Free Trade Agreements, argue that competition with other countries' cheaper labor forces will be intensified. Corporate suits against governments for denying permits or restricting the value of foreign investment will increase, and the Trans-Pacific Partnership will expand corporate efforts to manufacture in whichever country offers the lowest wages and the weakest environmental, health, and safety regulations.

Negotiations on the Trans-Pacific Partnership were completed in November 2015, and all member nations signed the Agreement in March 2016, after 7 years of secret negotiations. Member countries must now ratify the agreement. But opposition is fierce in many countries, including the United States. The ugly aftereffects of earlier Free Trade Agreements, and the consolidation of corporate control through the international extrajudicial legal system, has made many people and organizations in the signatory nations cautious of the Trans-Pacific Partnership, which is so much larger than earlier trade agreements, including the North American Free Trade Agreement.[52]

Most chillingly, the Trans-Pacific Partnership would roll back even the weak and poorly enforced Multilateral Environmental Agreements that have characterized all Free Trade Agreements to which the United States has been a party since 2007. The language is significantly weaker with respect to illegal trade in flora and fauna, illegal and unregulated fishing, and commercial whaling. Tellingly, the Trans-Pacific Partnership text does not even mention climate change or the United National Framework Convention on Climate Change to which all Trans-Pacific Partnership parties are subject. Moreover, the Trans-Pacific Partnership would require the United States to approve all exports of liquefied natural gas to all Partnership countries, thus greatly exacerbating both fracking and climate change.

The Trans-Pacific Partnership and Trade Tribunals

Even worse, the Trans-Pacific Partnership would at least double the number of corporate entities that could use the unaccountable trade tribunals to challenge the policies of the United States and other countries. Foreign firms would also gain new rights under these ISDS tribunals to sue against a country's financial regulatory policies. Under existing pacts, banks cannot bring suits in trade tribunals to demand taxpayer compensation for financial regulation.

Expanding the ISDS powers, foreign firms would be essentially granted the right to privately enforce the Trans-Pacific Partnership in the trade tribunals, challenging U.S. domestic laws on grounds not available in U.S. law. Even more nightmarishly, the corporations would be allowed to request—and receive—compensation for "expected future profits" the investor would have earned in the absence of the regulation or decision it is challenging.

More than 1,000 additional corporations in the Trans-Pacific Partnership nations, owning more than 9,200 subsidiaries in the United States, could use the trade tribunals against American interests. The Partnership would also grant new powers to more than 5,000 U.S. corporations to file trade tribunal cases against other signatory nations on behalf of the corporations' 19,000 subsidiaries in the signatory countries. Unfortunately, the trade tribunals would be empowered to hear cases on disagreements about natural resource extraction, infrastructure projects, and contracts for operation of utilities.

The ISDS tribunals under the Trans-Pacific Partnership, as negotiated, have no provision for outside appeal and no requirement to seek first redress in a country's own court system. And even if a government wins, it still must pay the tribunal's costs and legal fees. The attorneys who sit as tribunal judges are not required to be impartial and follow no conflict-of-interest rules. Conflict-of-interest guidelines are supposed to be devised as a side agreement, which has not yet occurred. The tribunal judges would even be allowed to rotate between being judges and advocates for investors bringing the cases. In other words, *they have every incentive to favor corporate compensation* in order to increase the number of cases that would benefit their clients, the tribunal system, and themselves.[53]

The Environmental Dark Side of International Trade

What of the effects of burgeoning international trade on the world's oceans? One immediate effect has been the dramatic increase in ship size, as well as the number of ships crossing the oceans. This led to the recent widening of the Panama Canal.

Larger Cargo Ships

The Canal was facing constraints, squeezed by the trend toward vast ships, which required more time to traverse the Canal just at the time of increasing demand for guaranteed levels of service and lower transit time in the locks and channels.

Several studies since the 1930s concluded that the best solution was building a new, third set of locks that were larger than the 1914 locks, to relieve the congestion resulting from the massive increases in international trade. Originally built for a maximum capacity of 80 million tons of shipping annually on "Panamax" ships (Panama Canal-Maximum-Sized Ship), by 2009, the Canal was handling 299 million tons, but as cargo ships expanded, less than half of them could traverse the Canal's narrow locks.[54]

The original Panamax ship was a mid-sized cargo vessel capable of passing through the lock chambers of the Panama Canal, meaning they were not to exceed the dimensions (length, width, and depth) of the lock chambers and the height of the Bridge of the Americas—965 long, 106 feet wide, and 39.5 feet in depth.

But cargo ships have been steadily growing in size. The first generation of these cargo ships, created in 1956, carried a maximum of 800 "twenty-foot-equivalent-unit containers" (20 feet long, by eight feet wide, by nine feet high). A trucking entrepreneur named Malcolm McLean invented modern cargo carriers by using truck trailers filled with products that could be lifted directly from a truck onto a ship. This eliminated the need to pack, transport, and unpack goods in barrels and kegs; whole containers could be moved efficiently onto ships. This process, called "intermodalism," revolutionized the shipping industry by creating containers that could be stacked on large decks, greatly lowering labor costs, and leading directly to the current boom in international trade.[55]

Cargo ships expanded in the 1970s to carrying up to 2,500 20-foot-equivalent-unit containers—that is, 10 gigantic containers across the deck, five high above deck, and four containers below.

To accommodate the larger ships, the Panama Canal Authority initiated the construction of the third lane of locks in 2009, with sufficient clearance for the larger vessels. The new lock dimensions are 1,400 feet in length, 180 feet in width, and 60 feet in depth, and the "New Panamax" ships are more than 1,300 feet long and carry a staggering 18,000 to 20,000 containers (Figure 16.4). The widening project, completed in June 2016, achieved the goal of doubling the capacity of the Canal by allowing passage to 79 percent of the cargo ships.[56]

FIGURE 16.4
Post-Panamax container ships *President Truman* and *President Kennedy*. Photograph by the U.S. National Oceanic and Atmospheric Administration (https://commons.wikimedia.org/wiki/File:Line0534.jpg).

The cascading effect of the expansion is likely to be a shift of some Asia-to-U.S. cargo ships to the U.S. east coast, which has led many east coast ports, such as Baltimore, Miami, New York/New Jersey, and Norfolk, into major projects to modernize their infrastructure and deepen the channels in anticipation of the shift.

One of the gigantic New Panamax ships visited the Port of Seattle for the first time in February 2016, somewhat to the worry of the National Oceanic and Atmospheric Administration, which is the country's first responder in spill-response planning. Cargo ships in this category have about 4.5 million gallons of fuel onboard. Large oil tankers, though smaller in size, have substantially more oil onboard, such as the infamous *Exxon Valdez*, which carried 55 million gallons of oil. But the new, enormous class of trading ships are set to become ever more common, if current trends continue; they will be plying many of the world's oceans and are therefore vulnerable to storms and coastal weather all over the world, greatly increasing the possibility of massive spills.[57]

Not only that, but since the 1980s, it is common for these colossal ships to carry hazardous chemicals, such as the 54 million tons of hazardous chemicals—with the potential to cause mass casualties—that passed through U.S. ports in 2003. Spills and their effects, not to mention "chemical terrorism," are of great concern.[58]

Containers Lost at Sea

Today's international shipping, which transports millions of containers around the world, has led to another serious problem—one that gets little publicity: the massive number of containers lost at sea. In 2013, scientists from the National Oceanic and Atmospheric Administration teamed up with the Monterey Bay Aquarium Research Institute to continue investigation of a single cargo container that fell off the ship *Med Taipei* in 2004 and sank in 4,000 feet of water in California's Monterey Bay National Marine Sanctuary, to find out how it affected marine ecosystems. Photos and data were compared from 2004, 2011, and 2013. Questions included (1) the effects of toxic paint on the container; (2) the effect on the marine environment of chemicals possibly leaching from inside the container, as the contents (steel-belted tires, in this case) begin to disintegrate; (3) the biological community growing on the container; and (4) whether heavy metals were present in animals living on or near the container.[59]

Estimates of how many containers are lost each year at sea vary wildly, as might be expected. Of the 120 million containers that crossed the oceans in 2013, how many made it to port? The World Shipping Council, after surveying the industry, estimated that between 2008 and 2013, 546 containers were lost at sea each year, if catastrophic events were not counted; 1,679 were lost each year if catastrophic events were included.[60]

Losses seem to be increasing as a result of catastrophic events. For example, in 2011, 2012, and 2013, the average annual loss of containers was 2,683, nearly a 300-fold increase from the World Shipping Council estimate. Why? The loss of the ship *Comfort* in the Indian Ocean that caused the loss of all 4,293 containers, and the *Rena* in 2011 off the coast of New Zealand, which caused the loss of 900 containers, caused the upsurge in container losses.[61]

However, the lack of accountability in this burgeoning area of marine pollution is disturbing. The European Parliament, among others, has raised questions as to whether the number of containers lost at sea is underreported, or whether accurate estimates of container losses even exist. Some industry sources say the losses may be as high as 10,000 containers a year. The main problem is that nobody knows for certain.[62]

Effects on Whales and Other Cetaceans from Commercial Ships

Noise pollution is one of society's growing problems because it is increasingly disrupting the population dynamics of marine animals, such as whales, dolphins, and porpoises (collectively = Cetaceans). Simply put, the growing noise of human activities is irreversibly shifting the composition, structure, and function of marine biophysical systems, as commercial ships have increased in number, travel throughout more of the world's oceans, have gotten progressively larger, and augmented an explosion of human activities in the marine realm.

Human contribution to noise pollution in the oceans has increased over the past five decades and is dominated by low-frequency sound (frequencies <1,000 Hz) from commercial shipping, oil and gas development, and military activities.[63] The ability of many sea creatures to seek food, find mates, protect their young, use their habitual routes of migration safely, and escape their predators is increasingly and severely compromised.

FIGURE 16.5
Blue whale. Photograph by T. Bjornstad, U.S. NOAA Fisheries (https://commons.wikimedia.org/wiki/File:Blue_Whale_001_body_bw.jpg).

The effects of underwater noise can be likened to being trapped in the center of an acoustic traffic jam, where the din comes simultaneously from all sides. In deep water, where marine animals rely on their sense of hearing, the noise is especially harmful. For example, high intensity anthropogenic sound damages the ears of fish.[64]

Noise from commercial ships, as well as other types of vessels and marine activities, scramble the communication signals used by whales, dolphins, and porpoises, which cause them to abandon traditional feeding areas and breeding grounds, change direction during migration, alter their calls, and blunder into fishing nets.[65]

In addition, dolphins and whales can no longer avoid colliding with ships on the open seas, where international shipping produces the most underwater noise pollution, with few regulations to control it. For example, both Cuvier's beaked whales and blue whales use sound to communicate with one another, as well as for hunting food. Cuvier's beaked whales produce clicks while diving up to a mile deep, as they hunt for food. The clicks produce echoes as they bounce off the bodies of the squid the whales are hunting—a phenomenon known as "echolocation." When the scientists played sonar sounds during experiments with tagged Cuvier's beaked whales, they stopped hunting as soon as the sound started, and swam rapidly, silently away. The scientists were using "active sonar," which is the purposeful emission of sound employed by fishing vessels, scientific vessels, submarines, commercial and military ships, and boats used for oil exploration, overwhelming areas of the ocean with manmade noise.

Contrary to Cuvier's beaked whales, however, blue whales showed almost no response to the emission of sonar while feeding on the surface, but those diving for krill reacted very differently. They stopped feeding just as soon as the sonar began and moved swiftly away from the source of the sound. These huge whales, which may be the largest animals that have ever lived, can scoop up a million calories' worth of krill in one gulp while diving, so disturbing their feeding deprives them of vast amounts of energy[66] (Figure 16.5). (For a more thorough discussion of environmental effects of ships on cetaceans, see *Interactions of Land, Ocean and Humans: A Global Perspective.*[67])

Management Considerations for a Rights of Nature System

The seas, estuaries, and land transportation networks of the world cannot sustain such an intense international trade system. Furthermore, the frenzied pace of international trade has led to extreme absurdities in every nation that show how unsustainable international trade has become, and how dramatic its effects are on local production, communities, and environments.

One December, I (Cameron) wanted to buy roasting chestnuts at my local chain grocery store in Salem, Oregon. Family farms growing these chestnuts lie within one hour's drive of the store. This local chain store has a policy of purchasing and showcasing local agricultural products, at which it has been very successful, pulling products from the Willamette Valley and farther afield in Oregon, thus supporting family farms. However, when I asked for the usual local roasting chestnuts, store personnel told me that it was cheaper to import them from Korea! Indeed, I only found Korean chestnuts for sale. It is a travesty of environmental economics—and a masterpiece of externalizing all the hidden costs onto the public—that it is cheaper to import roasting chestnuts from the other side of the world, when they grow on family farms an hour away.

Here are a few management guidelines for refashioning international trade into a sustainable, small-scale exchange of goods between unique nations within a Rights of Nature system that does not harm the environment or countries' own ability to protect their ecosystems and their people. In this instance especially, it will be necessary for all countries in the international community to adopt a Rights of Nature framework for trade policies. Otherwise, nations refusing to adhere to the restrictions and taking advantage of increased opportunities make the entire project unworkable. Cooperation among all nations is essential to restore the marine ecosystem to biophysical sustainability from the overuse it now suffers under the corporate international trade regime.

1. Implement through the United Nations a Rights of Nature statute as an enforceable environmental protocol binding on all nations. This statute must focus on granting and enforcing Earth's rights to clean air, pure water, and ecosystem integrity across the landscape. All international trade must be regulated under this protocol, tailored to recover and maintain the maximum sustainable productivity of ocean and land ecosystems.
2. Include in all international Rights of Nature legislation the rights of nongovernmental organizations to sue on behalf of the ecosystems of the world in international legal tribunals (if necessary, in addition to a specially created tribunal) both for ecocide and for actions that degrade the rights of the Earth to flourish free of pollution.
3. Set up an International Rights of Nature Tribunal along the lines of that already formed by the Global Alliance for the Rights of Nature. It must be paid for by all nations from general fund monies given to the United Nations, so there is no incentive to approve damaging projects in order to continue the Tribunal's funding. Appointed attorneys may only be from nonprofit entities (from different countries) with a strong track record of protecting Earth's ecosystems. It will hear testimony, produce reports, and make binding decisions on development proposals that threaten ecosystem integrity, sustainability, and productivity under Rights of Nature standards.
4. Cancel or drastically revise Free Trade Agreements and Bilateral Agreements worldwide so that trade between and among nations consists only of that nation's unique products. This includes The North American Free Trade Agreement, as well as smaller agreements. The Trans-Pacific Partnership, as negotiated, is extremely damaging to natural resources, ecosystem integrity, and national sovereignty, and must be discarded in full.
5. Encourage countries via international incentives to protect their own local economies and environments, as national sovereignties have the power to do. This protection includes making sustainable investments in their own economies, nurturing both local independence and unique goods for trading, and protecting and restoring environments.
6. Prohibit all corporate entities, large or small, from ability to sue any sovereign nation or any part thereof for the loss it has suffered, or may suffer, from that nation's polices or regulations, whether environmental, labor/safety, or tariffs. In addition, dismantle all ISDS tribunals.
7. Reduce international trade to a human scale. The risk of terrible sea tragedies from the New Panamax ships carrying 18,000 twenty-foot-equivalent-unit containers

or more is simply too great. No ships carrying more than a much smaller amount may be allowed. Fund research through the United Nations and coastal countries to determine the much smaller amount of international shipping, and size of ships, that the world's oceans can sustainably accommodate.

Endnotes

1. The following discussion of the Ecuadorian Constitution is based on: [Ecuadorian] Constitution, full language: http://pdba.georgetown.edu/Constitutions/Ecuador/english08.html (Title II, Ch. 7, Article 71) (accessed June 23, 2016).
2. *Ibid.* (Article 73).
3. *Ibid.* (Article 72).
4. *Ibid.* (Title II, Chapter 2, Section Two, Article 13).
5. *Ibid.*
6. *Ibid.* (Title II, Chapter 2, Section Two, Article 14).
7. *Ibid.* (Title II, Chapter Four, Article 57).
8. *Ibid.*
9. *Ibid.* (Title VI, Chapter 3, Article 281).
10. Language of the Bolivian law (compete text): http://www.worldfuturefund.org/Projects/Indicators/motherearthbolivia.html (accessed June 23, 2016).
11. *Ibid.* (Chapter III, Article 7, I(1)).
12. *Ibid.* (Article 7, I(7)).
13. *Ibid.* (Article 7, I(5)).
14. *Ibid.* (Chapter IV, Article 8).
15. *Ibid.* (Article 8, Section 2).
16. Universal Declaration of Rights of Mother Earth: Global Alliance for the Rights of Nature. https://therightsofnature.org/universal-declaration/ (accessed June 23, 2016).
17. *Ibid.* (Article 2 (1)).
18. *Ibid.*
19. *Ibid.* (Article 1 (7)).
20. *Ibid.* (Article 2).
21. Dave Hansford. Early Polynesians Sailed Thousands of Miles for Trade. http://news.nationalgeographic.com/news/2007/09/070927-polynesians-sailors.html (accessed June 19, 2016).
22. Mason Inman. Polynesians—And Their Chickens—Arrived in Americas Before Columbus. http://news.nationalgeographic.com/news/2007/06/070604-chickens.html (accessed June 19, 2016).
23. (1) Volga trade route. https://en.wikipedia.org/wiki/Volga_trade_route (accessed June 19, 2016); and (2) The Varangian Guard: The Vikings in Byzantium. http://www.soldiers-of-misfortune.com/history/varangian-guard.htm (accessed June 19, 2016).
24. Günther Stockinger. Abandoned Colony in Greenland: Archaeologists Find Clues to Viking Mystery. http://www.spiegel.de/international/zeitgeist/archaeologists-uncover-clues-to-why-vikings-abandoned-greenland-a-876626.html (accessed June 19, 2016).
25. Norwegian rule. https://en.wikipedia.org/wiki/Orkney: Norwegian_rule (accessed June 19, 2016).
26. Eugene Linden. The Vikings: A Memorable Visit to America. http://www.smithsonianmag.com/history/the-vikings-a-memorable-visit-to-america-98090935/?no-ist (accessed June 19, 2016).

27. Spanish colonization of the Americas: Mexico. https://en.wikipedia.org/wiki/Spanish_colonization_of_the_Americas#Mexico (accessed June 20, 2016).
28. Andrés de Urdaneta. https://en.wikipedia.org/wiki/Andrés_de_Urdaneta (accessed June 21, 2016).
29. New Spain. https://en.wikipedia.org/wiki/New_Spain (accessed June 29, 2016).
30. Frank Quimby. Miguel López de Legazpi. http://www.guampedia.com/miguel-lopez-de-legazpi/ (accessed June 29, 2016).
31. Potosí. https://en.wikipedia.org/wiki/Potos%C3%AD (accessed June 21, 2016).
32. Arturo Giraldez. *The Age of Trade: Manila Galleons and the Dawn of the Global Economy*. Rowman and Littlefield, Lanham, MD. (2015) pp. 146–148.
33. The foregoing three paragraphs are based on: (1) Arturo Giraldez. *The Age of Trade: Manila Galleons and the Dawn of the Global Economy*. Rowman and Littlefield, Lanham, MD. (2015) 257 pp; (2) Manila galleon. http://www.britannica.com/technology/Manila-galleon (accessed June 21, 2016); and (3) Manila galleon. https://en.wikipedia.org/wiki/Manila_galleon (accessed June 21, 2016).
34. The foregoing three paragraphs are based on: (1) Navigation Acts: United Kingdom. https://www.britannica.com/event/Navigation-Acts (accessed June 29, 2016); (2) Trent J. Bertrand. International trade. https://www.britannica.com/topic/international-trade (accessed June 29, 2016); and (3) General Agreement on Tariffs and Trade (GATT): International Relations. http://www.britannica.com/topic/General-36. The preceding discussion of the Panama Canal is based on: Panama Canal. https://en.wikipedia.org/wiki/Panama_Canal (accessed June 24, 2016).
35. General Agreement on Tariffs and Trade (GATT): International Relations. *op. cit.*
36. The preceding discussion of the Panama Canal is based on: Panama Canal. https://en.wikipedia.org/wiki/Panama_Canal (accessed June 24, 2016).
37. The preceding two paragraphs, including the quote by Hufbauer and Schott are based on: Mohammed Aly Sergie. NAFTA's Economic Impact. http://www.cfr.org/trade/naftas-economic-impact/p15790 (accessed June 24, 2016).
38. Julián Aguilar. Twenty Years Later, NAFTA Remains a Source of Tension. http://www.nytimes.com/2012/12/07/us/twenty-years-later-nafta-remains-a-source-of-tension.html?_r=0 (accessed June 24, 2016).
39. Mohammed Aly Sergie. NAFTA's Economic Impact. *op. cit.*
40. Julián Aguilar. Twenty Years Later, NAFTA Remains a Source of Tension. *op. cit.*
41. The foregoing two paragraphs are based on: (1) The Environment and NAFTA. http://www.globalization101.org/the-environment-and-nafta/ (accessed June 24, 2016); and (2) Russ Beaton and Chris Maser. Economics and Ecology: United for a Sustainable World. CRC Press, Boca Raton, FL. (2012) 191 pp.
42. The previous three paragraphs are based on: Mary E. Kelly. NAFTA and the Environment: Free Trade and the Politics of Toxic Waste. http://www.multinationalmonitor.org/hyper/issues/1993/10/mm1093_03.html (accessed June 24, 2016).
43. Stephen P. Mumme. The North American Free Trade Agreement's impact on the trinational environment remains controversial. http://fpif.org/nafta_and_environment/ (accessed June 24, 2016).
44. The preceding two paragraphs are based on: Sarah Anderson and Manuel Pérez-Rocha. Mining for Profits in International Tribunals: Lessons for the Trans-Pacific Partnership (p. 5). http://www.ips-dc.org/wp-content/uploads/2013/05/Mining-for-Profits-2013-ENGLISH.pdf (accessed June 25, 2016).
45. Nia Williams and Valerie Volcovici. TransCanada sues U.S. over Keystone XL pipeline rejection. http://www.reuters.com/article/us-transcanada-keystone-idUSKBN0UK2JG20160107 (accessed June 25, 2016).
46. (1) Manuel Pérez-Rocha and Stuart Trew. Corporate Rule. http://fpif.org/nafta-20-model-corporate-rule/ (accessed June 25, 2016); and (2) Deirdre Fulton. Amid TPP Fight, El Salvador Mining Case Shows Danger of Corporate Tribunals. http://www.commondreams.org/news/2015/05/11/amid-tpp-fight-el-salvador-mining-case-shows-danger-corporate-tribunals (accessed June 25, 2016).

47. Ian Austen. TransCanada Seeks $15 Billion from U.S. over Keystone XL Pipeline. http://www.nytimes.com/2016/01/07/business/international/transcanada-to-sue-us-for-blocking-keystone-xl-pipeline.html?_r=0 (accessed June 25, 2016).
48. Nia Williams and Valerie Volcovici. TransCanada sues U.S. over Keystone XL pipeline rejection. *op. cit.*
49. Sarah Anderson and Manuel Pérez-Rocha. Mining for Profits in International Tribunals: Lessons for the Trans-Pacific Partnership. *op. cit.* (p. 4).
50. The foregoing four paragraphs are based on: (1) *Ibid.* (pp. 1 and 2); (2) Deirdre Fulton. Amid TPP Fight, El Salvador Mining Case Shows Danger of Corporate Tribunals. *op. cit.*; (3) El Salvador Beats Mining Giant OceanaGold at World Bank Court. http://www.telesurtv.net/english/news/El-Salvador-Beats-Mining-Giant-OceanaGold-at-World-Bank-Court-20161013-0014.html (accessed November 28, 2016); and (4) Elisabeth Malkin. El Salvador Wins Dispute Over Denying a Mining Permit. http://www.nytimes.com/2016/10/15/world/americas/salvador-mining-dispute.html?_r=0 (accessed November 28, 2016).
51. Sarah Anderson and Manuel Pérez-Rocha. Mining for Profits in International Tribunals: Lessons for the Trans-Pacific Partnership. *op. cit.* (p. 6).
52. The preceding three paragraphs are based on: (1) TPP: What is it and why does it matter? http://www.bbc.com/news/business-32498715 (accessed June 25, 2016); and (2) Rebecca Howard. Trans-Pacific Partnership trade deal signed, but years of negotiations still to come. http://www.reuters.com/article/us-trade-tpp-idUSKCN0VD08S (accessed June 25, 2016).
53. The foregoing discussion is based on: Secret TPP Text Unveiled: It's Worse than We Thought. http://www.citizen.org/documents/analysis-tpp-text-november-2015.pdf (accessed June 25, 2016).
54. The preceding two paragraphs are based on: (1) Panama Canal expansion project. https://en.wikipedia.org/wiki/Panama_Canal_expansion_project (accessed June 25, 2016); (2) The Panama Canal Expands. http://www.wsj.com/articles/the-panama-canal-expands-1466378348 (accessed June 25, 2016); and (3) Panamax and New Panamax. http://maritime-connector.com/wiki/panamax/ (accessed June 30, 2016).
55. The Advent of the Modern-Day Shipping Container. http://oceanservice.noaa.gov/news/features/nov13/containers.html (accessed June 25, 2016).
56. The preceding two paragraphs are based on: (1) The Panama Canal Expands. *op. cit.*; and (2) Panamax and New Panamax. *op. cit.*
57. Doughelton. At the U.S.-Canadian Border, Surveying a World War II Shipwreck for History and Oil. https://usresponserestoration.wordpress.com/author/doughelton/ (accessed June 25, 2016).
58. (1) Emergency Responders Call Up NOAA's CAMEO. http://celebrating200years.noaa.gov/visions/spill_response/welcome.html (accessed June 25, 2016); and (2) Responding to Environmental Catastrophes: An Evolving History of NOAA's Involvement in Oil Spill Response. http://celebrating200years.noaa.gov/transformations/spill_response/welcome.html#intro (accessed June 25, 2016).
59. (1) Every day, millions of shipping containers crisscross the world's oceans. But what happens to the ones that get lost along the way? http://sanctuaries.noaa.gov/news/features/shippingcontainers.html (accessed June 25, 2016); (2) James Cave. Thousands of Containers Fall off Ships Every Year. What Happens to Them? http://www.huffingtonpost.com/2014/07/17/med-taipei-shipping-container-study_n_5593719.html (accessed June 25, 2016); and (3) Restoration Plan for The M/V Med Taipei ISO Container Discharge Incident. http://montereybay.noaa.gov/resourcepro/mt/welcome.html (accessed July 2, 2016).
60. Survey Results for Containers Lost at Sea—2014 Update. http://www.worldshipping.org/industry-issues/safety/Containers_Lost_at_Sea_-_2014_Update_Final_for_Dist.pdf (accessed June 25, 2016).
61. Mike Schuler. How Many Shipping Containers Are Really Lost at Sea? http://gcaptain.com/how-many-shipping-containers-lost-at-sea/ (accessed June 25, 2016).

62. (1) A British MEP has told the European Parliament that 2,000 containers are lost overboard from ships in EU waters every year. http://news.bbc.co.uk/democracylive/hi/europe/newsid_9112000/9112791.stm (accessed June 25, 2016); and (2) Tim Lister. Ship loses more than 500 containers in heavy seas. http://www.cnn.com/2014/02/21/world/container-ship-loses-containers/ (accessed June 25, 2016).
63. Donald A. Croll, Christopher W. Clark, John Calambokidis, and others. Effect of Anthropogenic Low-Frequency Noise on the Foraging Ecology of Balaenoptera Whales. *Animal Conservation*, 4 (2001):13–27.
64. R. D. McCauley, J. Fewtrell, and A. N. Popper. High intensity anthropogenic sound damages fish ears. *Journal of the Acoustical Society of America*, 113 (2003):638–642.
65. (1) Christine Erbe. Underwater noise of whale-watching boats and potential effects on killer whales (*Orcinus orca*), based on an acoustic impact model. *Marine Mammal Science*, 18 (2002):394–418; and (2) Sean Todd, Jon Lien, Fernanda Marques, and others. Behavioural effects of exposure to underwater explosions in humpback whales (*Megaptera novaeangliae*). *Canadian Journal of Zoology*, 74 (1996):1661–1672.
66. The preceding two paragraphs are based on: (1) Stacy L. DeRuiter, Brandon L. Southall, and John Calambokidis, and others. First direct measurements of behavioural responses by Cuvier's beaked whales to mid-frequency active sonar. *Biology Letters*, 9, doi: 10.1098/rsbl.2013.0223 (accessed July 4, 2013); (2) Jeremy A. Goldbogen, Brandon L. Southall, Stacy L. DeRuiter, and others. Blue whales respond to simulated mid-frequency military sonar. *Proceedings of The Royal Society B*, 280, doi:10.1098/rspb.2013.0657 (accessed July 4, 2013); (3) Victoria Gill. Blue and beaked whales affected by simulated navy sonar. http://www.bbc.co.uk/news/science-environment-23115939 (accessed July 4, 2013); (4) P.J.O. Miller, N. Biasson, A. Samuels, and P.L. Tyack. Whale Songs lengthen in response to sonar. *Nature*, 405 (2000):903; (5) K.C. Balcomb and D.E. Claridge. A mass stranding of cetaceans caused by naval sonar in the Bahamas. *Bahamas Journal of Science*, 8 (2001):1–12; (6) Ron Word. Whales vs. Navy: Fight Goes On. http://abcnews.go.com/Technology/wireStory?id=7074665 (accessed March 15, 2009); (7) Julie Watson and Alicia Chang. Navy Expands Sonar Testing Despite Troubling Sings. http://abcnews.go.com/Technology/wireStory/scientists-whales-flee-sonar-21225041 (accessed December 15, 2013); and (8) Wild Whales. Vancouver Aquarium, British Columbia, Canada. http://wildwhales.org/ (accessed December 22, 2013).
67. Chris Maser. *Interactions of Land, Ocean and Humans: A Global Perspective*. CRC Press, Boca Raton, FL. (2014) 308 pp.

17

Conclusion: Rights of Nature and Our Responsibility

Placing the Rights of Nature at the heart of a revised legal system, beginning with enshrining it in the federal Constitution and all state Constitutions, is ultimately the only way to ensure that humans retool our living systems to provide long-term, social-environmental sustainability. But as we have said, not even this, by itself, is enough. Nature must be the primary rights-holder, above all human-oriented rights. This is not because we advocate an "anti-human" legal system, far from it. Placing Nature's Rights first is the only way that human life can thrive sustainably. Human use of the global ecosystem must first allow Nature to flourish. Then, to ensure continuing social-environmental sustainability, all human activities must abide by Nature's Laws of Reciprocity. It is our responsibility to tackle this unprecedented problem for the sake of all generations.

Although the various ecological problems the world faces are becoming better known, there are undoubtedly further dismaying changes in store concerning the biosphere's ability to support life, both human and non-human. This portent is leading to an increasing sense of unease in American political, economic, and intellectual life.

Politically, there has been almost no sign of the bold and fearless leadership needed to adopt a new paradigm that will place America—and other nations following their own paths to sustainable futures—on the road to living within Nature's Laws of Reciprocity. The Rights of Nature is a strictly marginalized concept in the United States, even after its adoption into the Ecuadorian Constitution and Bolivian laws, and the innovative New Zealand statute protecting Te Urewera National Park. On the rare occasions that the Rights of Nature is mentioned in the United States, it is usually brushed aside with contempt, including the few courts that have had the opportunity to consider it. Communities that adopt a Rights of Nature ordinance have few, if any, protections under the current legal system.

Even if political officials demonstrated more than the most tepid leadership, the profit-oriented legal and economic structures, and the many corporate entities that benefit from continued resource overexploitation, funnel massive resources into defeating any proposals for true change toward a Rights of Nature system or even stringent, pollution-curbing laws.

Caught between the desperate need to fundamentally refashion and transform our human relationship with Nature, and the many forces of inertia militating against it, some scientists and researchers have begun publishing ecological doomsday scenarios. They seem to multiply daily, each more terrifying and punitive than the one before. All doomsday scenarios are based on the increasingly troubling data, especially on climate change, streaming in from every quarter showing that humans have been cavalierly and arrogantly living far outside Nature's boundaries, to say nothing of the drastic cumulative effects that have begun to unravel Earth's ecosystems as we know them.

The New Environmental Apocalypse Literature

Let us briefly sample three of these environmental apocalypses before moving on to discuss the final topics necessary to transform human behavior into sustainable harmony with the vast web of relationships we call Nature.

Environmental apocalypse 1: In the late 1960s, Dr. James Lovelock, a British scientist and inventor, began research on what ultimately became the most powerful new paradigm of environmental consciousness to emerge from the Western tradition: the Gaia Theory. "The Gaia Theory posits that the organic and inorganic components of Planet Earth have evolved together as a single living, self-regulating system. It suggests that this living system has automatically controlled global temperature, atmospheric content, ocean salinity, and other factors that maintain its own habitability. In a phrase, 'life maintains conditions suitable for its own survival.'"[1]

Lovelock was working at the Jet Propulsion Laboratory in California on methods of finding if life existed on Mars. After some journal articles in the early 1970s, Lovelock published a book called *Gaia: A New Look at Life on Earth* in 1979 that laid out the basic idea of the theory. American microbiologist Lynn Margulis contributed greatly to working out the scientific details of the theory via her knowledge of the myriad ways microbes affect the atmosphere and the planet. There has been continuing controversy over whether the Gaia Theory implies purposeful self-regulation or simple biophysical thermo-regulation. In 1990, Lovelock publicly stated that he does not hypothesize Gaia to be a living being with purpose.[2]

Regardless of this dispute—which involves questions of foresight, teleology, and ultimately of spiritual understanding—the Gaia Theory continues to be a tremendous catalyst for scientific research designed in a more holistic framework, rather than a strictly reductionist one. Initially, the Gaia Theory (or hypothesis) was just that—a theory. However, scientific research over the last quarter century has provided many exciting findings that verify and expand the original hypothesis. It now appears that it is not the life systems of Gaia alone that regulate life on Earth, but the entire system of organic and inorganic parts together—the relationship within and among living beings and the land, air, and the global seas.[3]

Predictably, some scientists refuted the Gaia Theory; one, paleontologist Peter Ward of the University of Washington at Seattle, coined an opposing term and theory: the Medea Hypothesis. The name is based on the Greek myth of Medea, a sorceress who, in some versions of the tale, killed her children. Ward, arguing from a strict Darwinist perspective, seeks to show that life is ultimately self-destructive, and thus far from maintaining a Gaian balance conducive to life. On the contrary, living creatures look only to ensure their own reproductive success. Past mass extinctions were based on microbes or plants using up all the atmospheric carbon dioxide, or pouring out too much oxygen or methane as a result of excessive species proliferation. Organic life itself is the culprit that has caused the biosphere to unravel several times on an evolutionary timescale and nearly extinguish all life permanently.[4]

Understanding Earth as Gaia leads to a more holistic understanding of the current problems facing the Earth, as a result of human abuse of the world's ecosystems and refusal to live in harmony with Nature's Laws of Reciprocity. But, as more problems become acute, some scientists are turning to sharper language, beginning with Lovelock himself. In 2006, and again in 2009, he wrote books with increasingly strident warnings and titles.

In his 2006 book, *Revenge of Gaia*, Lovelock predicted that after humans have increased the carbon dioxide in the atmosphere past the tipping point, global temperatures would spiral uncontrollably. Icecaps will disappear, raising temperatures faster and releasing methane and carbon dioxide trapped in the tundra, which will increase global warming yet further. This will disrupt dozens of critically important feedback loops essential for life. As the planet burns, humanity might not survive at all, or perhaps in severe poverty and greatly diminished numbers.[5]

In 2009, Lovelock reiterated this warning in *The Vanishing Face of Gaia*, saying it appears inevitable that the Earth will become at least five to six degrees hotter than it is now, a challenge humans can only meet with difficulty, especially as the Earth is overpopulated seven times more by humans than it can sustain for much longer. Climactic catastrophes will cause both ecosystem resilience and food supply to crash, leading to an inevitable die-off of the human population.[6]

As he stated in a 2009 news interview, he fears for the future of our overpopulated planet. "I am not a willing Cassandra and in the past have been publicly skeptical about doom stories, but this time we do have to take seriously the possibility that global heating might all but eliminate people from Earth."[7]

Environmental apocalypse 2: Another book laying out a gruesome doomsday hypothesis is *Under a Green Sky* by Peter Ward, the University of Washington earth and space sciences professor who put forth the Medea Hypothesis. Ward describes prior extinctions in Earth's ancient past that were caused by excessive carbon dioxide in the atmosphere, and warns that current human-caused global warming may cause a similar cataclysm. This grim past is the best guide to our future. As he says, "Thus this book, words tumbling out powered by rage and sorrow but mostly fear, not for us but for our children—and theirs."[8]

Under a world quickly warmed, perhaps in a century, by human emissions of carbon dioxide and methane, the ocean would eventually become so oxygen-starved that bacteria would begin producing toxic amounts of hydrogen sulfide. This gas would rise to the atmosphere, and break down the ozone layer. The resulting intensification of sunlight would then kill much of the green phytoplankton (minute, free-floating, aquatic plants, such as one-celled algae).

What would such a world look like—as it has in the past mass extinctions caused by carbon dioxide overload? Here is Ward's apocalypse:

> No wind in the 120-degree morning heat, and no trees for shade. There is some vegetation, but it is low, stunted, parched.... A scorpion, a spider, winged flies, and among the roots of the desert vegetation we see the burrows of some sort of small animals—the first mammals, perhaps.... Yet as sepulchral as the land is, it is the sea itself that is most frightening. Waves slowly lap on the quiet shore, slow-motion waves with the consistency of gelatine.... From shore to the horizon, there is but an unending purple colour—a vast, flat, oily purple, not looking at all like water, not looking anything of our world. No fish break its surface, no birds or any other kind of flying creatures dip down looking for food. The purple colour comes from vast concentrations of floating bacteria, for the oceans of Earth have all become covered with a hundred-foot-thick veneer of purple and green bacterial soup.... We look upward, to the sky. High, vastly high overhead there are thin clouds, clouds existing at an altitude far in excess of the highest clouds found on our Earth. They exist in a place that changes the very colour of the sky itself: We are under a pale green sky, and it has the smell of death and poison.[9]

Environmental apocalypse 3: Even more apocalyptic is the research recently published by Sergei Petrovskii, an applied mathematics professor at the University of Leicester in England. He lays out data showing that an unchecked rise in global temperatures might drastically reduce the amount of breathable oxygen in the atmosphere. Phytoplankton—the aforementioned minute plants responsible for producing about two-thirds (70 percent) of the world's oxygen—would, if the earth heated too far, simply stop producing oxygen via photosynthesis. The rate of oxygen production slowly changes with ocean warming, and if the rate of oxygen production becomes too low or too high, the system dynamics would alter swiftly, leading to the plankton's extinction and plummeting oxygen levels. Petrovskii hypothesized this apocalypse would herald its advance with few warning signs. He estimates Earth might reach this point of no return by 2100—less than a century.

For this disaster to occur, oceans would need to warm 6 degrees Celsius (42.8 degrees Fahrenheit). Climate scientists warn that to avoid disastrous, climatic effects, global temperatures must not rise more than 2 degrees Celsius (35.6 degrees Fahrenheit) above preindustrial levels—the goal set at the recent Paris Accords. On the other hand, staying below the 2 degrees Celsius limit is looking increasingly unlikely politically, with 2.7 degrees Celsius (36.9 degrees Fahrenheit), warmer than it currently is, seeming to be the most realistic goal. That is below Petrovskii's apocalypse threshold, but would still cause major damage to Earth's ecosystems and their ability to support life.[10]

What Is to Be Done?

What is to be done in the face of this apocalyptic literature? Clearly, the human community needs to dramatically transform its ways of living. But forcing apocalyptic scenarios down people's throats in the hope that, after swallowing the bitter medicine, they will see the light does not catalyze change. On the contrary, warnings of looming apocalypse tend to do one of three things: (1) paralyze people from taking any action, much less decisive, helpful action; (2) cause people to build a fortress around themselves and their immediate surroundings so they can continue living in what they consider a balanced, normal life; and (3) encourage desperate people to bargain with the immutable powers crushing human life, often by adopting symptomatic, band-aid solutions in the vain hope they will resolve the crisis.

People take fearless, altruistic, far-sighted action when love, caring, and hope inspire them. Humans live in relationship, not only with one another but also with all of Nature. Everything humans do now must work toward healing these relationships, and that is the reason we have laid out the scenario for a Rights of Nature legal and environmental framework. The actual effects of climate change on Gaia—to use Lovelock's term—are unknown. It may be, for example, that negative feedback loops will correct for human misuse of the climate cycles and maintain the chemical and biological balances necessary for human and other life to flourish. But one thing is certain: humans must now courageously accept responsibility and begin to change their many broken relationships with Nature. This task is within our grasp, and apocalyptic literature cannot guide us in commencing, pursuing, or accomplishing it.

Adopting the Rights of Nature Worldwide with Fearless Courage

The Rights of Nature blueprint we have laid out in this book is meant to provide a template for far-reaching transformation of human culture so that all societies may tailor themselves to live within Nature's Laws of Reciprocity. The United States, the nation of which we are citizens, has been our prime focus, as it is a highly industrialized nation with many serious and growing environmental problems and very little Rights of Nature focus.

However, in order for the Rights of Nature to create a true, worldwide paradigm shift, it will be necessary for all nations to adopt an equally strong framework and enforce it. Otherwise, the Rights of Nature restraint practiced by one country would simply be undermined by another's profiteering, thus fatally weakening the entire edifice. All nations have equal responsibility to restrain resource exploitation and place Nature first, lest all the gains of a few courageous countries be lost, to the detriment of all. There is no way around complete, communal responsibility among all nations to cooperatively retool their legal systems and enforce the Rights of Nature as the primary rights-holder. Any nation failing to take up the challenge would bear responsibility for destroying the hopes of all.

This will also require tackling what is perhaps the most difficult topic of all: uncontrolled population expansion. As some scientists, notably James Lovelock, have made clear, humans have overpopulated the earth—and continue to exacerbate the problem. Although there is no simple solution to this problem, there is a solution. Namely, grant women worldwide the same equality of choice that men unquestioningly expect and covet. It will be exceedingly difficult to design and enforce such a change, as well as the suite of ethical changes necessary to grapple successfully and humanely with the many issues. But our best minds, greatest caring, and wisest understanding must go into solving this and other problems successfully, or all the gains of a full, worldwide Rights of Nature system would be lost.

As an economy grows, natural capital such as air, soil, water, timber, and marine fisheries is reallocated to human use via the marketplace, where economic efficiency rules. The conflict between economic growth and the conservation and maintenance of natural resource systems is a clash between the economic ideals of efficiency and the realities of Nature's biophysical effectiveness. This economically driven divergence creates a conundrum because traditional forms of active conservation require money, which, in the United States, is highly correlated with income and wealth. The conservation and maintenance of biodiversity—in all its forms—will ultimately require the cessation of economic growth that is environmentally unsustainable.[11]

The number of threatened species is related to per capita Gross National Product (GNP) in five taxonomic groups in over 100 countries. Birds are the only taxonomic group in which numbers of threatened species have *decreased* throughout industrialized countries as prosperity increased. Plants, invertebrates, amphibians, reptiles, and mammals showed increasing numbers of threatened species with increasing prosperity. If these relationships hold, increasing numbers of species from several taxonomic groups are likely to be threatened with extinction as countries increase in affluence.[12]

Wealthy, industrialized nations, such as the United States, left such a large ecological footprint in the last four decades of the 20th century that the damage adds up to more than the poor, non-industrialized nations owe in debt. In fact, the rich nations have caused upward of \$2.5 trillion in environmental damage to poor countries. The well-off disproportionately affect the poor through such things as exacerbation of climate change (a negative outcome to which China is now also contributing), depletion of stratospheric ozone, agricultural intensification and expansion, deforestation, overfishing, and destruction of mangroves.[13]

Our worldwide intellectual and ethical challenge in implementing the Rights of Nature must include an understanding of these two critical aspects of uncertainty, both of which have often been denied, derided, or ignored:

1. Cumulative effects: No given factor can be singled out as the sole cause of anything. All things operate synergistically as cumulative effects that exhibit a lag period before fully manifesting themselves. Cumulative effects, which encompass many little, inherent novelties, cannot be understood statistically because ecological relationships are far more complex and far less predictable than our statistical models lead us to believe.

   At length, cumulative effects, gathering themselves below our level of conscious awareness, suddenly become visible. By then, it is too late to retract our decisions and actions even if the outcome is decidedly negative. So the cumulative effects of our activities multiply unnoticed until something in the environment shifts dramatically enough for us to see the effects through casual observation.

   Conservation under a Rights of Nature system must therefore accommodate both infrequent and unpredictable events and long-term trends by planning for the time scale of those events—without aiming to maintain the status quo.[14]

2. Monitoring: To correct our course, and discover how to adjust our actions in a Rights of Nature system—say, in a large-scale, ecological restoration project—it is necessary to monitor the results of human interventions. Good monitoring has seven steps: (1) crafting a vision and goals, (2) making a preliminary inventory, (3) modeling your understanding, (4) writing a caretaking plan with clearly defined objectives, (5) observing implementation, (6) verifying effectiveness, and (7) validating the outcome(s).

   Here, a caveat is necessary. Traditional uni-dimensional indicators, which measure the apparent health of a single condition (such as sunflowers in one prairie remnant bordering a corn field), ignore the complex relationships among soil, water, and air quality, and the relationship of the remnant to its surrounding landscape. When each component is viewed as a separate issue and monitored in isolation, measurements tend to become skewed and lead to ineffective decisions. Therefore, if an accurate assessment of repair and sustainability is to be achieved, viable indicators need to be multidimensional and must measure the *quality of relationships* in the form of biophysical feedback loops among the components being monitored.[15]

We are confident that Americans, and the nations of the world, will rise to the challenge of shouldering full responsibility for human overuse and exploitation of the living systems on which all life depends. Courage and altruism are two indomitable human traits. Another is deep resonance with relationship. That is the ultimate truth of embracing the Rights of Nature: it is repairing severed *relationships* with courage, responsibility, and a deep commitment to caring for one another and all life.

Endnotes

1. Gaia Theory: Model and Metaphor for the 21st Century. http://www.gaiatheory.org/overview/ (accessed August 23, 2016).
2. Gaia hypothesis. https://en.wikipedia.org/wiki/Gaia_hypothesis (accessed August 23, 2016).
3. (1) *Ibid.*; and (2) James Lovelock. *The Vanishing Face of Gaia*. Basic Books, New York. (2009) 286 pp.

4. (1) The Medea Hypothesis: Is Life on Earth Ultimately Self-Destructive? https://www.timeshighereducation.com/books/the-medea-hypothesis-is-life-on-earth-ultimately-self-destructive/408337.article?storyCode=408337&sectioncode=26 (accessed August 23, 2016); and (2) Moises Velasquez-Manoff. The Medea Hypothesis: A response to the Gaia hypothesis. http://www.csmonitor.com/Environment/Bright-Green/2010/0212/The-Medea-Hypothesis-A-response-to-the-Gaia-hypothesis/%28page%29/1 (accessed August 23, 2016).
5. Robin McKie. The Revenge of Gaia, by James Lovelock. https://www.theguardian.com/books/2006/feb/12/scienceandnature.features (accessed August 23, 2016).
6. Peter Forbes. The Revenge of Gaia, by James Lovelock. https://www.theguardian.com/culture/2009/feb/21/james-lovelock-gaia-book-review (accessed August 23, 2016).
7. Roger Highfield. The Vanishing Face of Gaia: A Final Warning by James Lovelock, review. http://www.telegraph.co.uk/culture/books/bookreviews/5017620/The-Vanishing-Face-of-Gaia-A-Final-Warning-by-James-Lovelock-review.html (accessed August 23, 2016).
8. Peter Ward. *Under a Green Sky: The Mass Extinctions of the Past, and What They Can Tell Us about Our Future.* Smithsonian Books, an imprint of Harper Collins, New York. (2007) p. xiv.
9. *Ibid.* (pp. 138–140).
10. (1) Julie M. Rodriguez. The world will run out of breathable air unless carbon emissions are cut. http://inhabitat.com/runaway-carbon-emissions-threaten-two-thirds-of-the-earths-oxygen-supply/?utm_content=buffer39638&utm_medium=social&utm_source=twitter.com&utm_campaign=buffer (accessed August 23, 2016); and (2) Yadigar Sekerci and Seregei Petrreovskii. Mathematical Modelling of Plankton–Oxygen Dynamics Under the Climate Change. http://link.springer.com/article/10.1007%2Fs11538-015-0126-0 (accessed August 23, 2016).
11. Russ Beaton and Chris Maser. *Economics and Ecology: United for a Sustainable World.* CRC Press, Boca Raton, FL. (2002) 191 pp.
12. (1) Robin Naidoo and Wiktor L. Adamowicz, Effects of economic prosperity on numbers of threatened species. *Conservation Biology,* 15 (2001):1021–1029; and (2) Oliver R.W. Pergams, Brian Czech, J. Christopher Haney, and Dennis Nyberg, Linkage of conservation activity to trends in the U.S. economy. *Conservation Biology,* 18 (2004):1617–1623.
13. U. Thara Srinivasan, Susan P. Carey, Eric Hallstein, and others, The debt of nations and the distribution of ecological impacts from human activities. *Proceedings of the National Academy of Sciences,* 104 (2008):1768–1773.
14. (1) Chris Maser. *Earth in Our Care: Ecology, Economy, and Sustainability.* Rutgers University Press, New Brunswick, NJ. (2009) 304 pp.; and (2) A.R.E. Sinclair, Simon A.R. Mduma, J. Grant, and others, Long-term ecosystem dynamics in the Serengeti: Lessons for conservation. *Conservation Biology,* 21 (2007):580–590.
15. Chris Maser. *Earth in Our Care: Ecology, Economy, and Sustainability. op. cit.*

Appendix: Common and Scientific Names of Plants and Animals

FUNGUS	
Fungus	Fungi
ALGAE	
Algae	Protista
Diatoms	Heterokontophyta
FERNS	
Bracken fern	*Pteridium aquilinum*
GRASSES AND GRASS-LIKE PLANTS	
Cheatgrass (aka downy brome)	*Bromus tectorum*
Crested wheatgrass	*Agropyron cristatum*
Grama grass	*Bouteloua* spp.
Grasses	Gramineae
Sedge	*Carex* spp.
FORBES TREES AND SHRUBS	
Acacia	*Acacia* spp.
Alders	*Alnus* spp.
Apple	*Malus domestica*
Ash	*Fraxinus* spp.
Birch	*Betula* spp.
Blue elderberry	*Sambucus cerulea*
Douglas fir	*Pseudotsuga menziesii*
English oak	*Quercus robur*
European beech	*Fagus sylvatica*
European hornbeam	*Carpinus betulus*
European linden	*Tilia europaea*
European maple	*Acer pseudoplantus*
Fir	*Abies* spp.
Hemlock	*Tsuga* spp.
Larch	*Larix* spp.
Maple	*Acer* spp.
Mesquite	*Prosopis* spp.
Noble fir	*Abies procera*
Norway spruce	*Picea abies*
Oak	*Quercus* spp.

Paperbark tea tree	*Melaleuca quinquenervia*
Pear	*Pyrus communis*
Pine	*Pinus* spp.
Piñion pine	*Pinus edulis*
Ponderosa pine	*Pinus ponderosa*
Quaking aspen	*Populus tremuloides*
Salal	*Gaultheria shallon*
Scots pine	*Pinus sylvestris*
Silver birch	*Betula pendula*
Sitka spruce	*Picea sitchensis*
Spruce	*Picea* spp.
Vine maple	*Acer circinatum*
Western hemlock	*Tsuga heterophylla*
Western redcedar	*Thuja plicata*
Western wild ginger	*Asarum caudatum*
Willow	*Salix* spp.
INVERTEBRATES	
BACTERIA	
Bacteria	Prokaryotes
Cyanobacteria	Cyanobacteria
Fecal coliform bacteria	*Escherichia coli*
VIRUS	
Infectious salmon anemia virus	*Isavirus* spp.
MOLLUSKS	
Clam	Bivalvia
Cockle	Cardioidae
Giant clam	*Tridacna gigas*
Oysters	Ostreidae
WORMS	
Earthworms	Annelida
ECHINODERMATA	
Sea cucumbers	Holothuroidae
CRUSTACEANS	
Barnacles	Crustacea
Krill	Euphausiacea
Sea lice	*Lepeophtheirus salmonis*
INSECTS	
Bees	Hymenoptera
Beetles	Coleoptera
Butterflies	Lepidoptera
Insects	Insecta

VERTEBRATES

FISH

Atlantic salmon	*Salmo salar*
Chum salmon	*Oncorhynchus keta*
Cod	*Gadus* spp.
Cutthroat trout	*Oncorhynchus clarki*
European perch	*European perch*
Fish	Actinopterygii
Marlin	Istiophoridae
Pink salmon	*Oncorhynchus gorbushca*
Sailfish	Istiophoridae
Salmon	Salmonidae
Sardines	Clupeidae
Sharks	Chondrichthyes
Steelhead trout	*Oncorhynchus mykiss*
Swordfish	Xiphiidae
Tuna	*Thunnus* spp.

REPTILES

American crocodile	*Crocodylus acutus*
Burmese python	*Python molurus bivittatus*
California glossy snake	*Arizona elegans*
Galápagos giant tortoise	*Chelonoidis nigra porteri*
Green iguana	*Iguana iguana*
Lizard	Lacertilia
Sea turtle	Chelonioidea

BIRDS

Balearic shearwater	*Puffinus mauretanicus*
Band-tailed pigeon	*Patagioenas fasciata*
Birds	Aves
Blue grouse	*Dendragapus obscurus*
Ducks	Anatidae
Golden eagle	*Aquila chrysaetos*
Great white egret	*Ardea alba*
Marbled murrelet	*Brachyramphus marmoratus*
Northern spotted owl	*Strix occidentalis*
Pileated woodpecker	*Dryocopus pileatus*
Reddish egret	*Egretta rufescens*
Ring-necked pheasant	*Phasianus colchicus*
Roseate spoonbill	*Platalea ajaja*
Ruffed grouse	*Bonasa umbellus*
Sage grouse	*Centrocercus urophasianus*
Steller's jay	*Cyanocitta stelleri*
Varied thrush	*Ixoreus naevius*
Wilson's warbler	*Wilsonia pusilla*
Winter wren	*Troglodytes troglodytes*
Wrens	Troglodytidae

MAMMALS

African lion	*Panthera leo*
Bears	Ursidae
Beaver	*Castor canadensis*
Beechy ground squirrel	*Spermophilus beecheyi*
Black bear	*Ursus americanius*
Black-footed ferret	*Mustela nigripes*
Black-tailed deer	*Odocoileus hemionus columbianus*
Blue whale	*Balaenoptera musculus*
Cuvier's beaked whale	*Ziphius cavirostris*
Deer mouse	*Peromyscus maniculatus*
Dolphins	Delphinidae
Dusky-footed woodrat	*Neotoma fuscipes*
Eastern cottontail	*Sylvilagus floridanus*
Erect man	*Homo erectus*
Florida panther	*Puma concolor*
Fox	*Vulpes* spp.
Gophers	Geomyidae
Horse	*Equus* spp.
Marten	*Martes Americana*
Mice	Rodentia
Mink	*Mustela vison*
Modern human	*Homo sapiens*
Moles	Talpidae
North American bison	*Bison bison*
North American elk	*Cervus elaphus*
Porpoises	Phocoenidae
Prairie dogs	*Cynomys* spp.
Raccoon	*Procyon lotor*
Red fox	*Vulpes vulpes*
Red tree mouse (vole)	*Arborimus longicaudus*
Reindeer	*Rangifer tarandus*
Rodents	Rodentia
Santa Rosa beach mouse	*Peromyscus polionotus leucocephalus*
Shrews	Soricidae
Spotted skunk	*Spilogale putorius*
Squirrels	Sciuridae
Vicuña	*Vicugna vicugna*
Whales	Cetacea

Index

Page numbers followed by f indicate figures.

D

T

V

Y

Z

For Product Safety Concerns and Information please contact our EU
representative GPSR@taylorandfrancis.com
Taylor & Francis Verlag GmbH, Kaufingerstraße 24, 80331 München, Germany

www.ingramcontent.com/pod-product-compliance
Ingram Content Group UK Ltd.
Pitfield, Milton Keynes, MK11 3LW, UK
UKHW050757180926
475701UK00007B/1090
* 9 7 8 0 3 6 7 8 8 4 3 5 2 *